Chalcogenide Nanophotonics

Chalcogenide Nanophotonics

Tun Cao
Ying Su
Yinghan Li
Hongze Gao

WILEY VCH

Authors

Dr. Tun Cao
Dalian University of Technology
116024 Dalian
China

Prof. Ying Su
Dalian University of Technology
116024 Dalian
China

Prof. Yinghan Li
Dalian University of Technology
116024 Dalian
China

Prof. Hongze Gao
Dalian University of Technology
116024 Dalian
China

Cover Design: Wiley
Cover Image: © Ischnura elegans (Vander Linden, 1820) – male

Typesetting: Straive, Chennai, India

Library of Congress Card No.: applied for

British Library Cataloguing-in-Publication Data A catalogue record for this book is available from the British Library.

Bibliographic information published by the Deutsche Nationalbibliothek The Deutsche Nationalbibliothek lists this publication in the Deutsche Nationalbibliografie; detailed bibliographic data are available on the Internet at <http://dnb.d-nb.de>.

Print ISBN: 978-3-527-35294-4
ePDF ISBN: 978-3-527-84455-5
ePub ISBN: 978-3-527-84456-2
oBook ISBN: 978-3-527-84457-9

Printed and bound by
CPI Group (UK) Ltd, Croydon, CR0 4YY

C9783527352944_210126

Contents

About the Authors

Tun Cao is a Professor at the School of Optoelectronic Engineering and Instrumentation Science, Dalian University of Technology, where he serves as the Dean. He is a recipient of the National High-level Talent Program, a member of the Discipline Review Group of the Academic Degrees Committee of the State Council, and a Fellow of *Optica*. His research interests lie in metamaterials, photonic crystals, and optoelectronic devices. He has led a broad portfolio of national and international research projects, including the National Key R&D Program, the Defense Basic Research Enhancement Program, international cooperation initiatives, National Natural Science Foundation of China programs, and industry-supported projects in the aerospace sector. He has received five ministerial- and provincial-level science and technology awards. Over the past five years, he has published more than 150 SCI-indexed papers in journals such as *Science Advances, Advanced Materials,* and *Physical Review Letters,* obtained over 50 granted national invention patents, and facilitated the transfer of two technological achievements.

Ying Su is an Associate Professor at the School of Optoelectronic Engineering and Instrumentation Science, Dalian University of Technology, and a member of Professor Tun Cao's research group. Her research focuses on opto-electro-thermal devices and their application technologies. She has presided over national and provincial research projects and contributed to major national R&D programs. She has published more than 20 SCI-indexed papers and holds over 10 invention patents. Her work has been recognized with the Second Prize of the Technological Invention Award from the Chinese Optical Engineering Society and the Second Prize of the Dalian Technological Invention Award.

Yinghan Li is an Assistant Professor at the School of Optoelectronic Engineering and Instrumentation Science, Dalian University of Technology. She received her BEng from Dalian University of Technology in 2015 and her PhD from the University of Science and Technology of China in 2024. She joined the research group of Professor Tun Cao at Dalian University of Technology in 2025. Her research focuses on optoelectronic display devices and thin-film technology, with specific interests in solution-based film formation kinetics, fabrication and modification of optoelectronic films, intelligent optical/electrical modulation, and functional coatings for thermal management.

Hongze Gao is an Assistant Professor at Dalian University of Technology, with a primary research focus on the optical operation of micro- and nanoparticles. He received his doctoral degree from Harbin Institute of Technology in 2023. Having published five SCI-indexed papers as the first or corresponding author, he has established a solid academic foundation in his research field. His current work centers on the sorting and identification of chiral nanoparticles by utilizing optofluidic technologies, with targeted applications in biosensing to enable the sensitive and accurate detection of biomarkers.

Preface

The rapid advancements in photonic technologies over recent decades have transformed light from a passive carrier of information into an active medium for computation, sensing, and energy conversion. At the core of this revolution lies the pursuit of materials that transcend traditional limits, enabling dynamic control over light at the nanoscale. Chalcogenide semiconductors, with their unique electronic and optical properties, have emerged as a pivotal material in this transformation. This book, *Chalcogenide Nanophotonics*, offers a comprehensive exploration of these remarkable materials, covering their fundamental physics, synthesis methods, optical phenomena, and cutting-edge applications in modern photonics.

Chalcogenides, comprising Group VI elements such as sulfur, selenium, and tellurium, have long fascinated researchers due to their tunable bandgaps, high refractive indices, and reversible phase transitions. These features make them indispensable for a wide range of applications, from ultrafast optical switches and nonvolatile photonic memories to infrared sensors and reconfigurable metamaterials. Their integration with silicon photonics and advanced nanofabrication techniques has bridged electronic and photonic functionalities, unlocking unprecedented opportunities in photonics.

This text is structured to guide readers from foundational principles to frontier innovations. Chapter 1 introduces the historical context and defining characteristics of chalcogenide semiconductors, emphasizing their role in advancing optoelectronic and thermoelectric technologies. Chapter 2 delves into the electronic band structure and material properties, elucidating how chemical bonding and lattice dynamics govern their optoelectronic behavior. Chapter 3 explores the intricate mechanisms of thin-film growth, offering insights into epitaxial techniques such as chemical vapor deposition and pulsed laser deposition, which are critical for achieving high-quality chalcogenide layers.

The optical properties of chalcogenides take center stage in Chapters 4–6, where we dissect dielectric functions, Raman spectroscopy, and emission mechanisms, while highlighting their applications in optical communications and integrated photonic devices. Chapters 7–8 venture into advanced topics, including chalcogenide-based photonic crystals and metamaterials, showcasing their potential for beam steering, perfect absorption, and chiral light manipulation.

Finally, Chapter 9 outlines future challenges and opportunities, from machine learning-driven material discovery to monolithic 3D integration for quantum photonics.

A distinctive feature of this book is its interdisciplinary approach, weaving together materials science, condensed matter physics, and photonic engineering. Each chapter integrates theoretical frameworks with practical case studies—such as phase-change memory devices leveraging GeSbTe alloys or GST-based metasurfaces for dynamic color displays—to illustrate the symbiotic relationship between material design and device performance. The inclusion of recent breakthroughs, such as van der Waals epitaxy for low-defect heterostructures and UV lithography for scalable metasurfaces, ensures relevance to both academic and industrial audiences.

We are deeply grateful to the many researchers whose pioneering work is cited throughout this text, and to our colleagues at Dalian University of Technology for their intellectual support. We also thank the funding agencies that enabled our exploration of chalcogenide nanophotonics, and extend special thanks to our students, whose curiosity and rigor continue to inspire innovation.

As we stand on the threshold of a new era in photonics—where intelligent systems seamlessly integrate light and matter—this book serves as both a roadmap and an invitation to researchers, engineers, and students alike, encouraging them to harness the infinite potential of chalcogenides. May it inspire new ideas, foster collaboration, and accelerate the development of transformative technologies.

Tun Cao, Ying Su, Yinghan Li, Hongze Gao
Dalian University of Technology
Liaoning Province, China

1 Introduction

As an important branch of research in two-dimensional semiconductor materials, chalcogenides have unique properties, diverse crystal structures, strong light–matter interactions, high carrier mobility, and other advantages. These excellent characteristics also make them candidate materials for solar cells, sensors, photodetectors, and other devices, which have aroused great research interest in the past few decades.

In this chapter, we define chalcogenide semiconductors and provide an overview of their historical development. The focus will be on the key properties, along with their applications across communication, sensing, and storage sectors. Through these introductions, the infinite potential of chalcogenide semiconductors in advancing electronic device technology is highlighted.

1.1 Definition of Chalcogenide Semiconductors

1.1.1 Basic Definition of Chalcogenide Semiconductors

Chalcogenide semiconductors are materials containing one or more chalcogen elements (sulfur, selenium, or tellurium) combined with other elements like metals or metalloids (e.g., germanium or arsenic). Their unique chemical and physical properties give them broad application prospects in semiconductor materials. These materials can usually be synthesized by simple chemical reactions or physical methods to form a wide range of crystal structures and morphologies. The main advantage of chalcogenide semiconductors is their excellent optoelectronic properties, including large nonlinear optical response, tunable bandgap, and good thermal stability. These properties enable them to show great potential for applications in nanophotonics, optical communications, optical storage, and other fields (Figure 1.1).

Since the beginning of the 20th century, chalcogenide semiconductors have gradually received attention from the scientific community. Early research mainly focused on the underlying physical properties of these materials. As technology advances, research is increasingly focusing on its performance in practical applications, particularly in the field of nanophotonics. Chalcogenide semiconductors have garnered attention due to their adjustable optical bandgap and high refractive index.

Chalcogenide Nanophotonics, First Edition. Tun Cao, Ying Su, Yinghan Li, and Hongze Gao.

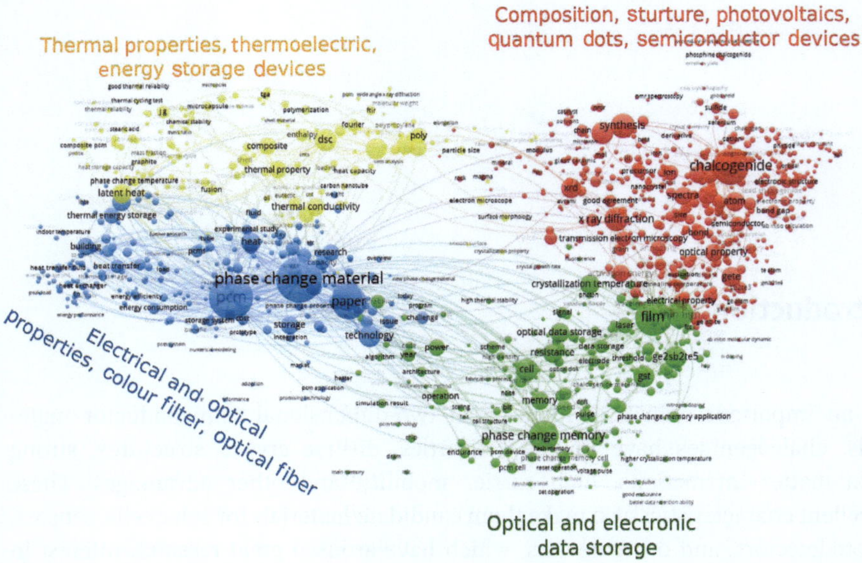

Figure 1.1 Semantic assessment of abstracts from the ISI Web of Science using the key word "chalcogenide". *Source:* Adapted from Cao et al. (2020).

Through bandgap engineering and phase transition modulation, these materials can realize the modulation of optical properties on a microscopic scale, which fully meets the special requirements of materials for modern optoelectronic devices. As an example, germanium-sulfur (GeS), a compound of germanium and sulfur, is a typical chalcogenide semiconductor with a low-energy bandgap suitable for photovoltaic conversion in the visible and infrared wavelengths. The composition of chalcogenides can be optimized by changing the ratio of metal elements to chalcogenides and by introducing other doping elements. This component modulation capability allows design flexibility for chalcogenides in optoelectronic devices. For example, lead sulfide (PbS), as an important narrow bandgap semiconductor material with excellent photoconductivity and thermoelectric properties, is widely used in fields such as photodetectors and thermoelectric converters. With the deepening of research, more and more novel ternary or quaternary sulfur compounds have been discovered, such as bismuth selenium (Bi_2Se_3) and bismuth alloy selenium ($Bi_2(SeTe)_3$). Chalcogenide materials possess excellent optical and nonstructural properties, making them applicable in emerging fields such as topological insulators and photoelectrocatalysis. Their complex composition lays a crucial foundation for investigating phase transitions, superconductivity, and nonlinear optical properties. Consequently, the study of chalcogenide semiconductors has become a significant area of focus in materials science and optoelectronic engineering.

In recent years, significant progress has been made in related research as methods for synthesizing chalcogenide semiconductors continue to mature. Currently, researchers have successfully applied chalcogenide semiconductors in a variety of optoelectronic devices, including optical switches, optical storage devices, and fiber-optic communication systems. Chalcogenide semiconductors have demonstrated excellent performance in cutting-edge applications such as optical field modulation and optical signal processing. Through an in-depth study of chalcogenide semiconductors, we can better understand their unique physicochemical properties and thus promote further development of related technologies.

1.1.2 Classification of Chalcogenides

Chalcogenides are compounds that have sulfur elements (e.g., sulfur, selenium, and tellurium) as their main components and are widely used in semiconductors, optoelectronic materials, and various types of new functional materials.

According to the valence state, chalcogenides can be roughly divided into the following categories: first, binary chalcogenides, such as molybdenum disulfide (MoS_2) and tungsten diselenide (WSe_2); second, ternary and multivariate chalcogenides, which contain not only sulfur but also other elements such as gallium, and indium, such as the ternary tin sulfide ($Sn_1-_xGe_xS$) and multivariate rare-earth sulfur compounds (RE_2S_3), which tend to have more complex structures and a wealth of physical properties. Among binary chalcogenides, the most common are transition metal disulfides (TMDs), which usually have a layered structure and can be prepared by mechanical stripping or chemical vapor deposition, etc., as single- or few-layer materials that exhibit excellent optoelectronic properties. For example, MoS_2, as a typical TMD, has become an important research object in the field of 2D materials due to its direct bandgap properties. WSe_2, on the other hand, implies excellent spin properties. In these materials, the interlayer interactions are relatively weak, allowing them to exhibit a wealth of physical phenomena such as superconductivity and photoconductivity when externally modulated. Ternary and multivariate chalcogenides have superior properties due to their compositional diversity. For example, Ge–Sb–Te system materials (GSTs) have been widely studied and applied in phase change memories, where they exhibit significant optical changes during the phase change process, enabling fast data access. Such alloys can also achieve bandgap modulation by adjusting the composition ratio, which provides new ideas for device performance optimization. Low-dimensional ternary chalcogenides are also gaining attention for thermoelectric, catalytic, and photodetector applications, where they typically exhibit high carrier mobility and excellent thermoelectric properties.

Chalcogenides can be classified into metal, nonmetallic, alloy, and composite chalcogenides based on their composition and structure. Metal chalcogenides, such as lead sulfide (PbS) and cadmium selenide (CdSe), typically exhibit excellent optoelectronic properties and superior conductivity and are widely used in optoelectronic devices and sensors. Nonmetallic sulfur compounds, such as hydrogen sulfide (H_2S) and boron sulfide (B_2S_3), excel in areas such as catalysis, gas sensing, and environmental monitoring. In addition, the study of alloy and composite chalcogenides provides a wide scope for the design of novel materials, such as modulating the electronic and optical properties of materials by changing the ratio of different compositions.

According to the crystal structure, chalcogenides can also be categorized into glassy and crystalline states. Glassy chalcogenides, including germanium sulfide

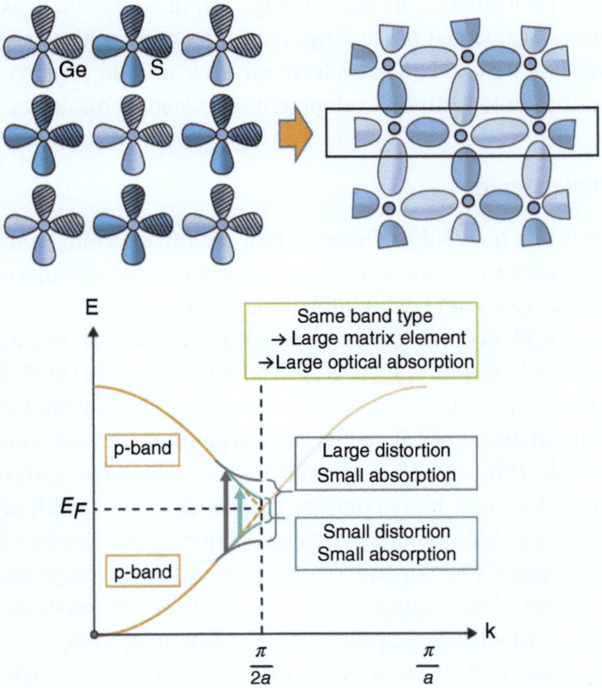

Figure 1.2 Visualization of bonding mechanisms in GeS and their impact on the electronic band structure. The upper section illustrates the atomic orbitals of Ge and S involved in bonding. σ-Bonds arise from overlapping p-orbitals, each occupied by approximately half an electron pair (ES ≈ 1), leading to a metallic band (represented by blue curves in the lower portion). Nevertheless, slight charge redistribution and/or minor Peierls distortions can induce a narrow bandgap.

(GeS) and arsenic sulfide (As_2S_3), as seen in Figure 1.2, are ideal for optical communication and storage devices because of their amorphous nature, which provides high optical transparency and nonlinear optical properties. In contrast, crystalline chalcogenides such as Sb_2S_3 and $Ge_2Sb_2Te_5$ exhibit exceptional optical and electronic properties due to their highly ordered structures. In nanophotonic applications, the phase transition and tunability of these materials make them key materials in novel photonic devices. For example, $Ge_2Sb_2Te_5$, as a phase-change material (PCM), can realize fast switching between solid and liquid states under different temperature and light conditions, providing the basis for high-performance memory and information processing devices.

Chalcogenides can be classified into various functional materials based on their application. These include photovoltaic materials, thermoelectric materials, photocatalysts, and more. For instance, chalcogenides used as photovoltaic materials have shown outstanding performance in solar cells and photodetectors, making them widely utilized in new energy technologies. Chalcogenides, such as bismuth selenide (Bi_2Se_3), are significant thermoelectric materials with vital applications in power generation and refrigeration due to their high thermoelectric efficiency.

Additionally, photocatalysts play an essential role in environmental remediation and energy conversion by facilitating chemical reactions using sunlight or other light sources, which helps achieve environmental purification and clean energy production. These categories provide a solid foundation for researching and developing new chalcogenides, paving the way for their applications in advanced fields such as nanophotonics.

In the realm of materials science and nanophotonics, chalcogenide semiconductors are of great importance due to their varied structures and extensive physical properties. A deeper exploration of the characterization and synthesis of different sulfur compounds, as well as the fundamental physical mechanisms behind them, is crucial for a more comprehensive grasp of their features and possible applications.

1.2 Basic Properties of Chalcogenide Semiconductors

1.2.1 Chemical Properties

Chalcogenides are compounds based on sulfur, selenium, tellurium, etc., which usually combine with other metal or semimetal elements to form diverse structures. These compounds are widely found in nature and have many chemical properties covering a wide range of aspects from atomic structure to chemical reactions. The stability and reactivity of chalcogenides are influenced by their chemical composition, crystal structure, and synthesis conditions. Many chalcogenide semiconductors demonstrate exceptional thermal stability and oxidation resistance under high temperatures and pressures, providing a theoretical foundation for their use in extreme environments. The reaction rate and mechanism of such materials can be optimized by adjusting the type or proportion of metals to meet the needs of industrial applications. The synthesis methods of chalcogenides, including solid-state reactions, solution methods, and vapor-phase deposition methods, provide multiple avenues to further enhance their properties. These chemical properties make chalcogenide semiconductors show great promise for future applications in nanotechnology research.

1.2.2 Physical Properties

The structural forms of chalcogenides—both amorphous and crystalline—play a crucial role in materials research, influencing not just their optoelectronic behavior but also offering distinct benefits across multiple applications.

Glassy chalcogenides are usually disordered during cooling and exhibit excellent nonlinear optical properties, making them suitable for fields such as optical communications and optical storage. In the glassy state, chalcogenides not only maintain high flexibility and plasticity but also have good optical transparency, which makes them have broad prospects in optical coatings and fiber applications.

Chalcogenides can form a variety of crystal structures, including cubic, hexagonal, and orthorhombic systems. These crystals typically exhibit high thermal conductivity and excellent electrical conductivity, which has led to increased interest in their use as thermoelectric materials. Specific chalcogenides, such as GeTe and Sb_2S_3, undergo phase changes at certain temperatures and pressures, significantly influencing their light response properties. The different crystal morphologies affect the bandgap and other optical properties of the material, which are essential for developing efficient optoelectronic devices. Additionally, structural changes during phase transitions enable these materials to display novel physical properties on the nanoscale, including the formation of defect states and the excitation of localized surface plasmon resonances (LSPR).

The mutual transformation between the glass and crystal morphology of chalcogenides is also an important research direction. By adjusting the temperature, pressure, and chemical composition, scientists can achieve a reversible transition from the crystalline to the glassy state. This transition affects the physical and chemical properties of the material, such as thermal conductivity and optical properties, and is therefore essential for understanding and utilizing the properties of sulfur compounds. In recent years, it has been shown that suitable doping and modulation can significantly increase the glass transition temperature and stability of chalcogenides, thus expanding their potential for applications in nanophotonics. Understanding the properties of glass and crystal forms of chalcogenides will provide a theoretical basis for the design and optimization of new materials and promote their further development in modern technology.

With the deepening of the research, the thermodynamic properties, photoelectric properties, and surface plasmon resonance of chalcogenides have been gradually revealed, which provides a theoretical basis for their expansion in different application fields. Especially for sensors and optical storage devices, chalcogenides are important candidates for emerging nano-optoelectronic devices due to their strong light absorption in specific wavelength bands and excellent electrical properties. By precisely controlling the physical properties of materials, scientists can develop more efficient and multifunctional optoelectronic devices to meet the growing market demand and technological challenges.

1.2.3 Optical Properties

Chalcogenides have garnered significant attention due to their unique optical properties, which show great potential for applications in nanophotonics. These materials are particularly notable for their nonlinear optical characteristics, enabling responses such as second harmonic generation (SHG) and self-focusing phenomena when subjected to external excitation. The nonlinear effects arise from their complex crystal structures and chemical compositions, making chalcogenides important in fields like optical frequency conversion and optical signal processing. The bandgap of chalcogenides can be tuned by altering the material's composition and structure. This bandgap engineering not only broadens their applications across a wide range of wavelengths but also allows for excellent light absorption and emission properties at specific wavelengths. The nonlinear optical properties and bandgap engineering of chalcogenide semiconductors will be explored in detail later.

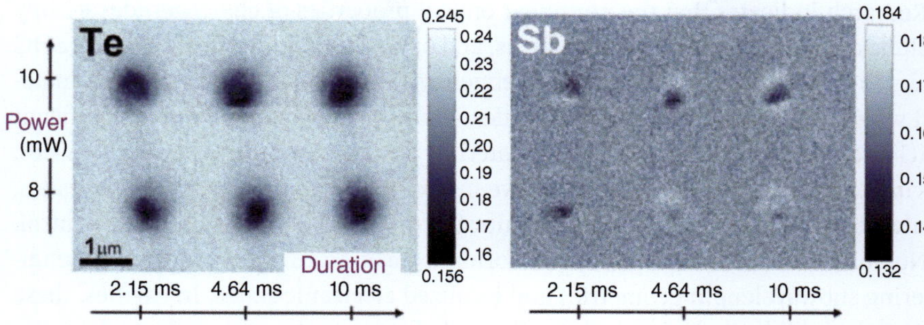

Figure 1.3 Auger electron spectroscopy with high resolution applied to Te-enriched GST during laser-driven crystallization. *Source:* Debunne et al. (2011) / IOP Publishing.

The phase transition behavior in chalcogenides also significantly affects their optical properties. When utilizing PCMs for photonics applications, chalcogenides can rapidly transition between different crystalline states, which in turn changes their optical refractive index. This feature provides a new implementation approach for optical storage, optical switches, and optical modulators. $Ge_2Sb_2Te_5$, for example, exhibits extremely high reflectance or transmittance variations during the phase transition process, which greatly enhances its applicability in optoelectronic devices. The high transparency and low optical loss of chalcogenides make them suitable for use in the manufacture of high-performance optical waveguides and optical fibers (Figure 1.3).

The crystal morphology (glassy and crystalline) of chalcogenides also has a significant impact on their optical properties. In the glassy state, chalcogenides typically have a wider spectral range and higher transparency, while in the crystalline state, their optical properties are affected by crystal orientation and defects, thus placing higher demands on the preparation process of the materials. By regulating the preparation conditions, chalcogenides with excellent optical properties can be obtained, thus promoting the realization and application of photonics-related technologies.

Chalcogenide semiconductors have garnered significant interest in nanophotonics owing to their exceptional nonlinear optical characteristics. These properties arise when a material's interaction with light depends nonlinearly on the incident intensity, typically scaling quadratically or beyond. In chalcogenides, such behavior originates from their large nonlinear optical susceptibilities and adjustable bandgaps, which position them as promising candidates for advanced optoelectronic applications. The nonlinear response of these materials relies on their unique crystal structures and chemical compositions, allowing them to exhibit several excellent properties in the field of optics. For example, binary or multivariate chalcogenides, such as GeTe and Sb_2S_3, can be tuned by adjusting the ratios of the constituent elements and the external conditions (e.g., temperature and pressure) to achieve the modulation of their nonlinear optical properties. This flexibility makes chalcogenides potentially practical in applications such as laser technology, light modulation, and frequency conversion.

Research indicates that the nonlinear optical properties of chalcogenides mainly include optical SHG, self-focusing effects, and optical switching effects. Optical SHG refers to the phenomenon where the frequency of light waves is doubled in a material when exposed to high light intensities. Many chalcogenides exhibit significant SHG effects, making them preferred materials for use in nonlinear optical devices. Owing to their large nonlinear refractive index and broad bandgap characteristics, chalcogenides show great potential for ultrafast optical pulse generation and control.

Notably, their nonlinear optical performance can be further improved by engineering subwavelength geometries and localized electronic states. In cavities, these materials exhibit strong optical nonlinear behavior, with potential mechanisms including multiphoton absorption, phase matching, and self-focusing phenomena. As an example, the nonlinear refractive index of chalcogenides increases significantly under intense laser irradiation, allowing for ultrafast optical switching and modulation. This phenomenon is attributed to the strong electron–phonon interactions in the material, as well as to the compounding and recombination processes in the excited states, which in turn lead to the enhancement of nonlinear optical effects. Therefore, these materials can find applications in lasers, ultrafast optics, and optical communications.

Utilizing the nonlinear optical properties of chalcogenides, the researchers have also developed novel perfect absorbers and optical filters. These devices enable efficient absorption and selective transmission of light at specific wavelengths, greatly improving the performance of photonic devices. For example, by employing structural modulation and chalcogenides of different compositions, it is possible to modulate the propagation characteristics of light in the design of the absorption peaks, thus realizing precise control of optical signals. With the in-depth study of the nonlinear optical properties of chalcogenides, the performance optimization and application prospects of these materials have been further expanded. For instance, recent studies have found that the nonlinear optical properties of these materials can be enhanced using different synthesis methods and post-processing techniques, thus broadening their application in modern photonic devices. These advancements not only enable performance optimization in optoelectronic devices but also offer viable pathways for implementing quantum computing architectures and photonic-integrated circuits. The demonstrated capabilities highlight chalcogenides' significant potential for nanophotonic applications, while establishing fundamental building blocks for next-generation ultrafast optical systems and advanced information processing platforms based on these materials.

Bandgap engineering involves the strategic modification of a material's composition, structure, or external parameters to achieve precise tuning of semiconductor bandgaps and their associated electronic properties. This technique is especially significant in chalcogenide semiconductors. The bandgap of chalcogenides is influenced not only by the elements that make up the material but also by factors such as crystal structure, orientation, and stress. By carefully regulating these parameters, it is possible to design the bandgap effectively.

The most common method in bandgap engineering is to change the constituent elements of the material. In the transition metal dichalcogenides MX_2, the bandgap properties can be significantly altered by substituting a different transition metal (M) or chalcogenide (X). For example, replacing rhenium (Re) in rhenium disulfide (ReS_2) with molybdenum (Mo) or tungsten (W) not only perturbs the atomic orbital interactions but also modifies the bandgap. Regulating the nonchemical composition of materials is an important approach, particularly in adjusting the number of layers present. The structure, whether it consists of single or multiple layers, has a significant impact on the bandgap of the material. For instance, single-layer MoS_2 acts as a direct bandgap semiconductor, while multilayer MoS_2 functions as an indirect bandgap semiconductor. Therefore, the bandgap can be precisely engineered through controlled modulation of material thickness and layer architecture.

In addition to the chemical composition and the number of layers, the engineering of the bandgap can be achieved equally well by the modulation of external conditions. Under high-pressure or low-temperature conditions, the bandgap of a material can change significantly due to stress- and temperature-induced changes in the crystal structure and adjustments in the electronic energy band structure. It has been found that under high pressure, certain sulfur compounds undergo a phase transition from semiconductors to metals and that the bandgap changes accompanying this phase transition allow them to exhibit different properties under different application conditions. Bandgap tuning through external stimuli (such as electric fields or optical excitation) enables dynamic control in photonic applications, particularly for optical modulators and high-speed switching devices. For example, by applying an electric field, the polarization of a material can be induced to modulate its energy band structure for rapidly controlling optical and conductive properties.

The research on the bandgap engineering not only improves the understanding of material properties but also provides a theoretical basis for the design of novel devices. In the field of optoelectronics and nanophotonics, the application of bandgap engineering has enabled these chalcogenide materials to exhibit good performance in optical communications, sensors, optical storage, and other areas. More importantly, developments in this area will drive future innovations in new, more efficient, and smarter optoelectronic devices. By further exploring the mechanisms involved in bandgap engineering, it will help to crack the limitations of the performance of existing materials and realize a wider range of material applications.

1.2.4 Optical Force

Optical forces refer to mechanical forces resulting from the interaction of light with matter, such as effects like optical tweezers and phototransformation. And the deformation mechanism of chalcogenide semiconductors under the action of optical forces is a hot spot of current research.

The main mechanisms of deformation of chalcogenide materials under optical forces include photo birefringence, optical torque, and photomobility. Photogenic birefringence is the phenomenon of birefringence that occurs when a material is exposed to light and stresses are generated within the material, resulting in a change in the refractive index. Light torque refers to the torque generated inside a material due to exposure to light, resulting in a rotational deformation of the material.

Photomobility means that light causes increased fluidity within the material, which results in deformation.

The application of chalcogenide semiconductors in optical force modulation is mainly based on their unique light–matter interaction mechanism and tunable optoelectronic properties. Taking phototropic phase transition–driven light field modulation as an example, chalcogenides (e.g., the Ge–Sb–Te system) can realize dynamic switching of refractive indices by amorphous–crystalline phase transitions in the presence of light. This characteristic, combined with the design of metasurfaces or photonic crystal microstructures, can generate a strong gradient electromagnetic field distribution in the local optical field, thereby inducing significant optical forces through the transfer of optical momentum. Reconfigurable optical tweezer systems developed with such materials enable subwavelength precision manipulation of nanoparticles. The lattice defects (e.g., sulfur vacancies) in two-dimensional transition metal chalcogenides (e.g., MoS_2, WS_2) significantly enhance the photoluminescence efficiency and amplify the light–matter interactions through the strong coupling of localized exciton states to the optical field. The local electromagnetic field intensity of defect sites may increase by more than 10 times, thereby enhancing the ability of optical forces to trap or drive micro- and nanostructures. In addition, photogenerated carriers generated by chalcogenide nanocrystals (e.g., Cu_2S–In_2S_3) in the presence of light can trigger lattice stress changes, leading to reversible deformations (e.g., bending or swelling) at the surface or interface of the material. This optomechanical coupling effect can directly convert light energy into mechanical displacement, providing a power source for optically driven actuators or microfluidic chips. Heterojunctions of chalcogenides (e.g., $Sb_{2-x}Bi_xS_3$) can dynamically adjust the direction and strength of the optical force by modulating the optical, thermal, and electrical multifield coupling. For example, in X-ray detection, the synergistic effect of photogenerated carriers and thermal expansion can optimize the sensitivity of the optical force in response to film deformation.

1.3 Application of Chalcogenide Semiconductors

1.3.1 Optical Communication

Optical communication technology is an advanced communication method that utilizes light as an information carrier for transmission, with significant advantages such as high speed, large capacity, and strong anti-interference ability. Chalcogenide semiconductors have shown their unique advantages in the field of optical communications, mainly due to their excellent optoelectronic properties and tunable bandgap. In recent years, the unprecedented growth in data transmission demands has revealed inherent limitations in conventional electronic communication systems, presenting critical scaling challenges. And optical communication has emerged as a key technology in addressing this issue, thanks to its high bandwidth and low latency. Chalcogenide materials, such as GeTe and Sb_2S_3, have unique advantages in this field due to their exceptional optical properties and tunable bandgap characteristics. These materials can effectively modulate their reflection, transmission, and absorption properties within specific wavelength ranges, enabling efficient modulation and demodulation of information transmission. Consequently, they can be utilized in the manufacture of essential devices, including optical modulators, optical switches, and optical waveguides.

Chalcogenide semiconductor materials offer new solutions to the challenges of conventional optical communication technologies in terms of spectral efficiency, transmission distance, and energy consumption. In optical communication systems, the application of chalcogenides is mainly reflected in their nonlinear optical properties and optical phase transition characteristics. The stimuli-responsive optical property modulation (temperature/pressure-dependent) in these materials significantly enhances gain while maintaining signal integrity during optical processing—essential characteristics for next-generation optical modulator design. For example, some chalcogenides undergo an insulator-to-metal transition during a phase transition, achieving effective switching of optical signals. In addition, by utilizing the phase transition properties of chalcogenides, efficient optical modulation can be achieved over a wider wavelength range, which means it can better adapt to the simultaneous transmission of multiple wavelength signals in optical communication systems. This characteristic makes them have great potential for application in high-speed optical communication, which can meet the bandwidth and speed requirements of future ultra-high-speed networks.

In addition, chalcogenide-based photodetectors offer significant sensitivity and range for optical communications. By adjusting the composition and structure of these materials, we can flexibly tune the bandgap, which allows for precise control of their optoelectronic properties. This enables efficient detection of optical signals at specific wavelengths. These enhancements not only improve the fidelity of information transmission and increase device compatibility but also broaden the application areas of optical communication networks, including fiber optic and wireless optical communications. In the future, as research on chalcogenide semiconductors deepens, we anticipate the development of more efficient and intelligent solutions for optical communications, further advancing communication technology.

1.3.2 Optical Storage

Optical storage technology, as an important part of the information storage field, relies on the optical properties of the material to realize the reading, writing, and storage of data. In the modern information society, with the explosive growth of data volume, traditional storage media face the double challenges of capacity and speed. Optical storage technology is gradually gaining attention due to its advantages of high density, high speed, and long-term stability. Chalcogenide semiconductors, as a new type of PCM, exhibit excellent optical properties and non-volatility, providing new possibilities for developing optical storage technology. These materials enable efficient modulation and storage of optical signals by changing their phase state, with advantages such as fast, reversible, and high-density storage. PCMs are characterized by their reversible structural transformations between distinct physical states under controlled external stimuli. A representative example is the laser-induced amorphous-to-crystalline phase transition, where the crystallization kinetics can be precisely modulated through optimization of laser parameters, including pulse duration and energy density. This controllable phase-switching behavior enables the development of non-volatile memory devices capable of reliable data storage operations.

Chalcogenide semiconductors, especially the Ge–Sb–Te (GST) series materials, are used in applications such as optical disks, photoresists, and high-performance two-dimensional materials due to their unique phase transition properties that give optical storage devices higher performance. These materials can rapidly change their crystal structure in the presence of light or an electric field, leading to the phenomenon of phase transitions, which enables the storage and reading of data. Laser irradiation can transform materials from an amorphous phase to a crystalline state, or vice versa, greatly improving the speed of data reading and writing. $Ge_2Sb_2Te_5$, for example, is a chalcogenide with excellent phase transition properties that is widely used in next-generation optical storage technologies. Due to its low-phase transition temperature, it can realize a fast-phase transition at room temperature, and thus the reaction speed in optical storage media is greatly improved. Thanks to their high thermal stability, chalcogenides exhibit excellent durability in information storage and are suitable for long-term storage applications, especially in optical discs, erasable optical discs, and flash memory technology. In addition, the tunability of chalcogenide semiconductors provides more scope for the design of optical storage devices. Researchers can optimize the optical properties of materials by changing their composition to meet the constantly increasing storage demands (Figure 1.4).

In recent years, the size of optical storage devices has been continuously reduced, greatly improving the density of data storage. Nanophotonic applications of chalcogenide semiconductors can further advance optical storage technology. Nanostructural modification provides an effective route to simultaneously engineer optical and electronic properties in functional materials, forming the physical basis for developing disruptive optical storage technologies. Through optical modulation technology, chalcogenide materials can realize reliable writing and reading of data in a shorter time, which effectively improves the response speed of the storage device. Although chalcogenides show promising applications in optical storage, they still face many challenges, such as material stability and durability. With further research, improving the stability, thermal conductivity, and light-absorbing ability of chalcogenide materials is expected to drive further development of optical storage technology. In the future, chalcogenide semiconductors will be more widely used in optical storage devices, providing strong support for the realization of faster and more efficient information storage.

1.3.3 Sensing Technology

The evolution of sensing technologies has enabled novel implementations of chalcogenide semiconductors, whose distinctive optoelectronic characteristics— including tunable bandgaps, pronounced thermo-optic effects, and phase-dependent

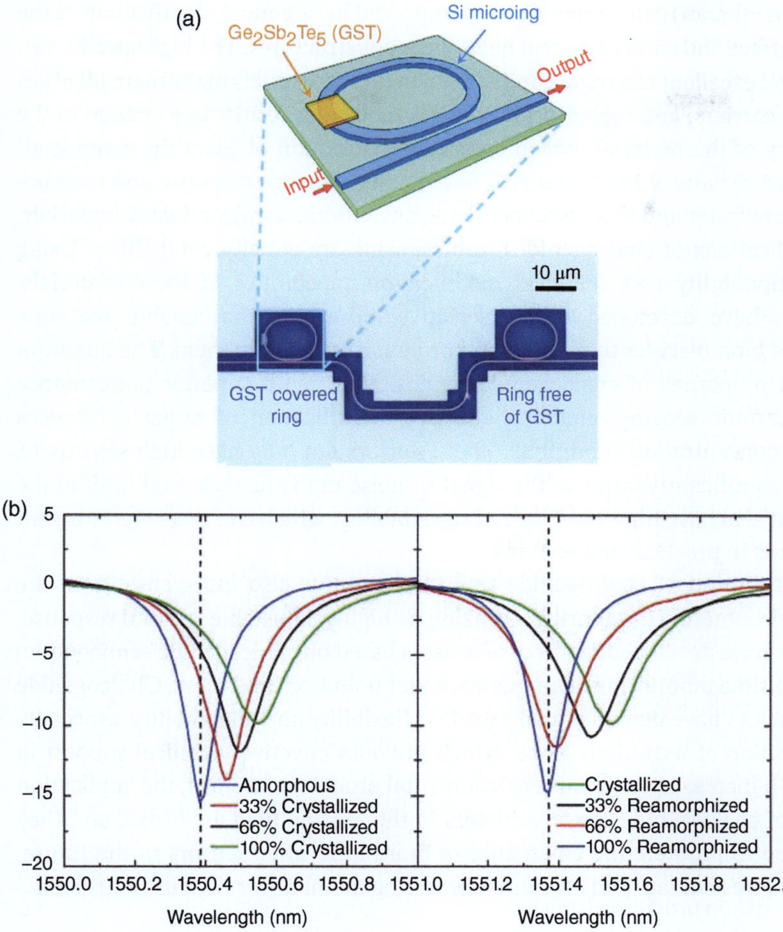

Figure 1.4 (a) A tunable photonic switch architecture comprising a Si racetrack resonator hybrid integrated with a $Ge_2Sb_2Te_5$ phase-change thin film, evanescently coupled to a Si waveguide. (b, c) Dynamically monitored transmission spectra demonstrating reversible optothermal switching characteristics during crystallization and reamorphization. *Source:* Rudé et al. (2013) / AIP Publishing LLC.

conductivity—make them ideal for advanced sensor designs. Nanophotonic engineering has particularly enhanced its detection capabilities, as evidenced by successful deployments in biochemical assays and ecological surveillance systems. The adjustable bandgap characteristics and nonlinear optical effects of these semiconductor materials give them superior sensitivity and response speed in low-concentration analysis and real-time monitoring processes. For example, two-dimensional chalcogenides such as molybdenum disulfide (MoS_2) and tungsten diselenide (WS_2) are widely used in gas sensors and biosensors due to their high mobility and excellent sensitivity.

Chalcogenide semiconductors have demonstrated exceptional performance in gas sensing. The selectivity and sensitivity to specific gases (e.g., ammonia, nitrogen dioxide, and ethanol) can be significantly improved by chemical modification of the material surface and construction of heterogeneous structures. The high specific surface area and excellent electrical conductivity of these materials make them ideal gas adsorption carriers, and the adsorption of gas molecules results in a change in the conductivity of the material, which enables the detection of gases. In many studies, modified sensors of MoS_2 and WS_2 have shown superior response and recovery times that are faster and more efficient than conventional semiconductor materials.

The applications of chalcogenides in biosensing are equally compelling. Using the biocompatibility and chemical modification capabilities of these materials, researchers have developed highly sensitive biosensors that enable real-time detection of biomolecules (e.g., DNA and proteins) by labeling them. The quantum dot optical properties of chalcogenides endow them with superior performance in optoelectronic sensing, enabling effective identification of target substances under low concentration conditions. These sensors not only have high sensitivity, which can significantly improve the signal-to-noise ratio and detection limit of the sensors, but also have high stability and repeatability, which can meet the stringent requirements in practical applications.

The contribution of chalcogenide semiconductors is also indispensable in the field of environmental monitoring. Utilizing its highly adjustable spectral response, researchers have developed a variety of sensors based on chalcogenide semiconductors for real-time monitoring of air quality, water pollution, and more. Chalcogenide semiconductors have demonstrated excellent flexibility and adaptability, especially in the detection of hazardous gases, which provides effective technical support in dealing with increasingly serious environmental problems. In short, the application prospects of chalcogenide semiconductors in the sensing field are broad, and they are expected to promote the emergence of more innovative sensors in the future, providing more accurate and efficient monitoring methods for various industries.

1.3.4 Other Applications

The applications of chalcogenide semiconductors in nanophotonics extend beyond conventional optical communications, data storage, and sensing systems. Their distinctive material characteristics facilitate diverse implementation possibilities across emerging technological frontiers. With the deepening of scientific research and the continuous development of related technologies, other applications of chalcogenides are gradually revealing their importance and market value. For example, in the field of chiral photonics, chalcogenide semiconductors also demonstrate enormous potential for applications. By constructing chiral nanostructures, chiral selective absorption and emission of light can be realized, which provides new ideas for the design of novel optical devices. In terms of achieving efficient chiral optical functionality, chalcogenide materials can optimize optical performance by adjusting their energy band structure on demand due to their bandgap engineering. This certainly offers more possibilities for realizing applications such as quantum communications, optical valves, and new laser sources.

Chalcogenide semiconductors have also received a lot of attention as excellent perfect absorbers. By modulating its microstructure and optical properties, it is possible to construct devices that can realize almost complete absorption of light of specific wavelengths, which are widely used in photothermal conversion, heat dissipation, and stealth technology. The design of such perfect absorbers not only improves the efficiency of energy utilization but also brings breakthroughs in areas such as photovoltaic conversion and energy storage. Some chalcogenide coatings can achieve nearly 100% light absorption efficiency across a wide spectral range, significantly enhancing the performance of photoelectric conversion devices. This advancement lays the groundwork for developing efficient photovoltaic equipment and infrared detectors. By combining sulfur compounds with other materials, it is possible to further improve their absorption properties. This approach can help meet the increasing demand for energy conversion and storage while also reducing production costs and material waste. In the future, these applications are expected to drive advancements in clean energy technology and contribute to sustainable development.

With the increasing demand for quantum computing and information processing, the prospect of chalcogenide semiconductors for applications in quantum optics cannot be underestimated. Based on its special electronic properties and tunable optical behavior, novel quantum information processing devices can be designed and developed. The performance of these devices depends on the quantum state manipulation of chalcogenides, which provides new solutions in areas such as quantum communication and quantum key distribution. It is expected to realize high-efficiency and low-energy consumption quantum computing platforms in the future, which will further promote the revolution in information technology and communication. The diverse applications of chalcogenide semiconductors are opening new directions and opportunities for the development of nanophotonics and materials science.

1.4 Research Progress of Chalcogenide Semiconductors

Owing to their special physical and chemical properties, chalcogenide semiconductors have shown great potential for applications in the field of nanophotonics, and numerous studies have been conducted around them.

On recent advances in chalcogenide semiconductors for nanophotonics, Tripathi et al. reviewed recent advancements in the chalcogenide PCMs in nanophotonics, covering various aspects from fundamental principles to practical applications, with focus on the advanced geometric engineering, the emergent functional capabilities, and the fundamental property modulation mechanisms governing light–matter interactions (Tripathi et al. 2023). Abdollahramezani et al. found that tunable nanophotonics can be achieved through chalcogenide PCMs. In the article, the authors provided an overview of the developments and trends in tunable metasurfaces and photonic integrated circuits using chalcogenide PCMs. The defining material properties, reversible structural transition processes, and pronounced thermo-optic coupling influence of widely studied categories of chalcogenide PCMs were outlined. Cen et al. (2019) provided an in-depth discussion on the fundamental principles and applications of phase transition material photonics of chalcogenides. Particularly, the developments of metal–chalcogenide–metal–trilayered plasmonic nanostructures are highlighted. The review pointed out the direction for the chalcogenide PCMs in growing areas of photonics. On the unique properties of semiconductors of chalcogenide semiconductors, Wuttig et al. revisited the correlations between the chemical bonding configurations and macroscopic properties, providing a theoretical basis for fundamentally optimizing the properties of the materials. Chew et al. focused on chalcogenide active photonics to provide new ideas for novel PCMs with enhanced spectral tunability across visible-to-near-infrared wavelengths through experimental and theoretical analysis. Martin-Monier et al. reviewed the durability of chalcogenide optical PCMs, discussed the factors influencing the crystallization and re-amorphization of several PCMs, examined their failure mechanisms, and formulated design rules for enhancing the cycling durability of these compounds. The review is critical for device stability and lifetime in practical applications.

In terms of related applications and preparation engineering, Cao et al. conducted a study on photonic GeSbTe phase transition metamaterials and their applications, revealing their unique advantages and potential application scenarios in the field of metamaterials. Piccinotti et al. optimized the properties of chalcogenide semiconductor alloys for nanophotonics applications by stoichiometrically engineering them to provide new strategies for material preparation.

The researchers have achieved remarkable results in the characterization of chalcogenide semiconductors and nanophotonic applications, and the multidimensional research from basic theory to practical applications has laid a solid foundation for the development of this field. However, there are still challenges, such as further optimization of material properties and expansion of new application areas, awaiting further exploration by researchers.

After understanding the history and conceptual framework of chalcogenide semiconductors and grasping their unique optoelectronic properties and relevance in modern photonics, the subsequent chapters of *Chalcogenide Nanophotonics* are structured to guide readers through a comprehensive progression from foundational principles to frontier innovations. Chapter 2 dissects electronic band theory, intrinsic material behaviors, and synthesis methodologies for chalcogenide thin films, providing critical insights for material optimization. Chapter 3 expands into deposition techniques, comparing chemical vapor deposition, thermal evaporation, and sputtering to address scalability and quality control. Chapter 4 transitions to optical phenomena, correlating macroscopic electrodynamics with advanced spectroscopic tools like Raman and emission spectroscopies. Chapters 5–8 pivot toward applied nanophotonics: optical communications (Chapter 5) analyze chalcogenide fibers and nonlinear effects; integrated photonic circuits (Chapter 6) explore memory devices and display technologies; photonic crystal platforms (Chapter 7) detail cavity and waveguide designs; and metamaterials (Chapter 8) showcase applications in optical switching, perfect lenses, and beam steering. By providing researchers with theoretical and practical strategies, we expect that chalcogenide semiconductors will occupy an increasingly important place in the future of nanophotonics.

2

Fundamentals of Chalcogenide Semiconductors

Chalcogenide semiconductors, composed of elements belonging to the Group VI family (O, S, Se, Te, and Po), represent a significant class of materials with unique physical and chemical properties. Due to these exceptional characteristics, chalcogenide semiconductor compounds have gained extensive attention and research in fields such as optoelectronics, thermoelectric materials, and phase-change memory devices. Chalcogenides have become indispensable materials in modern optoelectronic devices and next-generation storage technologies, owing to their outstanding electronic and optical properties, along with their highly tunable bandgaps (BGs). For example, in phase-change memory devices, the swift phase change from amorphous to crystalline states—accompanied by dramatic changes in conductivity—relies on modifications to the material's band structure. The amorphous phase typically exhibits localized electronic states and higher resistivity, while the crystalline phase arranges charge carriers in an orderly fashion, enabling efficient conduction. Moreover, the optical properties of chalcogenides further emphasize their uniqueness. Many of these compounds possess high refractive indices and wide transparency ranges, extending into the infrared spectrum, making them ideal candidates for photonic applications such as optical waveguides and infrared sensors. These characteristics are attributed to the polarization properties of heavy chalcogen atoms (e.g., tellurium), which enhance light–matter interactions.

This chapter will delve into the fundamental properties of chalcogenide semiconductors, with a focus on their electronic band structure, material characteristics, and the synthesis and characterization methods for chalcogenide thin films. A comprehensive grasp of these fundamental properties is essential for driving advancements in chalcogenide-based semiconductor technologies. Such insights not only offer theoretical understanding of their application potential but also provide practical strategies for material design and process optimization in research and development endeavors.

2.1 Theory of Electronic Band Structure

The electronic band structure of chalcogenide semiconductors plays a pivotal role in determining their electrical, optical, and thermal properties. The energy bands

Chalcogenide Nanophotonics, First Edition. Tun Cao, Ying Su, Yinghan Li, and Hongze Gao.

in these materials are influenced by their atomic composition, crystal structure, and the degree of hybridization between the orbitals of the constituent elements. Unlike conventional semiconductors such as silicon, chalcogenides exhibit a range of unique band structures, with many displaying direct or indirect BGs, depending on their composition and phase. Understanding the electronic band structure is essential for tailoring the material's properties for specific applications, such as optoelectronics and phase-change memory.

2.1.1 Band Theory

The development of band theory marks a transformative chapter in the history of solid-state physics. Emerging as the most profound application of quantum mechanics and quantum statistics in the study of solids, this theory revolutionized our understanding of electronic behavior in crystalline materials. Prior to its advent, the classical electron theory and Sommerfeld's free electron theory provided incomplete and often inconsistent explanations for the properties of metals. Band theory, however, introduced the concept of energy bands resulting from the periodic potential of atomic lattices, offering a comprehensive framework that not only addressed the shortcomings of earlier theories but also extended its applicability to semiconductors and insulators. This breakthrough paved the way for modern electronics, enabling the development of transistors, diodes, and other semiconductor devices. As such, band theory is not merely a theoretical cornerstone but a catalyst for technological innovation that continues to shape the modern world.

The formation of energy bands can be interpreted by analyzing the overlap of atomic orbitals in a solid. As isolated atoms combine to build a crystal, the electrons in their outer shells interact with the electrons of neighboring atoms. This interaction causes the specific energy levels of the isolated atoms to split into multiple sublevels. As more atoms are added to the crystal, the energy levels continue to split, forming energy bands. In a simple model, the energy levels of individual atoms are broad and continuous due to the overlap of atomic orbitals, creating a range of energies that electrons can occupy. For instance, in the case of semiconductor materials, the atomic energy levels of the constituent atoms split into allowed bands of energies, separated by a region where no allowed energy states exist—this region is called the BG.

In solids, the energy bands are divided into three primary groups according to their energy levels and the behavior of electrons within them:

- **Valence Band (VB):** Defined as the highest energy band entirely filled with electrons when the temperature is at absolute zero. The electrons present in the VB are bound to atoms and cannot move freely. These electrons are typically responsible for the chemical bonding between atoms in a solid.
- **Conduction Band (CB):** The energy level following the VB is known as the conduction band. Electrons occupying this band can move freely throughout the solid and are responsible for electrical conductivity. The CB contains a partial population of electrons in conductors and insulators, while in semiconductors, it may be empty or partially filled depending on temperature and doping conditions.

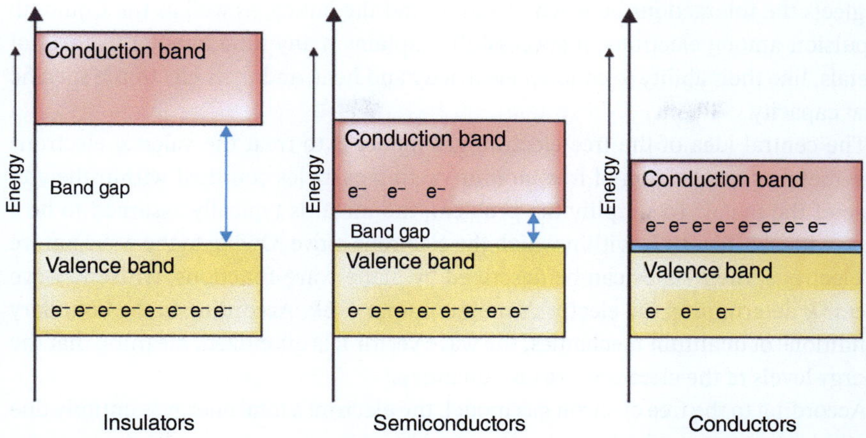

Figure 2.1 Band structure diagrams of insulators, semiconductors, and conductors.

- **Bandgap (BG):** The BG refers to an energy interval between the valence and conduction bands that contains no available electron states. The electron must absorb this energy to transition from the VB to the CB. The size of the BG is a critical factor that determines the electrical properties of a material.

The band structure of solid materials is fundamental to understanding their electrical properties, and it provides a clear explanation of the physical differences in electrical conductivity between conductors, semiconductors, and insulators. The conductivity of a material primarily depends on the number of electrons in its CB. When electrons gain enough energy to transition from the VB to the CB, they can move freely between energy bands, thus enabling electrical conduction.

In metallic conductors, the energy gap is extremely small (usually less than 1 eV), as shown in Figure 2.1a. Electrons are able to obtain enough thermal energy to enter the CB at room temperature, which results in excellent electrical conductivity. In contrast, insulating materials exhibit a large energy gap (typically greater than 9 eV), hindering electrons from acquiring sufficient energy to reach the CB at room temperature, thereby preventing electrical conduction. Semiconductors lie between conductors and insulators. Their BG is usually between 1 and 3 eV. This moderate BG allows semiconductors to exhibit unique and controllable electrical properties. By externally exciting the material (such as through light or thermal energy) or by adjusting the gap through methods like doping or applying an electric field, the conductivity of semiconductors can be effectively controlled, making them essential for a wide range of electronic applications.

2.1.1.1 Free Electron Gas Model

In the early development of solid-state physics, the free electron gas model provided a simple and intuitive framework for understanding the behavior of electrons in metals. This model assumes that the valence electrons in a metal are completely detached from the influence of atomic nuclei and can move freely throughout the crystal, akin to the motion of gas molecules in a container. Although this model neglects the interactions between electrons and the lattice, as well as the Coulomb repulsion among electrons, it successfully explains many fundamental features of metals, like their ability to conduct electricity and heat, and their electronic specific heat capacity.

The central idea of the free electron gas model is to treat the valence electrons in a metal as a collection of free, noninteracting particles confined within the volume of the metal. To simplify the problem, the metal is typically assumed to be a cube with side length L, within which the electrons move. Owing to the wave nature of electrons, their states can be described by plane wave functions, with the wave vector $\boldsymbol{k}$ determining the electron's momentum $p = \hbar\boldsymbol{k}$. According to the boundary conditions of quantum mechanics, the wave vector $\boldsymbol{k}$ is quantized, meaning that the energy levels of the electrons are also quantized.

According to the free electron gas model, the electron's total energy is entirely due to its kinetic energy, which can be described by

$$E(\boldsymbol{k}) = \frac{\hbar^2\boldsymbol{k}^2}{2m} \tag{2.1}$$

where $\hbar$ stands for the reduced Planck constant, m refers to the electron mass, and $\boldsymbol{k} = |\boldsymbol{k}|$ is the magnitude of the wave vector. This formula reveals that the electron's energy increases proportionally with the square of the wave vector, forming a parabolic energy band structure.

To describe the collective behavior of many electrons, statistical mechanics must be introduced. At absolute zero temperature, electrons fill the available states starting from the lowest energy level until all electrons are accommodated. The highest occupied energy level is called the Fermi energy E_F, and the corresponding wave vector is known as the Fermi wave vector $\boldsymbol{k}_F$. The Fermi energy is a critical parameter in the free electron gas model, as it determines many physical properties of metals. At zero temperature, the distribution of electrons follows the Fermi-Dirac distribution, meaning that all states with energies below E_F are occupied, while those above E_F are completely empty.

A key conclusion from the free electron gas model is the determination of the density of states (DOS), which quantifies how many electron states exist per unit energy. In three dimensions, this value scales with the square root of energy:

$$g(E) \propto \sqrt{E} \tag{2.2}$$

This relationship is particularly important in explaining the electronic specific heat of metals. At finite temperatures, only electrons near the Fermi energy can be thermally excited to participate in conduction, leading to the experimentally observed proportionality between electronic specific heat and temperature.

However, the free electron gas model has notable limitations. For example, it cannot explain the existence of insulators and semiconductors, nor can it describe the interactions between electrons and lattice vibrations (phonons). To more accurately describe the behavior of electrons in real materials, more sophisticated theories, such as band theory and density functional theory, must be employed. Nevertheless, as a cornerstone of solid-state physics, the free electron gas model has provided essential insights and a foundation for the development of subsequent theories.

2.1.1.2 Bloch's Theorem

In the annals of solid-state physics, one revolutionary theory stands out, remarkably first proposed by a young researcher. The year 1928 witnessed a paradigm shift when 23-year-old Felix Bloch, in his seminal doctoral thesis "Quantum Mechanics in Crystal Lattices," pioneered the energy band theory to explain electrical conduction in metals. This concept underwent significant advancement when Alan Wilson, in 1931, demonstrated that the critical difference between metals and insulators depends on the complete filling of energy bands, thereby establishing the theoretical framework for understanding semiconductor properties (Kane 1957). Through subsequent decades, the band theory has been meticulously developed and enhanced by generations of physicists, becoming an indispensable tool in materials science and condensed matter physics.

The dynamics of electrons in a crystalline lattice can, in principle, be reduced to solving the single-electron quantum wave equation under the effect of a periodic potential. However, obtaining explicit solutions is inherently difficult due to the complexity of the problem. Additionally, the periodic potentials in different crystals exhibit diverse forms and strengths, requiring tailored methods for each specific case. Bloch pioneered the study of single-electron wavefunctions in periodic potentials, establishing the general properties of these states. His work has served as a cornerstone for understanding electronic behavior in crystals and has provided essential guidance for solving practical problems in solid-state physics.

Bloch's Theorem: In a periodic potential, the electron's wave function maintains a plane wave structure, while its amplitude varies periodically in accordance with the lattice periodicity.

In an idealized, defect-free crystal, the atomic cores are assumed to be fixed in a periodic arrangement at their equilibrium positions. Within this framework, each electron is subjected to a potential field that arises from the combined effects of the average potential of all other electrons and the potential of the atomic cores. Under the periodic field approximation, the potential experienced by an electron is periodic in nature. This conceptual model is known as the periodic field model.

To describe how an electron behaves within a periodic potential, the corresponding Schrödinger equation is given by

$$\left[-\frac{\hbar^2}{2m}\nabla^2 + U(r)\right]\psi(r) = E\psi(r) \tag{2.3}$$

Here, $U(r) = U(r + \boldsymbol{R}_l)$ represents the periodic potential, and $\boldsymbol{R}_l = l_1 a_1 + l_2 a_2 + l_3 a_3$ denotes the lattice vector of the crystal.

Thus, the solution to the equation is expected to have the following form:

$$\psi_k(r) = e^{ik\cdot r} u_k(r) \tag{2.4}$$

where $\psi_k(r)$ is the electron wavefunction, $\boldsymbol{k}$ is the crystal wave vector, $e^{ik\cdot r}$ is a plane wave term, and $u_k(r)$ is a function that has the same periodicity as the crystal lattice, satisfying $u_k(r+\boldsymbol{R}) = u_k(r)$ for any lattice vector $\boldsymbol{R}$. This outcome, termed Bloch's theorem, characterizes the intrinsic nature of wavefunction solutions in the context of periodic potentials.

In other words, Bloch discovered that, regardless of the specific functional form of the periodic potential, the wave function of a single electron moving in such a potential is no longer a simple plane wave. Instead, it becomes a modulated plane wave, where the amplitude is no longer constant but varies periodically with the crystal's lattice period. Such a wave function, denoted as $\psi_{\boldsymbol{k}}(r) = e^{i\boldsymbol{k}\cdot r}u_{\boldsymbol{k}}(r)$, is termed a Bloch wave function or Bloch wave. The electron characterized by this wave function is known as a Bloch electron. This fundamental result is called Bloch's theorem. Alternatively, Bloch's theorem can be expressed as follows:

$$\psi_{\boldsymbol{k}}(\vec{r} + \overrightarrow{\boldsymbol{R}_n}) = e^{i(\vec{\boldsymbol{k}}\cdot\overrightarrow{\boldsymbol{R}_n})}\psi_{\boldsymbol{k}}(\vec{r}) \tag{2.5}$$

This result indicates that the wave functions at equivalent points in different unit cells differ solely by a phase factor $e^{i(\vec{\boldsymbol{k}}\cdot\overrightarrow{\boldsymbol{R}_n})}$. Since the phase factor does not alter the wave function's amplitude, the probability of locating an electron at corresponding points in different unit cells remains the same. This phenomenon is a manifestation of the crystal's inherent periodicity.

Bloch's theorem is rooted in the translational symmetry of the crystal lattice. Because the potential experienced by an electron is periodic, the Hamiltonian of the system is unchanged under translations by lattice vectors $\boldsymbol{R}$. This symmetry imposes a constraint on the form of the wave function: it must satisfy $\psi(r+\boldsymbol{R}) = e^{i\boldsymbol{k}\cdot\boldsymbol{R}}\psi(r)$, where $\boldsymbol{k}$ is a wave vector. This relationship ensures that the wave function's probability density is invariant under lattice translations, reflecting the crystal's periodicity. As a result, the wave function must be a plane wave whose amplitude is modulated by a function with the same periodicity as the lattice, consistent with Bloch's theorem.

Serving as a quantum number, the wave vector $\boldsymbol{k}$ characterizes electronic states in Bloch theory, where both the wavefunction and the energy levels vary as functions of $\boldsymbol{k}$. Each distinct value of $\boldsymbol{k}$ represents a unique electronic state.

For free electrons, physically, the wave vector $\boldsymbol{k}$ corresponds to the eigenvalue of momentum for a free electron. However, in the context of Bloch electrons moving in a periodic potential, the Bloch wave function is not an eigenfunction of the momentum operator. Instead, it is an eigenfunction of the Hamiltonian describing the electron's motion in the periodic potential. Thus, $\boldsymbol{k}$ does not represent the true momentum of the Bloch electron. Despite this, $\boldsymbol{k}$ possesses the dimensions of momentum and exhibits momentum-like behavior in various scenarios, such as the motion of electrons under external fields or their interactions with phonons and photons. Due to this behavior, $\boldsymbol{k}$ is commonly termed the "**crystal momentum**" or "**quasi-momentum**" of the electron.

In a periodic potential, the number of electron eigenstates is governed by the permitted values of $\boldsymbol{k}$. The periodicity of the crystal lattice and the application of periodic boundary conditions constrain k to discrete values within the first Brillouin zone. Specifically, $\boldsymbol{k}$ can take N distinct values, where N is the number of unit cells in the crystal. Consequently, each energy band can accommodate a maximum of $2N$ electrons, considering the spin degeneracy of the electronic states.

The plane wave factor $e^{i\boldsymbol{k}\cdot r}$ explicitly demonstrates that electrons propagating within a crystal lattice are no longer localized around individual atoms, but rather become delocalized throughout the entire crystalline structure. Such electrons are classified as itinerant electrons. Their motion adopts a form analogous to that of propagating plane waves. The periodic function $u_k(r)$ serves to modulate the amplitude of these waves, inducing periodic oscillations as the wave transitions from one unit cell to the next. However, this modulation does not alter the fundamental propagating wave characteristics inherent in the electronic states.

Electrons in crystals: $\psi_k(r) = e^{ik\cdot r}u_k(r)$
Free electron: $\psi_k(r) = Ae^{ik\cdot r}$
Isolated atom: $\psi(r) = Cu(r)$

The wave function of an electron in a crystalline solid can be understood as a hybrid state, intermediate between that of a completely free electron and that of an electron bound to an isolated atom. In the idealized case of an electron moving entirely freely within the crystal, $u_k(r) = A =$ const.; in contrast, for an electron entirely confined to a single atomic site, $e^{ik\cdot r} = C =$ const. However, the actual behavior of electrons in a crystal lies between these two extremes, resulting in a wave function of the form $\psi_k(r) = e^{ik\cdot r}u_k(r)$. The characteristics of the periodic function $u_k(r)$ are inherently linked to the strength of the electron–lattice interaction, serving as a key indicator of the degree of coupling between the electron and the periodic potential of the crystal.

From an energetic perspective, if an electron is confined solely to intra-atomic motion (as in the case of an isolated atom), its energy assumes discrete, quantized levels. Conversely, if an electron exhibits only collective, delocalized motion (as in the case of a free electron), its energy assumes a continuous spectrum. In crystalline solids, however, the behavior of electrons lies between these two extremes, incorporating both delocalized and localized motions. Consequently, the energy spectrum of electrons in crystals manifests as a band structure, characterized by alternating allowed energy bands and forbidden energy gaps. This unique structure arises from the interplay between the periodic potential of the lattice and the electron's dual nature of motion.

In the preceding discussion, we have merely outlined the general properties of electronic wavefunctions in solids through analytical considerations. To obtain precise and unambiguous results, it is imperative to solve the single-electron Schrödinger equation for the actual potential $V(r)$ of a specific solid of interest. However, even for relatively simple potentials, the process of solving the Schrödinger equation involves exceedingly intricate mathematical derivations. While such efforts are undoubtedly necessary to yield results that can be compared with experimental observations, they may not always be the most efficient approach for gaining a deeper understanding of how periodic potentials influence electron behavior. To focus more intently on the underlying physical principles, it is often advantageous to employ simplified potentials that allow for the solution of the Schrödinger equation with minimal mathematical complexity. This approach facilitates a clearer comprehension of the relevant physics without being bogged down by excessive computational details. Consequently, it becomes essential to consider models like the nearly-free electron and tight-binding approaches, which strike a balance between mathematical tractability and physical insight.

The nearly-free electron model operates on the assumption that the crystal potential is weak, allowing the electrons in the crystal to behave similarly to free electrons. In this scenario, the periodic potential acts as a small correction to the idealized free electron approximation. Using perturbation theory, the effects of the crystal potential are incrementally added to the unperturbed free electron model. This approach yields results that serve as a rough approximation for the valence bands of simple metals, including sodium (Na), potassium (K), and aluminum (Al). Although the model is a simplification, it provides valuable insights into the electronic properties of metals with weak crystal potentials and serves as a foundational concept in solid-state physics.

In the tight-binding model, the atomic potential is assumed to be so strong that electrons are effectively localized around individual atomic sites, with only minimal overlap between adjacent atoms. The weak interactions between neighboring atoms are treated as a perturbation to the localized electronic states. This approach provides a simplified yet insightful description of narrow inner-shell bands in solids, such as the 3D bands in transition metals. While the model's results are approximate, they offer a useful starting point for understanding the electronic structure of systems where strong atomic potentials dominate and interatomic interactions are relatively weak. The model's simplicity makes it particularly valuable for conceptual discussions and preliminary analyses of tightly bound electronic states.

2.1.2 Nearly-free Electron Model

We begin by considering a system of completely free electrons, described by the Hamiltonian

$$H_0 = \frac{p^2}{2m} \tag{2.6}$$

The eigenenergies of the plane wave eigenstates $|\boldsymbol{k}|$ are

$$\epsilon_0(\boldsymbol{k}) = \frac{\hbar^2|\boldsymbol{k}|^2}{2m} \tag{2.7}$$

We now consider the effect of adding a minor modification to the Hamiltonian in the form of a periodic potential

$$H = H_0 + V(r) \tag{2.8}$$

with V periodic, meaning

$$V(r) = V(r + \boldsymbol{R}) \tag{2.9}$$

where $\boldsymbol{R}$ represents an arbitrary lattice vector. The matrix elements of this potential correspond directly to its Fourier components:

$$\langle \boldsymbol{k}'|V|\boldsymbol{k}\rangle = \frac{1}{L^3}\int \mathrm{d}r e^{i(\boldsymbol{k}-\boldsymbol{k}')\cdot r}V(r) \tag{2.10}$$

This matrix element vanishes unless $\boldsymbol{k}' - \boldsymbol{k}$ equals a reciprocal lattice vector. Therefore, scattering between plane wave states $\boldsymbol{k}$ and $\boldsymbol{k}'$ can only occur if their difference corresponds to a reciprocal lattice vector.

Next, we employ perturbation theory. To first order in the perturbation, we obtain

$$\epsilon(\boldsymbol{k}) = \epsilon_0(\boldsymbol{k}) + \langle \boldsymbol{k}'|V|\boldsymbol{k}\rangle = \epsilon_0(\boldsymbol{k}) + V_0 \tag{2.11}$$

This simply results in a uniform constant energy shift for all eigenstates. In fact, it can be rigorously shown (to all orders in perturbation theory) that the sole impact of V_0 is to uniformly raise or lower the energies of all eigenstates by this constant amount. Henceforth, we will assume that $V_0 = 0$ for simplicity.

At second order in perturbation theory, we have

$$\epsilon(\boldsymbol{k}) = \epsilon_0(\boldsymbol{k}) + \sum_{\boldsymbol{k}'=\boldsymbol{k}+\boldsymbol{G}} \frac{|\langle \boldsymbol{k}'|V|\boldsymbol{k}\rangle|^2}{\epsilon_0(\boldsymbol{k}) - \epsilon_0(\boldsymbol{k}')} \tag{2.12}$$

where $'$ denotes that the summation is limited to cases where $\boldsymbol{G} \neq 0$ holds true. This summation requires careful treatment, as instances may occur where $\epsilon_0(\boldsymbol{k}) \approx (\boldsymbol{k}')$ for some $\boldsymbol{k}'$, leading to divergent terms that invalidate the perturbation expansion. These degenerate cases must be handled using degenerate perturbation theory, to be discussed next.

To identify when this degeneracy occurs, we solve the two equations:

$$\epsilon_0(\boldsymbol{k}) = \epsilon_0(\boldsymbol{k}') \tag{2.13}$$

$$\boldsymbol{k}' = \boldsymbol{k} + \boldsymbol{G} \tag{2.14}$$

To build our understanding, we first explore the simpler one-dimensional case. Given that $\epsilon(\boldsymbol{k}) \sim \boldsymbol{k}^2$, Eq. (2.13) has a solution only if $\boldsymbol{k}' = -\boldsymbol{k}$. This indicates that the system has solutions only when

$$\boldsymbol{k}' = -\boldsymbol{k} = \frac{n\pi}{a} \tag{2.15}$$

or exactly at the Brillouin zone boundaries (refer to Figure 2.2).

In higher dimensions, any zone boundary point $\boldsymbol{k}$ has a corresponding boundary point $\boldsymbol{k}'$ fulfilling Eqs. (2.13) and (2.14).

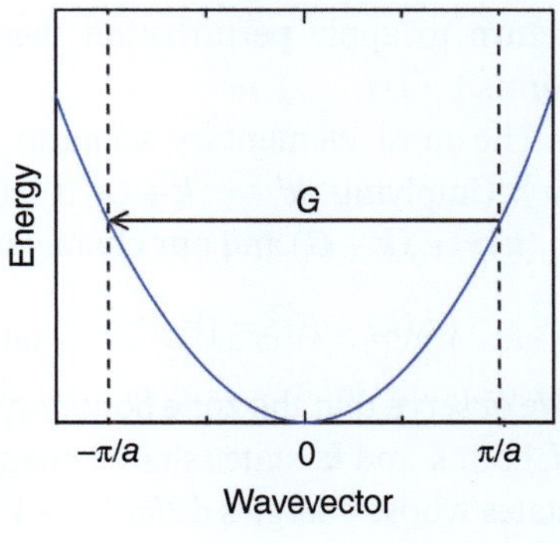

Figure 2.2 Scattering from the Brillouin zone boundary to the Brillouin zone boundary. States at opposite zone boundaries differ by $\boldsymbol{G}$ but are energetically degenerate, causing Eq. (2.11) to diverge when the identical energies yield a zero denominator.

In proximity to a zone boundary where Eq. (2.11) exhibits divergence, degenerate perturbation theory becomes necessary. This method involves initial diagonalization of the Hamiltonian within the degenerate subspace, followed by treatment of additional perturbations. Specifically, we perform exact diagonalization for states of equal energy that are coupled by the matrix elements.

2.1.2.1 Degenerate Perturbation Theory

When two plane wave states $|\boldsymbol{k}\rangle$ and $|\boldsymbol{k}'\rangle = |\boldsymbol{k}+\boldsymbol{G}\rangle$ possess nearly degenerate energies indicating that wavevectors $\boldsymbol{k}$ and $\boldsymbol{k}'$ lie near the zone boundaries, it is necessary to first diagonalize the matrix elements connecting these states. The relevant matrix is given by

$$\langle \boldsymbol{k}|V|\boldsymbol{k}\rangle = \epsilon_0(\boldsymbol{k}) \tag{2.16}$$

$$\langle \boldsymbol{k}'|V|\boldsymbol{k}'\rangle = \epsilon_0(\boldsymbol{k}') = \epsilon_0(\boldsymbol{k}+\boldsymbol{G}) \tag{2.17}$$

$$\langle \boldsymbol{k}|V|\boldsymbol{k}'\rangle = V_{\boldsymbol{k}-\boldsymbol{k}'} = V_{\boldsymbol{G}}^* \tag{2.18}$$

$$\langle \boldsymbol{k}'|V|\boldsymbol{k}\rangle = V_{\boldsymbol{k}'-\boldsymbol{k}} = V_{\boldsymbol{G}} \tag{2.19}$$

Here we have employed the definition of $V_{\boldsymbol{G}}$ from Eq. (2.10), while the relation $V_{-\boldsymbol{G}} = V_{\boldsymbol{G}}^*$ follows necessarily from the reality of $V(r)$.

In the context of this two-state system, we can represent an arbitrary wavefunction in the form:

$$|\Psi\rangle = \alpha|\boldsymbol{k}\rangle + \beta|\boldsymbol{k}'\rangle = \alpha|\boldsymbol{k}\rangle + \beta|\boldsymbol{k}+\boldsymbol{G}\rangle \tag{2.20}$$

Applying the variational principle to reduce the system's energy produces solutions equivalent to those found via the effective Schrödinger equation:

$$\begin{pmatrix} \epsilon_0(\boldsymbol{k}) & V_{\boldsymbol{G}}^* \\ V_{\boldsymbol{G}} & \epsilon_0(\boldsymbol{k}+\boldsymbol{G}) \end{pmatrix} \begin{pmatrix} \alpha \\ \beta \end{pmatrix} = E \begin{pmatrix} \alpha \\ \beta \end{pmatrix} \tag{2.21}$$

The eigenvalue equation governing the energy E takes the form:

$$(\epsilon_0(\boldsymbol{k}) - E)(\epsilon_0(\boldsymbol{k}+\boldsymbol{G}) - E) - |V_{\boldsymbol{G}}|^2 = 0 \tag{2.22}$$

(After completing the diagonalization within the degenerate subspace, one may return to apply perturbation theory to the remaining nondegenerate scattering terms.)

The most elementary scenario occurs when $\boldsymbol{k}$ lies exactly on a zone boundary (implying $\boldsymbol{k}' = \boldsymbol{k}+\boldsymbol{G}$ is similarly positioned). Under these conditions, $\epsilon_0(\boldsymbol{k}) = \epsilon_0(\boldsymbol{k}+\boldsymbol{G})$ and our characteristic equation reduces to

$$(\epsilon_0(\boldsymbol{k}) - E)^2 = |V_{\boldsymbol{G}}|^2 \tag{2.23}$$

We observe that the zone boundary develops an energy gap. Before adding potential d, both $\boldsymbol{k}$ and $\boldsymbol{k}'$ states shared energy $\epsilon_0(\boldsymbol{k})$. With $V_{\boldsymbol{G}}$ present, they mix to form new states whose energies differ by $\pm|V_{\boldsymbol{G}}|$.

In One Dimension Let us gain deeper insights by studying the simpler one-dimensional system. Suppose we have a periodic potential $V(x) = \tilde{V}\cos(2\pi x/a)$ with $\tilde{V} > 0$. The Brillouin zone boundaries are at $\boldsymbol{k} = \pi/a$ and $\boldsymbol{k}' = -\boldsymbol{k} = -\pi/a$ so that $\boldsymbol{k}' - \boldsymbol{k} = \boldsymbol{G} = -2\pi/a$ and $\epsilon_0(\boldsymbol{k}) = \epsilon_0(\boldsymbol{k}')$.

By examining equation (2.21), we find that under condition $\epsilon_0(\boldsymbol{k}) = \epsilon_0(\boldsymbol{k}')$, the physical solutions are given by $a = \pm\beta$, which corresponds to the quantum eigenstates of the system:

$$|\Psi_\pm\rangle = \frac{1}{\sqrt{2}}(|\boldsymbol{k}\rangle \pm |\boldsymbol{k}'\rangle) \tag{2.24}$$

associated with $E_\pm$ respectively. The wavefunctions $|\boldsymbol{k}\rangle$ in real space can be written as

$$|\boldsymbol{k}\rangle \to e^{ikx} = e^{ix\pi/a} \tag{2.25}$$

$$|\boldsymbol{k}'\rangle \to e^{-ik'x} = e^{-ix\pi/a} \tag{2.26}$$

we discover that the two eigenstates are given by

$$\psi_+ \sim e^{ix\pi/a} + e^{-ix\pi/a} \propto \cos(x\pi/a) \tag{2.27}$$

$$\psi_- \sim e^{ix\pi/a} - e^{-ix\pi/a} \propto \sin(x\pi/a) \tag{2.28}$$

Examining the probability densities $|\psi_\pm|^2$ corresponding to these eigenstates reveals a distinct spatial distribution: the higher-energy state ψ_+ exhibits maximal density at potential maxima V, while the lower-energy state ψ_- shows density accumulation at potential minima. This demonstrates the fundamental principle that periodic potentials mediate scattering between plane wave states $|\boldsymbol{k}\rangle$ and $|\boldsymbol{k}+\boldsymbol{G}\rangle$. When the two plane waves possess degenerate energies, their coupling becomes maximally strong. This leads to the formation of two hybridized states: an antibonding state with elevated energy (localized at potential maxima) and a bonding state with reduced energy (localized at potential minima).

2.1.2.2 k Not Quite on a Zone Boundary

In One Dimension Extension to near-zone-boundary $\boldsymbol{k}$ is straightforward. In 1D, we solve Eq. (2.22) for general $\boldsymbol{k}$ via zone-boundary expansion.

Consider the states located at the zone boundary $\boldsymbol{k} = \pm n\pi/a$, which are separated by reciprocal lattice vectors $\boldsymbol{G} = \pm 2\pi n/a$. As shown in Eq. (2.23), the energy gap that appears exactly at the zone boundary is $\pm|V_G|$. Now, take a plane wave with wavevector close to this boundary, $\boldsymbol{k} = n\pi/a + \delta$, δ is a small quantity, and n is an integer. Because of the periodic potential, this wavevector can be scattered into $\boldsymbol{k} = -n\pi/a + \delta$. We therefore obtain:

$$\epsilon_0(n\pi/a + \delta) = \frac{\hbar^2}{2m}[(n\pi/a)^2 + 2n\pi\delta/a + \delta^2] \tag{2.29}$$

$$\epsilon_0(-n\pi/a + \delta) = \frac{\hbar^2}{2m}[(n\pi/a)^2 - 2n\pi\delta/a + \delta^2] \tag{2.30}$$

The characteristic equation (Eq. (2.22)) is then

$$\left(\frac{\hbar^2}{2m}[(n\pi/a)^2+\delta^2]-E+\frac{\hbar^2}{2m}2n\pi\delta/a\right)\times\left(\frac{\hbar^2}{2m}[(n\pi/a)^2+\delta^2]-E-\frac{\hbar^2}{2m}2n\pi\delta/a\right)-|V_G|=0 \tag{2.31}$$

which simplifies to

$$\left(\frac{\hbar^2}{2m}[(n\pi/a)^2+\delta^2]-E\right)^2=\left(\frac{\hbar^2}{2m}2n\pi\delta/a\right)^2+|V_G|^2 \tag{2.32}$$

or

$$E_{\pm}=\frac{\hbar^2}{2m}[(n\pi/a)^2+\delta^2]\pm\sqrt{\left(\frac{\hbar^2}{2m}2n\pi\delta/a\right)^2+|V_G|^2} \tag{2.33}$$

Expanding the square root for small δ we obtain

$$E_{\pm}=\frac{\hbar^2(n\pi/a)^2}{2m}\pm|V_G|+\frac{\hbar^2\delta^2}{2m}\left[1\pm\frac{\hbar^2(n\pi/a)^2}{m}\frac{1}{|V_G|}\right] \tag{2.34}$$

Under the condition of weak perturbation (the regime of interest here), the second bracketed term dominates unity in magnitude. Consequently, for one of the two solutions, the quantity within brackets becomes negative.

Near the BG at the Brillouin zone boundary, the energy dispersion follows a quadratic dependence on δ, as illustrated in Figure 2.3(a). Meanwhile, Figure 2.3(b) reveals that small gaps emerge at these boundaries within an otherwise parabolic band structure when using the repeated zone scheme. (This representation aligns with the reduced zone scheme if only a single zone is considered.)

This structure is strikingly similar to the tight-binding model: energy bands correspond to allowed eigenstates, while gaps separate forbidden energy regions. Additionally, the spectrum repeats periodically across the Brillouin zone (see Figure 2.3(b)), mirroring tight-binding behavior.

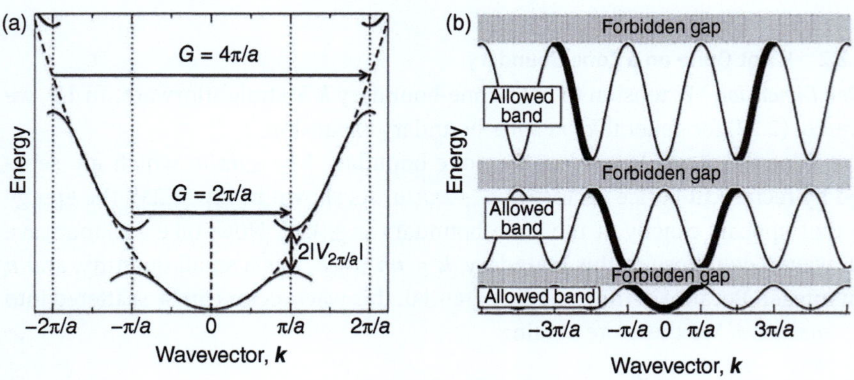

Figure 2.3 (a) The dispersion relation in the nearly free electron model is parabolic, with BGs opening at the Brillouin zone boundaries. (b) Similarly, the spectrum remains mostly parabolic, but small gaps appear at the zone edges due to periodic potential effects.

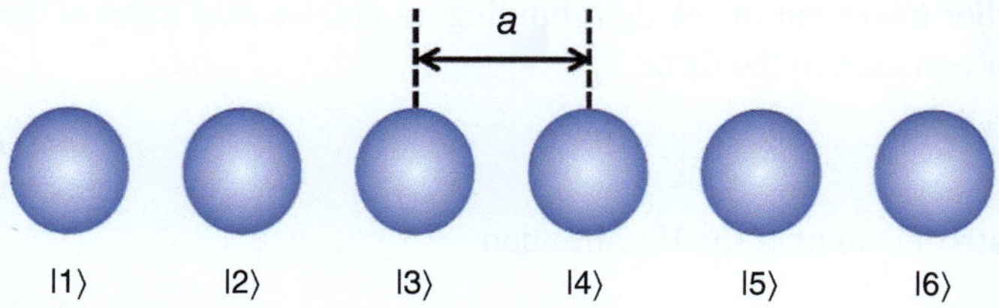

Figure 2.4 Representation of a one-dimensional tight-binding system where each lattice site hosts a single atomic orbital, permitting electron transfer between adjacent sites.

In Two and Three Dimensions The fundamental behavior of the nearly free electron model remains consistent across two and three dimensions. When approaching Brillouin zone boundaries, BGs emerge as a consequence of scattering from reciprocal lattice vectors. Notably, electronic states with energies just above the zone boundary crossing become elevated, while those slightly below experience a downward energy shift.

The dimensional dependence of zone boundary conditions reveals an important difference: While in one dimension each zone boundary $\boldsymbol{k}$ pairs with exactly one other $\boldsymbol{k}$ that simultaneously satisfies the reciprocal lattice vector requirement (such that $\boldsymbol{k} - \boldsymbol{k}' = \boldsymbol{G}$) and the boundary matching conditions (such that $\epsilon(\boldsymbol{k}') = \epsilon(\boldsymbol{k})$) specified in Eqs. 2.13 and 2.14, this unique correspondence does not generally hold in higher dimensions. While the one-dimensional case involves simple mixing between two degenerate plane waves (see Eq. (2.20)) leading to gap opening, higher dimensions present a more complex scenario. Here, numerous $\boldsymbol{k}'$ vectors, each connected to $\boldsymbol{k}$ through reciprocal lattice vectors and sharing the same unperturbed energy, may simultaneously satisfy the boundary conditions. The proper eigenstates must therefore incorporate all these degenerate plane wave contributions. This phenomenon is clearly demonstrated in a two-dimensional square lattice, where the four high-symmetry points labeled $(\pm\pi/a, \pm\pi/a)$ exhibit complete energy degeneracy in the unperturbed system and can be transformed into one another via appropriate reciprocal lattice vectors.

2.1.3 Tight-binding Model in One Dimension

In this analysis, we model orbitals in a macroscopic one-dimensional solid system as an extended chain of these molecular orbitals, as illustrated in Figure 2.4.

In this framework, each atom nn hosts a single orbital denoted by $|n\rangle$. For simplicity, we consider the system under periodic boundary conditions, meaning there are N sites with site N identified as site 0. Additionally, we assume that all orbitals are mutually orthogonal:

$$\langle n | m \rangle = \delta_{n,m} \tag{2.35}$$

Consider now a generalized trial wavefunction expressed as

$$|\Psi\rangle = \sum_{n} \phi_n \, |n\rangle \tag{2.36}$$

Following our earlier treatment of the tight-binding model, we may express the effective Schrödinger equation in the form:

$$\sum_m H_{nm}\phi_m = E\phi_n \tag{2.37}$$

where H_{nm} is the matrix element of the Hamiltonian

$$H_{nm} = \langle n|H|m\rangle \tag{2.38}$$

As established in our earlier examination of the molecular orbital approach, this Schrödinger equation formulation represents a variational approximation. Specifically, rather than determining the exact ground state solution, it identifies the optimal ground state configuration achievable within the constrained orbital basis of our model.

The accuracy of the variational method can be systematically improved through Hilbert space expansion by incorporating additional orbitals into the theoretical framework. For instance, rather than limiting to a single orbital $|n\rangle$ at each site, one can introduce multiple orbitals $|n, \alpha\rangle$ with α ranging from 1 to some integer p. As the number p grows, the method gains accuracy and ultimately approaches exactness. The linear combination of atomic orbitals method employs tight-binding-like orbitals to systematically improve approximations to the exact Schrödinger equation. A significant practical consideration arises when expanding the orbital basis: the orthogonality condition ($\langle n, \alpha\,|\,m, \beta\rangle = \delta_{nm}\delta_{\alpha\beta}$) generally no longer holds. While this renders the effective Schrödinger equation mathematically more complex, the underlying physical principles remain unchanged.

The system's Hamiltonian is expressed in the form:

$$H = K + \sum_j V_j \tag{2.39}$$

where $K = p^2/(2m)$ is the kinetic energy and V_j is the Coulomb interaction of the electron at position $\boldsymbol{r}$ with the nucleus at site j,

$$V_j = V(r - R_j) \tag{2.40}$$

where R_j is the position of the jth nucleus.

With these definitions, we have

$$H\mid m\rangle = (K + V_m)\mid m\rangle + \sum_{j\neq m} V_j\mid m\rangle \tag{2.41}$$

At this point, we identify $K + V_m$ as representing the Hamiltonian for an isolated system containing only the mth nucleus. Consequently, when we select the tight-binding orbitals to correspond to the atomic orbitals $|m\rangle$, we obtain

$$(K + V_m)\mid m\rangle = \epsilon_{atomic}\mid m\rangle \tag{2.42}$$

Here, ϵ_{atomic} represents the energy eigenvalue for an electron localized at the mth nucleus in complete isolation. This allows us to express the Hamiltonian as

$$H_{n,m} = \langle n\mid H\mid m\rangle = \epsilon_{atomic}\delta_{n,m} + \sum_{j\neq m}\langle n\mid V_j\mid m\rangle \tag{2.43}$$

We now have to analyze the final term in this equation. Physically, this term represents the probability amplitude for electron transfer (or "hopping") from the mth to the nth atomic site, mediated by interaction with other nuclei in the system. In most practical cases, such transitions are only significant between nearest-neighbor sites, leading us to express this term as

$$\sum_{j \neq m} \langle n \mid V_j | m \rangle = \begin{cases} V_0 & n = m \\ -t & n = m \pm 1 \\ 0, & \text{otherwise} \end{cases} \tag{2.44}$$

which defines both V_0 and t. (The V_0 term represents an on-site energy shift rather than an inter-site hopping process.) We emphasize that the system's translational invariance requires the energy to depend solely on the relative coordinate $n - m$, a condition explicitly satisfied by this formulation.

By applying the matrix elements established above, we obtain

$$H_{n,m} = \epsilon_0 \delta_{n,m} - t(\delta_{n+1,m} + \delta_{n-1,m}) \tag{2.45}$$

where we have now introduced the formal definition:

$$\epsilon_0 = \epsilon_{atomic} + V_0 \tag{2.46}$$

The Hamiltonian presented here represents the extensively studied tight-binding chain model. The parameter t, possessing dimensions of energy, serves as the hopping integral that facilitates electron transitions between adjacent lattice sites within the time evolution operator. The magnitude of t exhibits a strong dependence on interatomic spacing, demonstrating an exponential decay with increasing orbital separation while attaining maximal values at minimal distances.

2.1.4 Nanoelectronics: Superlattices and Heterostructures

In 1970, Esaki and Tsu proposed a novel semiconductor BG engineering methodology based on periodic heterostructures. By vertically stacking alternating semiconductor layers to form an artificial superlattice (a binary composite lattice), the electronic band structure could be deliberately tailored. The energy dispersion relation for electrons in this configuration appears in Figure 2.5. Originally speculative due to material constraints, such superlattices are now routinely manufactured via molecular beam epitaxy, employing alternating layers of small BG (such as GaAs) and large BG (such as GaAlAs) semiconductor compounds.

The defining structural characteristic of semiconductor superlattices lies in their sub-mean-free-path layer thicknesses, enabling ballistic electron transport across multiple small-BG regions within a single subband. This unique architecture yields dramatically enhanced carrier mobility compared to bulk materials, making superlattices particularly valuable for high-frequency microwave applications. A remarkable quantum mechanical consequence emerges from this elevated mobility and the artificially reduced Brillouin zone dimensions—electrons can be electrically accelerated to zone-boundary energies where they undergo Bragg reflection. This realization of Bloch's theoretical prediction (originally conceived for bulk crystals)

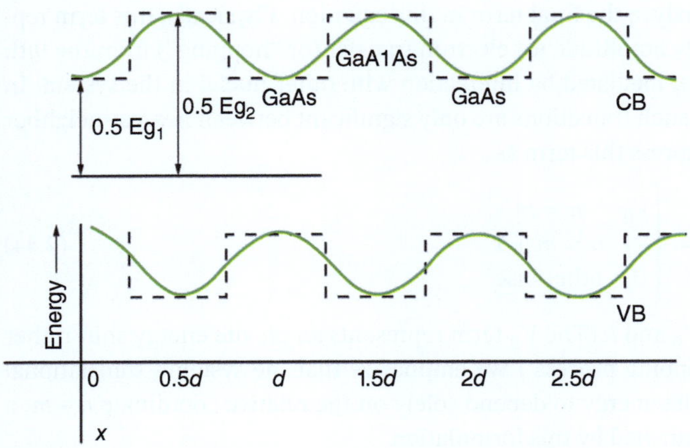

Figure 2.5 Variation of electron energy in the conduction and valence bands in a superlattice.

represents a significant breakthrough; while elusive in natural semiconductors, these characteristic conductivity oscillations manifest clearly in superlattice structures. The resulting negative differential resistance effect creates inherent electrical instability that can be strategically employed in oscillator and switching devices through careful engineering of the non-monotonic I–V characteristics.

In semiconductor laser design, layered heterostructures serve as effective traps for excited carriers. Unlike superlattices, where carriers propagate through subbands, this configuration employs a thin low-BG layer enclosed by higher-BG materials. At cryogenic temperatures and under weak electric fields, quantum confinement arises from the perpendicular wave vector quantization, enabling complete electron localization within the narrow-gap region. The smallest allowed wave vector magnitude along this direction is $2\pi/d$, which guarantees that at low temperatures and under weak electric fields, electron movement is effectively restricted to the low-BG plane. When an external magnetic field B is applied perpendicular to this plane, the electrons undergo circular orbits around the field direction, with their angular velocity determined by the cyclotron frequency:

$$\omega_c = \frac{eB}{m} \tag{2.47}$$

The occupancy of these cyclotron orbits is governed by the in-plane band structure. This leads to discrete quantization steps in the Hall resistance (the transverse resistance measured perpendicular to the applied electric field) when plotted against electron energy. The fundamental resistance quantum, now known as the von Klitzing constant, was precisely measured in these experiments—a discovery that earned Klaus von Klitzing the 1985 Nobel Prize in Physics (Klitzing 1986). While similar effects had been theoretically predicted, von Klitzing's breakthrough revealed that the Hall resistance in two-dimensional electron systems follows exact quantization according to

$$R_H = \frac{h}{ne^2} \tag{2.48}$$

Consequently, the quantum Hall effect has emerged as the gold standard for determining the ratio of e to h.

2.2 Basic Material Properties

Chalcogenide-based materials constitute a unique class, widely used in semiconductors, optoelectronics, and thermoelectrics. Examining the intrinsic properties of such materials is crucial for their optimization and design in practical applications. This section provides a detailed introduction to the basic material properties of chalcogenides, including structure properties, electrical properties of noncrystalline semiconductors, temperature dependence of d.c. conductivity, drift mobility and photoconduction, optical absorption, thermal conductivity, and specific heat. Through an in-depth exploration of these properties, we can gain a better understanding of the behavior of chalcogenide compounds and establish a theoretical foundation for future material design and applications (Zakery and Elliott 2003; Sahoo and Naik 2022b).

2.2.1 Structure Properties

Chalcogenide thin films, composed of elements from the chalcogen group (such as sulfur, selenium, and tellurium), exhibit a rich variety of structural properties that are intrinsically linked to their functional performance in a wide range of applications. These materials are characterized by their unique atomic arrangements, which can range from highly ordered crystalline structures to disordered amorphous networks. The structural characteristics of chalcogenide thin films—such as crystallinity, phase composition, surface morphology, and stability—play a pivotal role in determining their optical, electrical, and thermal behaviors. A comprehensive understanding of these properties is essential for optimizing their performance in advanced technologies, including phase-change memory, infrared optics, and photovoltaics. This section delves into the key structural aspects of chalcogenide thin films, supported by advanced characterization techniques that provide detailed insights into their atomic and microstructural features.

2.2.1.1 Crystallinity and Phase Structure

The crystallinity of chalcogenide thin films is one of the most critical structural properties, as it directly influences their functional characteristics. X-ray diffraction (XRD) is a primary tool for analyzing the crystalline phases and lattice parameters of these materials. For instance, amorphous chalcogenide films, which lack long-range atomic order, are known for their high optical transparency in the infrared region, making them ideal for applications in infrared optics and sensors. On the other hand, crystalline phases, such as those found in Ge–Sb–Te (GST) alloys, serve as key materials for phase-change memory devices owing to their ability to undergo rapid and reversible transitions between amorphous and crystalline states. The degree of crystallinity can be precisely controlled through deposition parameters, such as substrate temperature and deposition rate, as well as post-deposition annealing processes. This tunability allows for the optimization of material properties to meet specific application requirements.

2.2.1.2 Surface Morphology and Roughness

Surface morphology is another crucial structural property that significantly impacts the functionality of chalcogenide thin films. Methods like scanning electron microscopy (SEM) and atomic force microscopy (AFM) are utilized to analyze the film's topography at micro- and nanoscales. A smooth and uniform surface is often desirable for optical coatings and photonic devices, as it minimizes scattering losses and enhances light transmission. Conversely, controlled surface roughness can be advantageous in applications such as solar cells, where it improves light trapping and absorption efficiency. The deposition method—whether thermal evaporation, sputtering, or chemical vapor deposition (CVD)—plays a significant role in determining the film's morphological characteristics. For example, sputtered films tend to exhibit denser and smoother surfaces compared to thermally evaporated films, which may have higher roughness due to columnar growth.

2.2.1.3 Phase Composition and Structural Units

The phase composition of chalcogenide thin films is a key structural property that governs their functionality. Raman spectroscopy is a powerful technique for identifying different phases and their distribution within the film. In systems such as As–S or As–Se, the presence of specific structural units, such as As_4Se_4 or As_2S_3 pyramids, can significantly influence the material's optical BG and mechanical stability. Phase separation or the formation of secondary phases, often induced by thermal or laser annealing, can further modulate the film's properties. For example, in Ge–Se systems, the formation of $GeSe_2$ tetrahedra can enhance the material's thermal stability and optical properties. Understanding and controlling phase composition is therefore critical for tailoring the material's performance for specific applications.

2.2.1.4 Structural Stability and Environmental Durability

The structural stability of chalcogenide thin films under varying environmental conditions is a vital consideration for their practical deployment. Thermal analysis techniques, such as differential scanning calorimetry, are used to study phase transitions and thermal degradation processes. For instance, the glass transition temperature (T_g) and crystallization temperature (T_x) are key parameters that determine the thermal stability of amorphous chalcogenide films. Understanding these properties is essential for applications in harsh environments, such as infrared sensors or space optics, where the material must withstand extreme temperatures and radiation. Additionally, the mechanical properties of chalcogenide films, including hardness and adhesion, are evaluated using nanoindentation and scratch tests. These tests ensure the film's durability and reliability in device integration, particularly in applications where mechanical stress or wear resistance is critical.

A comprehensive theoretical understanding of an amorphous solid's properties requires detailed knowledge of its structural configuration. Even when constrained by interatomic forces between individual atoms and the tendency toward short-range order, there is an infinite number of allowed structures for any amorphous material. However, in view of the importance of short-range order in determining many physical properties, it is valuable to determine the nature and extent of this as far as possible. Techniques used for structural investigations include electron, X-ray, and neutron diffraction, extended X-ray absorption fine structure, and infrared and Raman spectroscopy, as well as several less direct methods.

The diffraction pattern characteristic of an amorphous material exhibits broad, diffuse halos or rings that diminish in intensity with increasing scattering angle. Crucially, these patterns lack the discrete spots or well-defined concentric rings that would signify the presence of crystalline order within the sample. From the angular dependence of the scattered intensity $I(s)$, one determines a diffraction function $F(s)$ given by

$$F(s) = s\left\{\frac{I(s)}{f^2(s)} - 1\right\} \tag{2.49}$$

where $s = 4\pi \sin \theta$ and $f(s)$ is the atomic scattering factor. The radial distribution function $J(r)$ is defined in terms of the atomic density $\rho(r)$ at distance r from any chosen atom:

$$J(r) = 4\pi r^2 \rho(r) = 4\pi r^2 \rho_0 + r\boldsymbol{G}(r) \tag{2.50}$$

where ρ_0 is the average atomic density and

$$\boldsymbol{G}(r) = \frac{2}{\pi}\int_0^\infty F(s) \sin s\, r ds \tag{2.51}$$

The radial distribution function (RDF), obtained through Fourier transformation of the scattering intensity profile, provides a one-dimensional representation of atomic spatial arrangement. When properly normalized, the RDF exhibits characteristic peaks at specific radial distances r, corresponding to mean interatomic separations between first, second, and higher-order neighbor atoms. Following correction for thermal effects, peak widths reflect the statistical variation in bond lengths, while integrated peak areas quantify the coordination number for each atomic shell. In the case of germanium shown here, the first two coordination numbers are 4 and 12 as in the crystal, and the average first interatomic distance is, within experimental error (about $\pm 1\%$), unchanged. The second-neighbor distance, however, shows a distribution about the crystalline value owing to bond-angle distortions. The curve is distributed about a parabola representing the RDF of a hypothetical amorphous solid of the same density ρ_0 but with matter uniformly distributed in space. The decreasing amplitude of the oscillations with r is a consequence of the lack of long-range order.

There are several problems involved in obtaining a reliable RDF, not the least of which is concerned with termination errors in the Fourier integral which can introduce spurious ripples into $J(r)$. There are various procedures for eliminating these, but considerable care is necessary. In amorphous solids containing more than one type of atom, the situation is more complicated and, for a complete study, partial interference functions are required. These can sometimes be derived by using electron diffraction in conjunction with XRD, since the atomic scattering factor f depends on the atomic number Z as $Z^{1/3}$ for the former and as Z^2 for the latter. However, inelastically scattered electrons have to be eliminated, and other corrections are necessary.

2.2.2 Electrical Properties

In this section, we discuss the density of electron states, mobility edges, and a variety of electrical measurements on noncrystalline semiconductors. Generally speaking, the experimental techniques are the same as or similar to those employed for crystalline materials; however, the models used in this book for interpreting data are somewhat different.

While our discussion focuses on microscopically homogeneous systems—a condition generally valid for conventional melt-quenched glasses (excepting SiO_2 containing Na_2O in a certain composition range) and properly deposited thin films—real materials often display local inhomogeneities. These microscopic variations in density and composition generate spatially dependent property fluctuations, including the characteristic BG variations illustrated in Figure 2.6(a). Complementary to structural inhomogeneities, electrostatic potential fluctuations (Figure 2.6(b)) may arise from spatial charge density variations. A representative case occurs in vapor-deposited Ge and Si films containing void structures, where incomplete orbital reconstruction can lead to localized charge accumulation. When the characteristic length scale of these potential variations significantly exceeds the quantum tunneling distance, charge transport through conduction and valence bands may be more accurately described by classical percolation theory rather than mobility edge models. The concept of variable-range hopping by electrons with energies at the Fermi energy *EF* must remain valid at low temperatures if *N(EF)* is finite, and so parts of this chapter may be applied even to such inhomogeneous films. For glow-discharge-deposited Si and Ge and for the majority of chalcogenide films and glasses, we think that any long-range potential fluctuations are small in magnitude, and indeed most data can be interpreted without any need to consider them.

2.2.2.1 Density of States

The electronic DOS constitutes a well-defined physical quantity applicable to both ordered and disordered materials, with its energy dependence measurable through

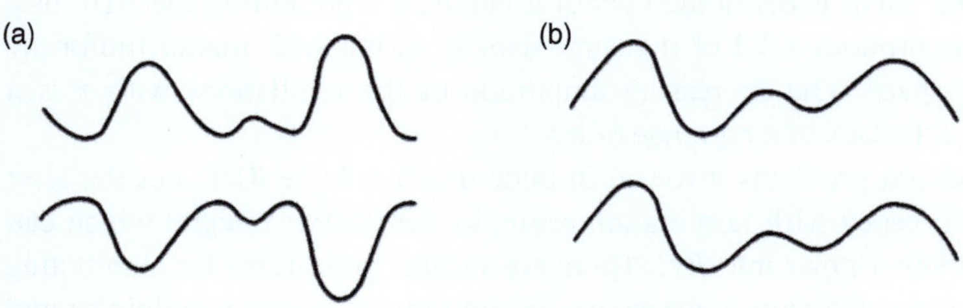

Figure 2.6 Possible forms of spatial fluctuation in the band edges of amorphous semiconductors: (a) elastic and (b) electrostatic.

multiple experimental techniques. Similar to crystalline systems, amorphous materials exhibit characteristic energy bands separated by forbidden gaps, where defect-related states may emerge within these gaps. Importantly, preservation of fundamental short-range structural parameters—including characteristic interatomic distances, bonding angles, and coordination numbers—ensures that the overall DOS profile maintains key features observed in crystalline counterparts, albeit with certain disorder-induced modifications. For the most part, however, it is the similarities between crystals and non-crystals that were, in the past, the most surprising feature so long as the band structure of crystals was thought to depend essentially on long-range order. It has not yet been possible, however, to calculate, except for certain idealized potentials, the form of the density near band edges, which frequently determines transport properties.

A particularly intriguing and characteristic feature of chalcogenides and many other amorphous semiconductors is the unusual positioning of the Fermi level. Unlike crystalline semiconductors, where doping can readily shift the Fermi energy toward the conduction or VB, in these materials, the Fermi level tends to remain pinned near mid-gap over a wide temperature range. This behavior, however, is not universal. For instance, Spear and Le Comber (1976) demonstrated that heavily doped amorphous silicon, prepared via glow discharge decomposition of SiH_4 with PH_3 or BH_3, exhibits a Fermi energy that shifts significantly—approaching or even reaching the conduction or VB. In contrast, impurities in amorphous silicon and germanium (whether in bulk glasses or evaporated/sputtered films) typically fail to produce such shifts, likely due to the strong tendency for bond saturation. Despite this, amorphous semiconductors are not truly "intrinsic" in the conventional sense, where the Fermi level is determined solely by the balance between valence and CB carriers. Instead, as summarized by Fritzsche (1973, 1974), multiple lines of evidence—including the persistence of a linear $\log(\sigma)$ versus $1/T$ relationship at low temperatures, thermopower measurements, and field effect studies—consistently indicate that the Fermi energy remains pinned near mid-gap, suggesting a unique electronic stabilization mechanism in these disordered systems.

2.2.2.2 Temperature Dependence of the d.c. Conductivity

Based on the aforementioned models characterizing electronic states and mobility edges in noncrystalline semiconductors, we can identify three principal charge transport mechanisms. Each conduction process becomes predominant within particular temperature ranges, reflecting the fundamental physics of disordered materials.

a) Charge transport occurs through carriers excited into delocalized states above the mobility edges (*Ec* or *Ev*). The electronic conductivity is given by

$$\sigma = \sigma_{min} \exp\left(-\frac{E_C - E_F}{kT}\right) \tag{2.52}$$

A plot of $\ln \sigma$ *versus* l/T will yield a straight line if $E_C - E_F$ is a linear function of T over the temperature range measured. If so, we write

$$E_C - E_F = E(0) - \gamma T \tag{2.53}$$

and the slope of such a plot will be $E(0)/k$, and the intercept on the σ axis will be $\sigma_{min}\exp(\gamma/k)$.

b) Charge transport mediated by carriers thermally activated into localized states near band edges, followed by hopping conduction in the vicinity of mobility edges E_A or E_B. For electron-dominated conduction, the conductivity expression takes the form:

$$\sigma = \sigma_1 \exp\left\{-\frac{E_A - E_F + w_1}{kT}\right\} \tag{2.54}$$

where w_1 represents the hopping activation energy, we observe characteristic temperature-dependent behavior. As established in the preceding analysis, the variable-range hopping nature dictates that w_1 decreases with lowering temperature. Nevertheless, since the dominant temperature dependence arises from the carrier activation term, the conductivity follows an approximately linear relationship when plotting lnσ versus $1/T$.

An estimate of σ_1 is not easy to make, but it is expected to be several decades smaller than σ_{min} partly because of a lower effective DOS near E_A compared with E_C and also because of a lower mobility as discussed in Section 6.4.2. In addition, the experimental slope and intercept of ln σ *versus* $1/T$ at $1/T = 0$ will be affected by any temperature dependence of $E_A - E_F$ analogous to that affecting the slope and intercept associated with process (a).

c) When the DOS at an energy level E_F is nonzero, an additional conduction channel emerges from carriers near E_F undergoing hopping transitions between localized states. This mechanism bears a strong analogy to impurity conduction in heavily doped crystalline semiconductors. The corresponding conductivity contribution can be expressed as follows:

$$\sigma = \sigma_2 \exp(-w_2/kT) \tag{2.55}$$

where $\sigma_2 \leq \sigma_1$ and w_2 is the hopping energy, of the order of half the width of the band of states.

2.2.2.3 Drift Mobility and Photoconduction

The phenomena described so far in this chapter depend on "states in the gap," only insofar as these determine the position of the Fermi energy E_F and, if $N(E_F)$ is finite, hopping conduction by electrons with energies near E_F. Measurements of drift mobility and still more of photoconduction, however, must often be interpreted in terms of deep traps and recombination centers not necessarily at E_F; this is why we consider them together.

Direct measurements of the drift mobility have proved very useful in determining the mechanism of charge transport of electrons or holes. Charge carriers are introduced at a specific location within the sample, and their transit time t_t over a distance d is measured while subjected to an electric field F. The drift mobility is subsequently calculated as

$$\mu_D = d/Ft_t \tag{2.56}$$

In the experiments described below, the materials have high resistivity, typically greater than 10^7 Ωcm, and the dielectric relaxation time is large compared with t_t; the excess carriers remain unscreened by others, similar to what is observed in drift mobility measurements of minority carriers in crystalline semiconductors. Both hole and electron transits can thus be observed.

In amorphous materials, the main interest is in the effect of localized states (traps). In a crystalline material, one could argue as follows. If there are *Nt* traps with discrete energy ε_t, below the CB, and if μ_0 is the conductivity mobility, then the drift mobility μ_D is given by

$$\mu_D = \mu_0 \left(\frac{n_0}{n_0 + n_t} \right) \tag{2.57}$$

where n_0 and n_t are the carrier densities in the band and in the traps, respectively. Assuming a thermal equilibrium distribution between the free and trapped carriers during transit, this ratio can be expressed in terms of the trap parameters, so that

$$\mu_D = \mu_0 \left\{ 1 + \frac{N_t}{N_c} \exp\left(\frac{\varepsilon_t}{kT} \right) \right\}^{-1} \tag{2.58}$$

which approximates to

$$\mu_D = \mu_0 \left(\frac{N_c}{N_t} \right) \exp\left(-\frac{\varepsilon_t}{kT} \right) \tag{2.59}$$

apart from high temperatures, where thermal emission becomes more likely. In this context, N_c represents the effective DOS at the edge of the CB. These quantities are frequently not known, but $e\mu_0 N_c$ is the pre-exponential factor in the dark conductivity, and if conduction is in extended states, this should be $\sigma_{min} \exp(\gamma/k)$. Thus, if μ_D is plotted against $1/T$, the slope yields the trap depth, and, if μ_0 and N_c are known, the intercept yields the trap density. The formula can be generalized to situations in which there is a continuous distribution of traps or, in amorphous semiconductors, to a continuous range of localized states below E_c, as will be shown below.

Normally, for unambiguous interpretation, it is important to ascertain that a well-defined transit time exists and that it scales correctly with field and sample thickness. Some spreading in the sheet of charge during transit owing to diffusion is inevitable, but more commonly, a spectrum of arrival times results from a statistical spread in trapping and release time. The assumption that there is thermal equilibrium between electrons in the CB and the traps will not be valid if it is too large. If the thermal release time is the reciprocal of

$$\nu_{ph} \exp\left(-\frac{\varepsilon_t}{kT} \right) \tag{2.60}$$

and t_p is the duration of the pulse, the condition that a trap will release electrons during the passage of the pulse is

$$\nu_{ph} t_p \exp\left(-\frac{\varepsilon_t}{kT} \right) > 1 \tag{2.61}$$

Deeper traps lead to a tail in the current pulse or to a lack of saturation in the charge collected. Under these conditions, one can still determine a time t_t' from which a drift mobility can be determined, but one is then probing only those traps with release times less than t_t'.

Observations of transients under varying conditions of field, temperature, and thickness can be used to determine the carrier lifetime τ or the time for a carrier to become lost by a deep trapping event. The product $\mu_D\tau$ is the range per unit applied electric field, and is an important material parameter (in the commercial process of xerography, for example). Measurement of the total charge collected under conditions where the range is much larger than the sample thickness enables a direct measurement of quantum efficiency to be made.

2.2.3 Optical Absorption

This section is concerned with those absorption processes that occur at the higher end of the frequency spectrum, particularly those associated with interband electronic transitions.

The critical difference in amorphous semiconductors' optical absorption stems from the breakdown of crystalline selection rules, particularly the k conservation requirement. This relaxation enables optical transitions that would be forbidden in ordered crystalline structures, significantly altering the absorption spectrum.

To establish a foundation for understanding optical absorption in amorphous semiconductors, we first examine the well-characterized absorption edges in crystalline materials. Crystalline semiconductors exhibit two distinct classes of optical transitions at their fundamental absorption edge: direct and indirect transitions. These processes both originate from photon–electron interactions that promote VB electrons to the CB, but differ fundamentally in their momentum conservation mechanisms. Indirect transitions require additional phonon participation to accommodate the necessary momentum transfer during the electronic transition, since the photon itself contributes negligible momentum to the system.

When excitonic effects are disregarded, the functional form of the absorption coefficient is principally determined by the energy-dependent DOS $N(E)$ in the participating bands. Considering direct transitions between simple parabolic bands ($N(E) \propto E^{1/2}$), the absorption characteristics can be derived as follows:

$$\alpha n_0 \hbar\omega \sim (\hbar\omega - E_0)^n \tag{2.62}$$

where $n = 1/2$ or $3/2$, determined by whether the optical transition is quantum-mechanically allowed or forbidden. E_0 denotes the optical BG, and n_0 is the refractive index. This absorption behavior remains largely unaffected by temperature, except for possible changes in E_0. For indirect transitions

$$\alpha n_0 \hbar\omega = \frac{(\hbar\omega - E_0 + h\nu_{ph})^n}{\exp(h\nu_{ph}/kT) - 1} + \frac{(\hbar\omega - E_0 - h\nu_{ph})^n}{1 - \exp(-h\nu_{ph}/kT)} \tag{2.63}$$

These two distinct terms correspond to phonon-assisted absorption and emission processes, characterized by different proportionality coefficients and thermal dependencies. Quantum selection rules dictate the exponent values: $n = 2$ for allowed transitions and $n = 3$ for forbidden transitions. Higher-order phonon processes may introduce additional term pairs with modified temperature dependences.

In general, both direct and indirect transitions can occur in a crystalline semiconductor. However, in materials in which the smallest gap is a direct one, indirect transitions, which are associated with smaller absorption coefficients, are not observed. An exception to this is the special case of vertical transitions involving absorption of a phonon of very small wavevector.

All the above types of optical transitions are modified when the electron–hole interaction is not ignored. The mutual attraction allows bound states of the electron and hole, namely excitons, to be formed with energy less than that of the free pair. For direct allowed transitions, absorption into these exciton states is in the form of a hydrogenic spectrum below the continuum at E_0. Moreover, the continuum itself is modified in shape. For direct forbidden transitions, the $n = 1$ line is missing, and the continuum absorption, although increased above the edge, starts at $\alpha = 0$.

For allowed and forbidden indirect transitions, inclusion of electron–hole interaction does not lead to a series of absorption lines; transitions into bound exciton states give rise to a continuous absorption. However, additional terms similar to those given in the equation above will occur, corresponding to the transitions into bound exciton states. For these, the power n is 1/2 and 3/2 for allowed and forbidden transitions, respectively. For transitions into the continuum, the corresponding power laws are 3/2 and 5/2.

Examples of most of the above relationships have been observed in crystalline semiconductors, and investigations on certain materials have yielded detailed information concerning the electronic structure at the band extrema.

Materials including alkali halides, Cadmium Sulfide (CdS), and trigonal selenium, present a fundamentally different absorption edge profile. The absorption coefficient in these systems rises exponentially with photon energc01-bib-0325y, creating a broad absorption tail extending over several electron volts—markedly different from the sharp, confined absorption onset seen in direct BG materials. This characteristic Urbach edge typically conforms to the empirical relationship:

$$\alpha = \alpha_0 \exp\left\{-\frac{\gamma'(E_0 - \hbar\omega)}{kT}\right\} \tag{2.64}$$

where γ' is a constant and T represents the absolute temperature, decreasing to a threshold value T_0 and remaining constant at T_0 for lower temperatures. As a result, the edge broadens progressively with increasing temperature above T_0. We shall briefly summarize the various models that have been proposed.

a) **Bound Exciton Interaction with Lattice Vibrations:** Toyozawa (1959, 1964) first proposed that the typical Gaussian profile of an exciton line transitions to an exponential form at its leading edge when quadratic terms in its interaction with phonons are considered. This theory seems to account for the exponential tails observed in alkali halides particularly well. The difficulty seems to be in justifying the fact that the quadratic terms outweigh the linear terms in the interaction. For other theoretical treatments based on the same ideas, the reader should see Mahr (1963), Mahan (1966), and Keil (1966).

b) **Electric-field Broadening of the Absorption Edge:** At photon energies below the onset of the field-free absorption, an approximately exponential tail is introduced, and at higher energies, the absorption coefficient oscillates. The "red shift" at low energies is the Franz–Keldysh effect and is due to a finite probability for tunneling of the band Bloch states into the energy gap. Clearly, if this explanation is invoked to explain Urbach's behavior, then the origin of the electric field has to be considered. In principle, this could arise from charged impurity states (Redfield 1963). However, the temperature dependence of the slope must be taken into account in any proposed model. Dexter (1967) introduces this by postulating that the electric field arises from the vibrating atoms in the material. In ionic solids, optical phonons are involved; in the more covalent materials, instantaneous changes in electronic charge clouds have to be invoked. The magnitude of the effect depends on the electron–phonon interaction, which is strong in piezoelectric solids like CdS and trigonal Se.

c) **Electric-field Broadening of an Exciton Line**: Dow and Redfield (1970) analyzed the absorption behavior of direct excitonic transitions under a uniform electric field. A numerical study of the shape of the absorption showed that, in contrast with the Franz–Keldysh results, its variation with photon energy is accurately exponential. Therefore, these researchers suggest that the "spectral" Urbach rule results from the broadening of an exciton line caused by an electric field. The parameter f represents the electric field strength, defined as the ratio between the potential energy change across the exciton radius caused by the field and the exciton binding energy. The Bohr radius a of the exciton ground state and the Rydberg constant R are expressed as follows:

$$a = \frac{\hbar^2 \kappa}{m^* e^2} \quad R = \frac{e^2}{2\kappa a} = \frac{m^* e^4}{2\kappa^2 \hbar^2} \tag{2.65}$$

where m^* is the reduced mass of the electron–hole pair. Thus

$$f = \frac{F|e|a}{R} = 3.89 \times 10^{-10} \left(\frac{m}{m^*}\right)^2 \kappa^3 F \tag{2.66}$$

where m is the free-electron mass and F is in V/cm. For $m^* = m$ and $\kappa = 6$, the line marked $f = 0.6$ corresponds to a field $\sim 7 \times 10^6$ V/cm. It is interesting that a classical criterion for direct field ionization of an exciton ($F_C = \kappa R^2/4e^3$) corresponds to $f = 0.125$. As seen from the curves, a noticeable exciton peak persists even when the field strength is nearly 10 times greater than this. The influence of the electron–hole interaction on the shape of absorption curves can be seen by comparison, in which the Coulomb interaction has been "turned off," giving the normal Franz–Keldysh result:

$$\alpha = \alpha_0 \exp\left\{-C(E_0 - \hbar\omega)^{3/2}\right\} \tag{2.67}$$

If Dow and Redfield's model is a correct description of Urbach's rule, it is, of course, still necessary to consider the origin of the internal electric field and the temperature dependence of the slope. This point will be considered later.

As mentioned above, the primary absorption edge in most amorphous semiconductors obeys an exponential relationship, i.e., In a is proportional to $\hbar\omega$. At the time of writing, only certain thin films of Ge and Si and perhaps InSb stand out as notable exceptions. Due to the challenges encountered in experimental procedures related to measuring small values of the absorption constant in thin films, it is not easy to test for exponential behavior in those amorphous semiconductors that can be prepared only in this form. Verification of exponential behavior is thus established with certainty only for those amorphous semiconductors that can be obtained as a glass by melt quenching, even though, of course, thin films of these materials are used for transmission measurements at high values of a.

2.2.4 Other Measurements

2.2.4.1 Thermal Conductivity and Specific Heat

Assessments of the temperature-driven behavior of thermal conductivity and specific heat of several noncrystalline materials have revealed a low-temperature behavior that differs considerably from that in crystalline materials. The T^3 dependence observed in crystalline SiO_2 (quartz) and common to other crystalline materials arises when the phonon mean free path l is constant and equal to the sample size, the thermal conductivity K then having the same Debye dependence on T as the specific heat C_v:

$$K = \frac{1}{3}C_v l v \quad C_v = 234NkT^3/\rho\Theta^3 \tag{2.68}$$

where v is the velocity of sound, N is the number density of atoms, p is the density, and Θ is the Debye temperature. As T is raised above 10 K, phonon scattering reduces l and K falls. It should also be mentioned that the absolute values of the thermal conductivity of crystalline materials vary considerably according to the degree of their chemical and structural perfection.

Returning now to the noncrystalline materials, K is several orders of magnitude lower than values generally found in crystals, is similar for all glasses measured to date, and exhibits a weaker temperature dependence ($K \propto T^n$ where $1.8 < n < 2$) than found in crystals.

The temperature dependence of the specific heat in several glasses is compared with that in quartz. The behavior in glasses follows

$$C_v = AT + BT^3 \tag{2.69}$$

Not only is there an additional linear term in the specific heat, which is of comparable magnitude in many noncrystalline materials, but the normal T^3 term is usually larger than that calculated from the elastic constants.

Several models have been proposed to explain the linear temperature term in the specific heat. A purely electronic mechanism in which the dependence arises from a finite density of one-electron states at the Fermi level is possible in principle, but the required density ($\sim 10^{20-21}$ cm^{-3}) would not be compatible with the optical transparency. This latter objection is perhaps overcome in the model of two-electron (bipolaron) states proposed by Anderson (1975) (and also in its adaptation by Mott, Davis, and Street (1975) to specific defects); the optical energy for one-electron excitation can be considerably greater than the thermal excitation energy for the two-electron system. However, according to Phillips (1976), the relaxation (hopping) time for bipolaron excitation is calculated to be too long except perhaps for a small fraction of the states that are considerably closer than the average.

2.2.4.2 Photoemission and Density of States

Photoemission is one of the most valuable experimental techniques available in the study of the electronic band structures of semiconductors. This technique entails the photoexcitation of charge carriers from either valence or core levels into the vacuum, followed by an examination of the emitted electrons' energy distribution in relation to the incident photon energy. The experiments are relatively difficult to perform properly, and skill is needed to interpret the energy distribution curves (EDCs).

For photons of energy $\hbar\omega$, the energy distribution of emitted electrons is governed by the relation:

$$N(E,\hbar\omega)\mathrm{d}E = \left\{\frac{K(\hbar\omega)}{\alpha(\hbar\omega)}\right\} T(E)S(E,\hbar\omega)N_c(E)N_v(E-\hbar\omega)\mathrm{d}E \tag{2.70}$$

Here, K represents a scaling factor that incorporates the parameters of the experimental setup, α denotes the absorption coefficient, T stands for the escape function, and S signifies the proportion of excited electrons with energy E that are lost due to scattering. Similar to optical absorption techniques, photoemission spectroscopy provides information on the combined (or joint) DOS in both the valence and conduction bands. This formulation operates under the assumption that the selection rule for the k conservation is negligible.

There are, however, two major advantages that photoemission measurements have over optical absorption. The first is that the energy at which a peak in N_cN_v occurs can be related to some fixed point in the band structure, such as the top of the VB. In optical absorption experiments, only energy differences are determined. In photoemission, a determination of absolute energies can, in principle, be made with a knowledge of the electron affinity of the semiconductor, and if a retarding potential method is used to measure the energy distributions of the collector. In practice, however, the measured retarding potential can be related to the energy at the top of the VB by noting the maximum value of the retarding potential for a given ha) at which electrons appear in the energy distributions.

If the photon energy is raised above, say, 20 eV, the structure associated with the CB disappears and the EDCs then reproduce the energy dependence of the density of valence states with some modification due to slowly varying photoionization cross-sections. At even higher energies, core-level spectra are obtained. Photoemission experiments using high photon energies are referred to as X-ray photoemission spectroscopy (XPS) or electron spectroscopy for chemical analysis (ESCA).

2.3 Synthesis and Characterization of Chalcogenide Film

The fabrication of chalcogenide thin films is a critical step in harnessing their potential for various applications. Techniques such as thermal evaporation, sputtering, CVD, and pulsed laser deposition (PLD) are commonly employed to deposit these films onto substrates. Each method offers distinct advantages in terms of film uniformity, stoichiometry control, and scalability, but also presents challenges related to defects, stress, and interfacial properties. Understanding the relationship between deposition parameters and film characteristics is essential for optimizing performance in specific applications.

Characterization of chalcogenide thin films involves a multidisciplinary approach, combining structural, compositional, optical, and electrical analyses. Techniques such as XRD, Raman spectroscopy, AFM, and ellipsometry (ELL) are routinely used to assess crystallinity, phase composition, surface morphology, and optical constants. Electrical measurements, including resistivity and switching behavior, provide insights into the functional performance of these films in devices. Advanced characterization methods, such as in situ transmission electron microscopy (TEM) and XPS, further enable the study of dynamic processes and interfacial phenomena at the nanoscale.

2.3.1 Synthesis of Chalcogenide Film

In the field of thin film fabrication of chalcogenide compounds, four predominant manufacturing techniques are widely employed: CVD, thermal evaporation (TE), PLD, and sputtering (SP). These methods have become the cornerstone of thin film technology due to their unique advantages in controlling film composition, thickness, and structural properties. CVD offers precise stoichiometric control through gas-phase reactions, while TE provides high-purity films through vacuum sublimation. PLD enables the growth of complex multicomponent films through laser ablation of targets, and sputtering facilitates excellent film uniformity through momentum transfer in a plasma environment.[6, 7] As illustrated in Figure 2.7, the current synthesis methods commonly employed for chalcogenide compounds are presented.

2.3.1.1 Chemical Vapor Deposition

Deposition, also known as thin-film deposition, is a critical process in semiconductor manufacturing that involves the formation of a thin layer of material on a wafer substrate. This process is fundamental to the fabrication of integrated circuits, enabling the creation of various functional layers required for device performance. The primary techniques for thin-film deposition are CVD and physical vapor deposition, each offering distinct mechanisms and applications.

CVD is a sophisticated technique that utilizes thermal energy, glow discharge plasma, ultraviolet radiation, laser irradiation, or other energy sources to induce chemical reactions among gaseous precursors. These reactions occur on the heated surface of a solid substrate, resulting in the formation of stable solid materials that are deposited as thin films on the wafer surface. CVD is renowned for its versatility and precision, making it a cornerstone of modern semiconductor fabrication.

In the realm of integrated circuit manufacturing, CVD is employed to deposit a wide array of thin-film materials, with the exception of certain metal films that may require alternative methods due to specific material properties or process constraints. Among the dielectric materials commonly deposited via CVD are

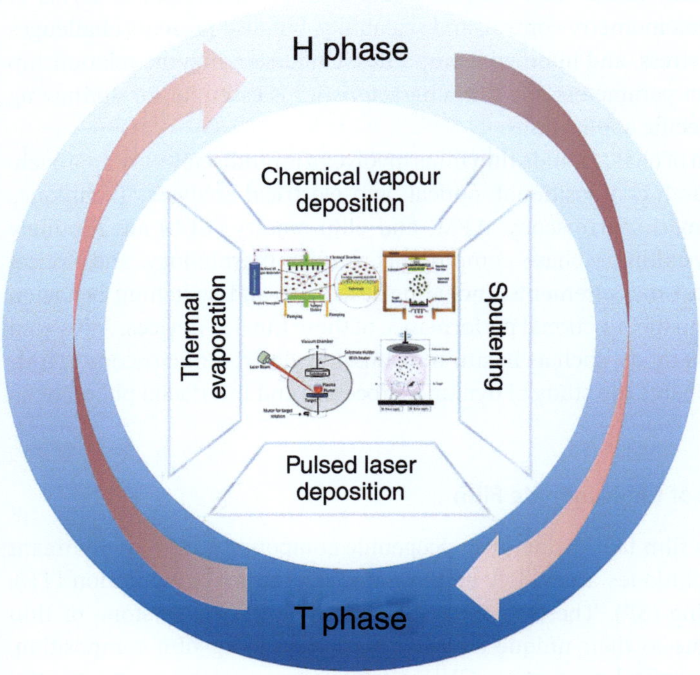

Figure 2.7 Common synthesis methods for chalcogenide thin films.

silicon dioxide (SiO_2), silicon nitride (Si_3N_4), borophosphosilicate glass (BPSG), and phosphosilicate glass (PSG). Conductive materials such as tungsten silicide (WSi_x), tungsten (W), and polysilicon are also routinely deposited using CVD techniques. Additionally, semiconductor materials like silicon (Si), gallium arsenide (GaAs), and gallium phosphide (GaP) can be effectively formed through CVD processes.

Thin film deposition via CVD is essentially a physicochemical process characterized by crystal growth from the gas phase. In a system with continuous gas flow, the growth mechanism can be divided into the following consecutive stages:

1) **Transport of Reactant Gases to the Substrate Surface:** The mixture of reactant gases is delivered to the vicinity of the substrate surface through controlled gas flow within the reaction chamber. This step ensures that the reactants are uniformly distributed and reach the deposition zone.
2) **Diffusion of Reactant Molecules to the Substrate Surface:** The reactant molecules diffuse from the main gas stream to the substrate surface. This diffusion process is influenced by factors such as gas flow dynamics, temperature gradients, and the geometry of the reaction chamber.
3) **Adsorption of Reactant Molecules onto the Substrate Surface:** Upon reaching the substrate, the reactant molecules adsorb onto its surface. The adsorption process depends on the surface temperature, the chemical affinity between the reactants and the substrate, and the surface morphology.

4) **Surface Chemical Reactions and Deposition:** The adsorbed molecules undergo chemical reactions, either with other adsorbed species or with molecules from the gas phase. These reactions result in the formation of solid material, which deposits onto the substrate surface, forming the thin film.
5) **Desorption of Reaction By-products:** The gaseous by-products generated during the chemical reactions desorb from the substrate surface. Efficient desorption is crucial to prevent contamination and ensure the purity of the deposited film.
6) **Diffusion and Removal of By-products:** The desorbed by-products diffuse away from the substrate surface and reenter the main gas stream. They are then transported out of the deposition zone and expelled from the reaction chamber.

These steps occur continuously and sequentially during the CVD process. However, the rate at which each step proceeds can vary significantly. The overall deposition rate depends on the slowest stage in the process, referred to as the **rate-limiting step**. Identifying and optimizing this step is critical for enhancing the efficiency and quality of the CVD process. Factors such as temperature, pressure, gas flow rates, and reactant concentrations must be carefully controlled to ensure that the rate-limiting step does not hinder the overall deposition process. Through comprehension and control of these stages, engineers are able to customize the CVD process to precisely regulate film characteristics—including thickness, uniformity, and composition—across diverse applications in semiconductor manufacturing, optics, and advanced materials science.

At atmospheric pressure, the rate of silicon thin-film deposition using various silicon sources is closely related to temperature. The deposition rate exhibits distinct behaviors in different temperature regimes:

1) **High-Temperature Regime:** In the high-temperature region, the deposition rate shows little dependence on changes in temperature. Here, the rate is primarily determined by the diffusion rate of reactant molecules to the substrate surface, corresponding to Step 2 in Section 2.3.1.1 in the CVD process. The diffusion of reactants through the boundary layer above the substrate becomes the rate-limiting step, as the chemical reactions occur rapidly once the reactants reach the surface.
2) **Low-Temperature Regime:** In the low-temperature region, the deposition rate follows an exponential relationship with temperature. In this regime, the rate is governed by the kinetics of the surface chemical reactions, corresponding to Step 4 in Section 2.3.1.1. The activation energy of the reactions plays a significant role, making the deposition rate highly temperature-dependent.

For a CVD reaction to be viable, it must satisfy three key volatility criteria:

1) **Volatility of Reactants:** At the deposition temperature, the reactants must possess sufficiently high vapor pressure to allow their introduction into the reaction chamber at a practical rate. If the reactants are gaseous at room temperature, the reaction system can be simplified. However, if the reactants have low volatility at room temperature, a carrier gas is required to transport them into the reaction

chamber. In such cases, the gas delivery lines must be heated to prevent condensation of the reactants.

2) **Volatility of By-products:** The reaction by-products, excluding the deposited material, must be volatile. This ensures that they can be easily desorbed from the substrate surface and removed from the reaction chamber, preventing contamination and interference with the deposition process.
3) **Low Volatility of Deposited Material:** The deposited material itself must have a sufficiently low vapor pressure at the deposition temperature to ensure that it remains on the heated substrate. This criterion is essential for the formation of a stable and continuous thin film.

These criteria collectively ensure the efficiency and controllability of the CVD process. By carefully selecting reactants and optimizing process conditions, such as temperature, pressure, and gas flow rates, engineers can achieve high-quality thin films with desired properties for applications in semiconductor manufacturing, optics, and advanced materials science. Understanding the temperature-dependent behavior of deposition rates and adhering to the volatility criteria are critical for the successful design and implementation of CVD processes.

2.3.1.2 Thermal Evaporation

TE operates by applying an electric current to a refractory filament, which supports the material to be evaporated—typically metals like gold or palladium or pure carbon. The entire setup is housed within a vacuum chamber where the sample is positioned. Resistive heating occurs as a high current flows through the filament, subsequently raising the temperature of the evaporation source. Once the source material reaches its vaporization temperature, an atomic flux is emitted into the chamber. This stream of metal or carbon atoms deposits onto all exposed surfaces, including the sample. To achieve a uniform coating, the sample stage is often rotated and tilted—a method sometimes referred to as rotary evaporation. Conversely, in specialized applications, nonuniform deposition can be intentionally induced for shadowing effects. By fixing the sample at an oblique angle relative to the evaporation beam, protruding surface features cast shadows, creating contrast that accentuates topographic details.

Sputter coating operates through the ejection of atoms from a target material via bombardment by high-energy particles, followed by their deposition onto the specimen surface. Unlike TE techniques, this process functions at higher pressures and does not require direct line-of-sight between source and sample. Current sputtering systems are categorized into five principal variants based on particle generation methodology: plasma-based, ion beam, radio frequency, Penning, and magnetron configurations. Comprehensive descriptions of these systems' operational principles and designs are well-documented in specialized literature. Modern applications of sputter deposition have expanded significantly beyond conventional metal/carbon coatings for microscopy applications. Modern sputtering technology has evolved beyond simple metal deposition, now enabling the precise coating of complex compounds for advanced electronic applications. Commercial systems offer sophisticated target materials and configurations. A fundamental SEM-compatible sputter coater typically comprises: (i) a vacuum chamber housing a target material cathode, (ii) an anode, and (iii) a sample stage. The process initiates with inert gas (Ar/N_2) introduction, ionized via glow discharge plasma. Bombardment ejects target atoms that undergo gas-phase collisions, resulting in omnidirectional dispersion. While net deposition favors the anode region, stochastic atomic motion ensures comprehensive surface coverage, even on topographically complex substrates. This isotropic deposition characteristic facilitates uniform coating even on rough surfaces at the microscale.

2.3.1.3 Pulsed Laser Deposition

PLD is a newer technique developed in the last decade that has so far been used largely in research settings. It uses energy from a pulsed laser to ablate material in a vacuum from a target, leading to a plasma plume consisting of uncharged particles, ions, and electrons in which the material is transported at velocities of over 1 m/s to a substrate. The deposition occurs in three phases: laser ablation/plasma creation, transport to the substrate, and condensation/film creation.

The key driver for the development of this technique is the superior retention of stoichiometric composition of complex materials during deposition. Materials such as YBCO, Y-stabilized BaZrO, and others that are being used in electroceramic applications have significantly benefited from this technique when transitioning these materials from their bulk, sintered form to thin films (Zhang et al. 2020; Sahoo and Naik 2022a; Schrader and Moore 1997). In addition, the high deposition rates greater than 10 nm/min, and the fact that the energy source is outside the vacuum chamber, eliminating contamination from, for example, heated filaments, are strengths of this technique. Laser energy and pulse rate, substrate temperature, and residual pressure impact the film properties, requiring optimization, which in the past was largely done empirically due to a lack of complete understanding of the processes. Challenges include the high directionality and small coverage of the plasma plume, requiring rastering/scanning the laser plume or substrate movement with respect to the plume to obtain uniform deposition rates and surface coverage. In addition, the plasma often includes macroscopic molten particles that impact film growth. Mechanical velocity filters, as well as off-axis deposition and collision between two plasmas, have been used to remove these macroscopic particles (Long 2002). Efforts are underway to better understand and model the laser ablation and deposition process, which may constitute the basis for future industrial/larger-scale use of the technology. Optical spectroscopy and laser-induced fluorescence can be used to characterize the plasma for research purposes.

2.3.1.4 Sputtering

Sputtering is a plasma-based technique in which noble gas ions—most commonly argon ions (Ar^+)—are accelerated toward a target (cathode), causing the ejection of material particles. These ejected particles form a vapor cloud that subsequently condenses onto the substrate. This method is widely used for depositing metallization layers in microelectronics and microelectromechanical systems (MEMS) devices. The sputtering process can be divided into four main stages:

1) Ion generation via collisions between inert gas atoms (Ar) and electrons, followed by ion acceleration toward the target.
2) Ejection of target atoms caused by ion impacts on the target surface.
3) Movement of these liberated atoms through the gas phase toward the substrate.
4) Condensation of the atoms onto the substrate surface.

Typically, the substrate-to-target distance ranges from 5 to 10 cm, extending up to 35 cm for certain sputtering systems, optimizing the condensation efficiency of sputtered atoms. The formation of seeds and clusters during sputtering occurs similarly to evaporation. However, atoms arriving at the substrate during sputtering possess higher kinetic energy (3–10 eV) compared to evaporation (~0.2 eV). Additionally, collisions between target atoms and gas atoms cause deposition from multiple directions, making the process largely isotropic and resulting in superior step coverage compared to evaporation. While sputtering causes more intense particle bombardment on the substrate surface—potentially damaging the crystal structure—it also enables in situ annealing of the deposited films. The heating of the substrate (e.g., wafer) due to the high-energy particle bombardment during the process has to be taken into account, in particular when considering deposition on polymer materials, such as during lift-off processes or for polymer MEMS devices. The layer properties can be influenced using a bias voltage. The key differences between sputtering and evaporation are as follows:

- Sputtered atoms and molecules have a higher impact energy (3–10 eV, compared to 0.2–0.26 eV for evaporation).
- Due to collisions with gas particles (Ar), the free target atoms reach the substrate surface from various directions, leading to good step coverage.
- The substrate has a higher exposure to gas than in evaporation.
- Applying a bias voltage can influence the layer parameters (e.g., planarization of the surface).
- Changing the polarity of the substrate and the target, the substrate can be exposed to backsputtering.

Backsputtering can be used for

- Cleaning of the substrate surface prior to deposition.
- In situ annealing of layers in sandwich layer structures (i.e., layers are successively deposited and partially backsputtered, thereby annealing the layer through physical impact and heat).
- Sputter etching of layers that are otherwise difficult or impossible to pattern.

2.3.2 Structural and Compositional Characterization

Thin films of chalcogenide compounds are critical in countless applications ranging from phase-change memory devices to infrared optics. In order to fully understand their structure, composition, and optical properties, it is necessary to adopt relevant characterization techniques. As shown in Figure 2.8, it is currently a commonly used characterization technology for chalcogenide thin films.

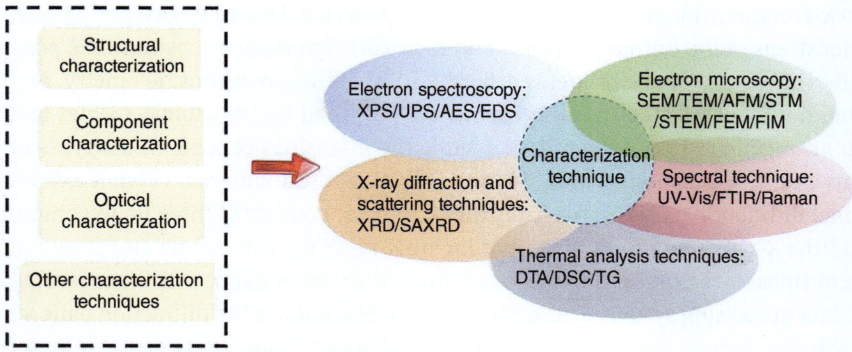

Figure 2.8 Common characterization technology for chalcogenide thin films.

2.3.2.1 Structural Characterization

Currently, the commonly used structural characterization methods for chalcogenide compounds include the following:

1) **X-ray diffraction:** This technique is employed to analyze the crystal structure and phase composition of the materials.
2) **Transmission electron microscopy:** TEM is utilized to observe the microstructure and identify crystal defects at a high resolution.
3) **Scanning electron microscopy:** SEM is used to examine the surface morphology and analyze the distribution of elements across the surface.

These methods provide comprehensive insights into the structural properties of chalcogenide compounds, facilitating a deeper understanding of their material characteristics.

X-ray Diffraction XRD is a fundamental technique used to analyze the phase composition and crystalline structure of materials. During diffraction analysis, a sample—whether crystalline or amorphous—is exposed to X-rays, producing diffraction effects of varying intensities. The resulting diffraction pattern is uniquely influenced by factors such as the material's composition, crystal structure, internal bonding, molecular arrangement, and overall conformation.

XRD technology offers numerous advantages, including nondestructiveness, contamination-free operation, rapid analysis, high measurement accuracy, and the ability to provide extensive information on crystal integrity. As a contemporary analytical technique for investigating material structure and composition, XRD is extensively utilized in both scientific research and industrial production across multiple fields. The core principle of XRD involves the interaction between a monochromatic X-ray beam and the crystal lattice. Since crystals consist of unit cells with atoms arranged periodically, the distances between atoms are comparable to the wavelength of the incoming X-rays. This leads to constructive interference of X-rays scattered by the atoms, resulting in pronounced diffraction peaks at certain angles. The positions and intensities of these peaks provide detailed information about the crystal's internal structure.

Based on this principle, a crystal's diffraction pattern is defined by two key aspects: the locations of the diffraction peaks and their corresponding intensities. The positions of these peaks are influenced by the unit cell's dimensions, geometry, and orientation, whereas the intensity values are governed by the atomic species and their arrangement inside the unit cell while the intensities depend on the types of atoms and their positions within the unit cell. As a result, different crystals exhibit distinct diffraction patterns. An X-ray diffractometer consists of three main components: the X-ray generation system (producing X-rays), the goniometer and detection system (measuring the 2θ angle and acquiring diffraction data), and the recording and data processing system. These parts collectively produce the diffraction pattern. Notably, the goniometer serves as the central element, featuring a complex design that directly affects the accuracy of the measured results.

Transmission Electron Microscopy TEM operates by directing a focused electron beam through an extremely thin specimen, typically under 100 nm in thickness or suspended on a specialized grid. As electrons interact with and penetrate the sample, they generate contrast patterns that form a projected image. This electron-transmitted image undergoes subsequent magnification before being captured by various detection systems, including fluorescent screens, photographic emulsions, or modern digital detectors (e.g., scintillator-coupled charge-coupled devices (CCDs) or direct electron detectors).

Transmission electron microscopes achieve substantially superior resolution as compared to optical microscopes due to the extremely short de Broglie wavelengths of accelerated electrons. This quantum mechanical property permits visualization of structural features at atomic-scale dimensions—including individual atomic columns—representing several orders of magnitude improvement over the diffraction-limited resolution of photon-based microscopy. As such, TEM has become an indispensable analytical tool across multidisciplinary research domains, including materials science, chemistry, and structural biology. TEM serves as a critical analytical tool across diverse scientific disciplines. In biomedical research, it facilitates groundbreaking studies in oncology and viral pathogenesis. Materials scientists employ TEM for nanoscale characterization of advanced materials, while environmental researchers utilize it for particulate pollution analysis. The technique also proves invaluable in semiconductor development and nanotechnology innovation. Beyond these applications, TEM contributes significantly to paleontological investigations and palynological studies, demonstrating its remarkable interdisciplinary utility.

Modern transmission electron microscopes offer diverse operational modalities that extend beyond basic imaging capabilities. These include scanning TEM (STEM) for high-resolution Z-contrast imaging, electron diffraction for crystallographic analysis, and spectroscopic techniques for elemental characterization. The system's versatility allows for a synergistic combination of these modalities. Notably, even in conventional imaging mode, TEM employs multiple distinct contrast generation mechanisms—including mass-thickness, phase, and diffraction contrast—each providing unique structural information about the specimen under investigation.

TEM contrast mechanisms originate from various specimen-electron interactions. Mass-thickness contrast results from local variations in material density or physical thickness. Z-contrast (atomic number-dependent) arises from differential electron scattering by elements. Crystalline materials generate diffraction contrast through Bragg scattering variations. Phase contrast emerges from quantum mechanical wave interactions with atomic potentials. Spectrum imaging captures energy loss characteristics, while additional contrast modes exist for specialized applications. Each mechanism provides distinct structural or compositional information about the specimen. Each contrast mechanism provides distinct structural and chemical information about the specimen, with the specific data obtained being highly dependent on both the selected imaging mode and the precise configuration of the microscope's optical components (lens excitations, aperture selections, and detector settings). This operational versatility enables TEM to deliver an unparalleled range of atomic-scale characterization capabilities. Under optimal conditions, modern TEMs can simultaneously determine atomic positions, elemental identities, and even bonding configurations. Consequently, TEM has become an indispensable characterization platform for nanoscale research across both hard and soft matter sciences.

The pioneering work of Max Knoll and Ernst Ruska in 1931 marked the first successful demonstration of TEM. Their research team achieved a milestone in 1933 by developing a TEM prototype surpassing the resolution limits of optical microscopy, followed by the introduction of the first commercial instrument in 1939. Ruska's fundamental contributions to electron optics were ultimately recognized with the 1986 Nobel Prize in Physics, cementing TEM's status as one of the most significant scientific advancements of the 20th century.

In TEM, image formation relies on the electron wavefront emerging from the specimen. The projector lens system precisely maps this exit wave distribution onto the detection plane. The resultant image intensity I can be mathematically described as proportional to the temporal average of the squared modulus of the electron wavefunction Ψ, representing the exit beams' quantum mechanical amplitude. This relationship holds under optimal imaging conditions with a high-performance detection system:

$$I(x) = \frac{k}{t_1 - t_0} \int_{t_0}^{t_1} \Psi\Psi^* \mathrm{d}t \tag{2.71}$$

Various imaging techniques aim to manipulate the electron waves emerging from the specimen to extract meaningful data about either the sample or the beam. As derived from the preceding equation, the resulting image is influenced not only by the beam's amplitude but also by the phase of the electrons. However, at lower magnification levels, phase-related effects are frequently negligible.

For higher-resolution imaging, thinner samples and higher electron beam energies are necessary. In this regime, the sample no longer absorbs electrons (as predicted by Beer's law) but instead acts as a phase-altering medium, shifting the phase of the incident electron wave while leaving its amplitude unchanged. This type of sample is termed a pure phase object. For extremely thin specimens, phase effects primarily govern image formation, making direct interpretation of intensity variations challenging. To enhance image contrast, the TEM can be intentionally defocused, leveraging the contrast transfer function (CTF) of the microscope. While the CTF typically reduces contrast in most cases, this approach proves beneficial when imaging weak phase objects, where phase-to-intensity conversion improves visibility.

1) In imaging mode, the objective aperture is placed at the back focal plane of the objective lens, the location where diffraction patterns appear. By adjusting the aperture to allow only the central (undiffracted) beam to pass, a bright-field image is obtained, since scattered electrons are filtered out. Conversely, if a diffracted beam is permitted to pass through the aperture, a dark-field image is generated. The selected electron signal is then magnified by the intermediate and projector lenses before being projected onto a viewing screen or camera, ultimately forming a resolved image of the specimen.
2) In diffraction mode, a selected area aperture is used to accurately specify the sample region generating the observed diffraction pattern. The diffraction pattern is then magnified and displayed on the viewing screen by tuning the intermediate lens current. This technique serves as an essential analytical tool for both cellular structure reconstruction and crystallographic orientation analysis, providing critical structural information at the atomic scale.

Scanning Electron Microscopy The first functional SEM was developed by Manfred von Ardenne in 1937 as an innovative alternative to TEMs. This pioneering instrument employed a finely focused electron beam to scan a small raster area, achieving superior resolution while simultaneously addressing the chromatic aberration limitations inherent in TEM systems. Subsequent advancements by research institutions, notably the Cambridge Scientific Instrument Company, culminated in the 1965 commercialization of a complete SEM system—the Stereoscan. Modern SEM systems represent a significant capital investment, with current market prices averaging approximately $1 million USD per unit.

The SEM represents a specialized class of electron microscopy that employs a low-energy electron beam for surface scanning and specimen analysis. This technology emerged as a solution to the inherent limitations of optical microscopy, particularly the resolution constraints imposed by visible light wavelengths. Unlike conventional light microscopes, electron microscopes utilize significantly shorter electron wavelengths, thereby achieving superior resolving power and enabling detailed examination of microscopic structures at nanometer scales.

The SEM fundamentally differs from TEM in its operational mechanism. While TEM relies on transmitted electrons passing through a specimen, SEM utilizes secondary and backscattered electrons emitted from the sample surface. This imaging technique operates by directing a focused electron beam that interacts with the sample, converting the kinetic energy of incident electrons into various detectable signals, including secondary electrons, backscattered electrons, and characteristic X-rays. Secondary electrons, generated through inelastic scattering, primarily reveal surface topography with high spatial resolution. Backscattered electrons, resulting from elastic scattering, produce atomic number contrast that reflects compositional variation. Diffracted backscattered electrons enable crystallographic analysis through electron backscatter diffraction patterns, while emitted photons permit elemental characterization via X-ray spectroscopy.

How does the SEM work?

- The electron emission source and electromagnetic lens system, positioned at the column's apex, utilize tungsten filament technology comparable to that employed in transmission electron microscopes. This configuration ensures stable electron beam generation and precise focusing capabilities essential for high-resolution imaging.
- Upon application of thermal energy to the electron source, electrons are liberated and subsequently accelerated toward the positively charged anode. This acceleration results in high-velocity electron motion essential for microscope operation.
- The incident electron beam induces dual emission phenomena from the specimen surface: high-energy primary electrons and low-energy secondary electrons. These interaction products carry complementary information, with secondary electrons revealing nanoscale surface topography while primary electrons provide compositional data through atomic number contrast.
- The specimen can be visualized in the SEM without special treatment—air-dried samples suffice. However, microbial specimens require fixation, dehydration, and drying to maintain cell structure and prevent collapse in the microscope's vacuum.
- The specimens are mounted and sputter-coated with a thin layer of heavy metals, which facilitates charge dissipation across the sample surface, thereby enhancing image resolution and clarity.
- The microscope achieves scanning by systematically sweeping a focused electron beam across a thinly prepared sample. As the electrons interact with the specimen's surface, they induce the emission of low-energy secondary electrons. These emitted particles are subsequently captured and analyzed by a dedicated detector.
- Upon entering the detector, the secondary electrons collide with a scintillator—a luminescent material that emits light when struck by charged particles. These light flashes are then transformed into an electrical signal via a photomultiplier, which relays the data to a cathode ray tube. The resulting image, resembling a television display, can be observed directly or captured photographically.
- The number of secondary electrons captured by the detector depends largely on the specimen's topography. Elevated surfaces tend to emit more electrons, resulting in stronger detector signals, while recessed areas produce fewer electrons, leading to weaker detection.
- As a result, elevated features on the specimen's surface produce a brighter display on the monitor, while recessed areas create darker contrast, forming a topographical map of the sample's structure.

SEM presents several inherent limitations that constrain its applications.

- The purchase cost is very high.
- They are cumbersome to transport.

- Their operation requires vibration-free and electromagnetically shielded environments.
- Stable voltage supply is necessary for their maintenance.
- Continuous cooling system access is essential for proper upkeep.

The scanning-transmission electron microscope (STEM) emerged as an innovative hybrid technology, marrying the fundamental operating principles of both SEM and TEM. In this sophisticated instrument, a finely focused convergent electron beam forms a nanometer-scale probe that systematically raster-scans across the specimen surface.

Raman Spectroscopy The Raman scattering intensity in thin-film multilayers exhibits complex thickness dependence, displaying non-monotonic behavior as layer thickness varies. This arises from optical interference effects between incident and Raman-scattered photons propagating through the film stack. Researchers can leverage this interference phenomenon for dual purposes: amplifying weak Raman signals through constructive interference, while simultaneously extracting valuable information about film thickness and optical constants through careful analysis of the intensity variations.

As an inelastic light scattering phenomenon, Raman scattering occurs when incident photons interact with matter, resulting in emitted photons with shifted frequencies. This distinguishes it from elastic scattering processes like Rayleigh scattering, where the scattered radiation maintains the original frequency. The energy difference between incident and scattered photons corresponds to vibrational or rotational transitions in the molecular or crystalline system. The observed energy shift in Raman scattering directly reflects the quantized vibrational or rotational states of molecular or crystalline systems. This unique feature establishes Raman spectroscopy as a complementary technique to infrared absorption for vibrational analysis. Under monochromatic excitation, the scattered radiation forms distinct spectral features: Stokes lines (lower energy) originate from ground-state scattering, while anti-Stokes lines (higher energy) arise from thermally populated excited states. The intensity ratio between these features, governed by Boltzmann statistics, serves as a reliable thermometer for probing the system's vibrational temperature (Weber 1979).

The Raman effect originates from light-induced charge separation in molecules, where the applied electromagnetic field displaces electrons relative to nuclei, generating an induced dipole moment $\mu = \alpha E$. This polarizability tensor α must exhibit first-order dependence on the normal mode coordinate q for the vibration to be Raman-active. Group theory dictates that only vibrational modes transforming as the irreducible representations of the molecule's point group that match the polarizability tensor components will appear in the Raman spectrum, establishing rigorous selection rules distinct from infrared spectroscopy.

In contrast, infrared (IR) absorption spectroscopy operates on the principle of direct resonance between the incident IR radiation frequency and the natural vibrational frequency of a specific molecular mode. The intensity of absorption in IR spectroscopy is determined by the variation in the dipole moment during vibrational motion. Upon photon absorption, the molecule transitions to an excited vibrational state, and the photon is fully absorbed. Only vibrational modes that alter the molecular dipole moment are considered IR-active. Notably, some molecular symmetries may exhibit activity in Raman spectroscopy but remain silent in IR spectroscopy, and vice versa. This complementary relationship between Raman and IR techniques enhances their utility in vibrational analysis, offering both unique and overlapping insights into molecular structure and symmetry.

Optical interference induces a non-monotonic thickness dependence in Raman signals, facilitating both enhanced detection and thin-film property assessment. Raman scattering's inelastic nature and polarizability sensitivity reveal vibrational and rotational information, while IR spectroscopy responds to dipole moment changes. Their combined application forms a robust analytical strategy for molecular and crystalline systems, delivering a holistic view of vibrational dynamics and structural symmetries.

2.3.2.2 Component Characterization

X-ray photoelectron spectroscopy XPS is a surface analysis technique that involves exposing a sample to X-rays and measuring the kinetic energy of the emitted electrons to assess its surface characteristics. One of the key strengths of XPS, also referred to as ESCA, is its surface sensitivity and ability to provide chemical state information about the sample's constituent elements. This makes it a highly effective analytical method for determining the atomic composition and chemical properties of materials. When monochromatic X-rays interact with a sample, they eject photoelectrons with distinct energy levels characteristic of the elements present. The XPS spectrum is generated by measuring the distribution of these electrons across different binding energies.

During the 1950s and 1960s, physicist Kai Siegbahn developed modern XPS while working at Sweden's Uppsala University. Recognizing the technique's ability to provide valuable chemical data, he named it ESCA. He was awarded the Nobel Prize in Physics in 1981 for his pioneering research. In XPS analysis, soft X-ray irradiation causes atomic ionization and core-electron emission. The method's surface sensitivity (sampling depth <10 nm) arises from the finite mean free path of emitted photoelectrons, which depends on their kinetic energy. The instrument detects and measures photoelectrons to produce a spectrum plotting emission intensity against binding energy. Since each element exhibits unique binding energy signatures, XPS enables precise surface elemental identification. With detection capabilities extending to all elements except hydrogen and helium, this technique has been successfully applied to diverse materials, including polymers, textiles, geological samples, and electronic materials.

X-ray photoelectron spectroscopy performs surface analysis by exposing samples to monochromatic, low-energy X-rays and measuring the kinetic energy of emitted electrons. Common X-ray sources include Mg Kα (1.2536 keV) and Al Kα (1.4866 keV) radiation, which penetrate solids only to micron-scale depths. This shallow penetration ensures photon interactions primarily with surface atoms, inducing photoelectric electron emission. When core electrons are ejected by X-ray irradiation, their kinetic energy (K.E.) can be determined through energy analysis:

$$K.E. = h\nu - B.E. - \varphi_s \tag{2.72}$$

In this expression, $h\nu$ represents the photon energy, B.E. denotes the binding energy of the atomic orbital from which the electron is emitted, and φ_s stands for the spectrometer's work function. To determine the binding energy of an electron, Eq. (2.1) can be rearranged to obtain Eq. (2.2), where the terms on the right are either known ($h\nu$ and Φ_{spec}) or measured in the XPS experiment (KE):

$$BE = h\nu - KE - \varphi_s \tag{2.73}$$

The binding energy (BE) of core electrons serves as a unique fingerprint for each element, forming the fundamental basis of XPS analysis. Since each element possesses characteristic BE values, this enables both identification and quantification of surface constituents. The technique can detect any electrons with binding energies below the excitation source's energy. Notably, an electron's BE is an intrinsic material property independent of the X-ray source. Chemical shifts—variations in BE caused by differences in chemical environment and electron density—provide crucial information about the sample's chemical state.

Secondary ion mass spectrometry As a powerful surface analysis tool, secondary ion mass spectrometry (SIMS) enables detailed investigation of the elemental distribution and stratification in solid materials. The principle involves bombarding the sample with energetic particles and subsequently detecting the charged secondary particles emitted from the surface.

SIMS stands out due to its exceptional sensitivity, capable of detecting nearly all elements in the periodic table—even the lightest ones. Additionally, it offers unique capabilities such as isotope discrimination, excellent depth resolution, and a broad dynamic range spanning over six orders of magnitude. These advantages make SIMS a powerful analytical tool for surface and thin-film characterization. However, a key limitation lies in its quantitative accuracy, as SIMS only measures ionized species, whose fraction is often challenging to determine precisely.

SIMS plays a crucial role in advanced material research, particularly in electronics, optics, and mechanical engineering. It is widely used to analyze thin films, including hard coatings for cutting tools, superconducting layers, amorphous semiconductors, and optical coatings. Beyond these applications, its versatility extends to metallurgy, ceramics, geology, organic chemistry, and biology. However, optimal performance in each case requires careful selection of experimental parameters, such as vacuum conditions, primary beam current/energy, mass resolution, and sensitivity.

SIMS instruments exist in various configurations, each specifically optimized for different analytical applications. These system variations arise from adaptations in design to meet particular measurement requirements.

SIMS currently stands as the most sensitive microanalytical technique available. This exceptional sensitivity stems from two key factors: the technique's ability to extract abundant particles from minute sample volumes, coupled with modern detectors capable of detecting single ions.

The sensitivity of an analytical measurement is influenced by both the sputter rate Y and the ionization probability $\alpha_x^{\pm}$, which together determine the spectrometer's minimum detectable concentration of an element. Additionally, in SIMS, the detection system's transmission efficiency β plays a crucial role. Typically, quadrupole mass spectrometers exhibit a transmission value around 10^{-3}, whereas time-of-flight systems achieve a significantly higher transmission of 10^{-1}.

The detection limit of an element refers to its lowest measurable concentration in an analytical procedure. For SIMS, typical detection limits are around 1 ppm, but this value can vary significantly based on the specific element and its host matrix.

Ion Beam Spectrochemical Analysis Ion beam spectrochemical analysis (IBSCA) is a surface analysis technique based on sputtering, closely related to SIMS/SNMS. Unlike these methods, IBSCA detects optical emissions from excited sputtered neutrals or ions rather than analyzing the particles themselves. Developed concurrently with SIMS in the 1960s and early 1970s, this method is also referred to as surface composition by analysis of neutral and ion impact radiation or beam-induced light emission.

The practical application of IBSCA as a routine analytical tool, similar to SIMS, faces a fundamental limitation: the requirement for measurable, reproducible signals. While analyses of metallic and semiconductor samples demonstrate pronounced oxygen-dependent variations in photon yield, oxide materials such as glasses exhibit minimal matrix effects, making IBSCA suitable for precise stoichiometric measurements in these systems. Consequently, IBSCA is primarily employed as a complementary technique alongside methods like SIMS, particularly for analyzing highly insulating surfaces and thin films. Its most common applications involve depth profiling of multilayer glass substrates and examining near-surface regions of glass samples, where environmental interactions frequently create stoichiometric variations distinct from the bulk material.

The technique follows the fundamental principles of sputter-based erosion methods, where a beam of noble gas ions (commonly Ar^+) bombards the sample surface. In non-elemental solids, this bombardment triggers a cascade of physical and chemical phenomena: initial stoichiometric changes in the surface region from ion–atom interactions, subsequent atomic redistribution within the solid, and finally surface material removal via secondary particle emission. The ion bombardment process generates various emission products, including electrons and surface particles (atoms or molecules) that may exist in neutral, charged, or excited states. Several theoretical models have been proposed to explain the excitation mechanisms of sputtered particles, such as the electron-transfer model, level-crossing model, and ITE model. Since the flux of ejected particles reflects the elemental composition of the near-surface region, monitoring this emission allows for quantitative determination of concentration-depth profiles.

The sputter erosion process typically employs energetic ions (0.5–10 keV) with current densities ranging from 10–100 uA/cm^2. Excited states of sputtered ions and neutrals exhibit extremely short lifetimes (10^{-13}–10^{-14} s), resulting in de-excitation phenomena occurring within 2 cm of the sample surface. Photon emission in the visible spectrum (250–900 nm) is collected using fused silica transfer optics for detection and analysis. Ultraviolet (UV) line detection necessitates specialized optical components (e.g., CaF_2 lenses) and detection systems (including high-vacuum optical spectrometers). Both excited-state emissions from sputtered atoms/ions/molecules and solid-state luminescence (particularly in glass samples) can be characterized using optical multichannel analyzers or CCD-based detectors. Elemental identification is achieved by correlating observed emission peaks with known transition wavelengths documented in spectral databases.

2.3.2.3 Optical Characterization

FTIR Spectroscopy Fourier transform infrared spectroscopy (FTIR) represents the predominant vibrational spectroscopy technique today. This method utilizes Fourier transform mathematics to simultaneously capture infrared spectra across all wavenumbers, contrasting with traditional dispersive spectroscopy that sequentially measures individual wavenumbers. The superior signal-to-noise ratio of FTIR has rendered dispersive methods largely obsolete in modern spectroscopic analysis (Schrader 1995).

At the heart of FTIR systems lies the Michelson interferometer, which processes incoming infrared radiation through precise optical manipulation. This critical component consists of a beamsplitter and two mirrors (one fixed, one movable). The beamsplitter divides the IR beam into two equal components, directing them toward their respective mirrors. Upon reflection, these beams recombine at the beamsplitter, creating an interference pattern that ultimately irradiates the sample prior to detection.

The moving mirror's primary function is to create variable optical path lengths, enabling interference between the two beam paths. When positioned equidistant from the beamsplitter (matching the fixed mirror's position), the system exhibits zero path difference ($\delta = 0$). Controlled translation of the moving mirror introduces precise optical path differences, which are essential for generating the interferogram. The varying optical-path difference creates interference phenomena analogous to XRD in crystals. As δ changes continuously through mirror displacement, the recombined beams undergo periodic constructive interference (when $\delta = n\lambda$, where n is an integer) and destructive interference (when $\delta = (1/2 + n)$). This controlled modulation of path length, achieved by precise mirror positioning (Figure 2.9), generates the characteristic interference pattern essential for Fourier transformation.

An interferogram graphically represents how light interference intensity varies with optical path difference. In FTIR systems, this pattern emerges from the superposition of countless wavelengths, exhibiting three key features: First, a strong central peak at zero path difference (maximum constructive interference). Second, symmetric wings are diminishing rapidly. Third, composite sinusoidal patterns from all constituent wavelengths. Although the detector measures these interference patterns after sample interaction, the actual infrared spectrum only emerges after mathematical processing of this raw interferometric data.

Fourier transformation is crucial for converting an interferogram into an infrared spectrum. This method is based on the concept that any mathematical function

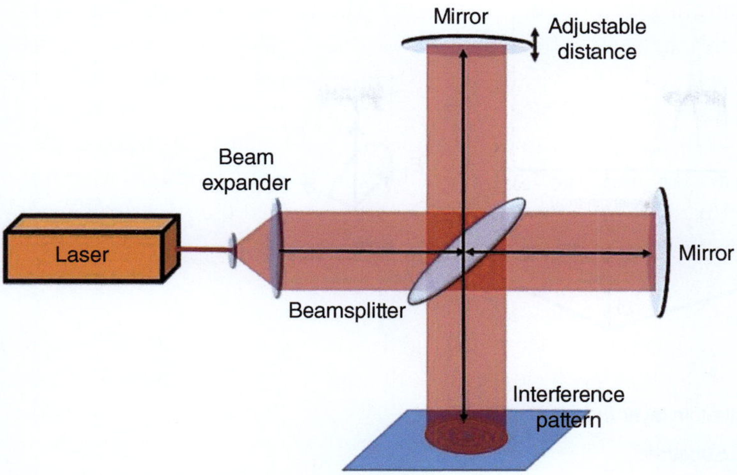

Figure 2.9 Optical diagram of a Michelson interferometer in FTIR.

can be broken down into a sum of sinusoidal components. The resulting spectrum, which plots light intensity against wave number, contains all the spectral information encoded in these waves. Modern FTIR instruments utilize a computational technique known as the fast Fourier transform to efficiently perform this conversion, significantly accelerating the analysis.

The Fourier transform is highlighted here as a key technique for translating a function between the time domain (t) and the frequency domain (ω). This conversion facilitates the interpretation of signals in terms of their spectral characteristics rather than their time-dependent variations:

$$F(\omega) = \frac{1}{\sqrt{2\pi}} \int_{-\infty}^{\infty} f(t)e^{-i\omega t}\mathrm{d}t \tag{2.74}$$

In an FTIR instrument, the Fourier transform converts the intensity versus optical path difference to the intensity versus wave number. The optical path difference can be considered as in the time domain because it is obtained by multiplying time with the speed of a moving mirror. The velocity of the moving mirror determines the wave number, which can be treated as a frequency-domain quantity since it equals the frequency divided by the speed of light.

UV-Vis-IR Ellipsometry Ellipsometry **(ELL)** serves as a highly sensitive analytical technique for characterizing thin films, multilayer structures, and substrate materials by precisely determining their thickness, refractive index, and extinction coefficient. Beyond simple layer analysis, it can effectively evaluate rough surfaces, interfacial properties, material gradients, and heterogeneous material compositions (Röseler 1993).

ELL can measure film thicknesses spanning an exceptionally wide range from sub-nanometer (0.1 nm) to macroscopic scales (100 μm). The measurable thickness range depends on both the spectral range employed and the uniformity of

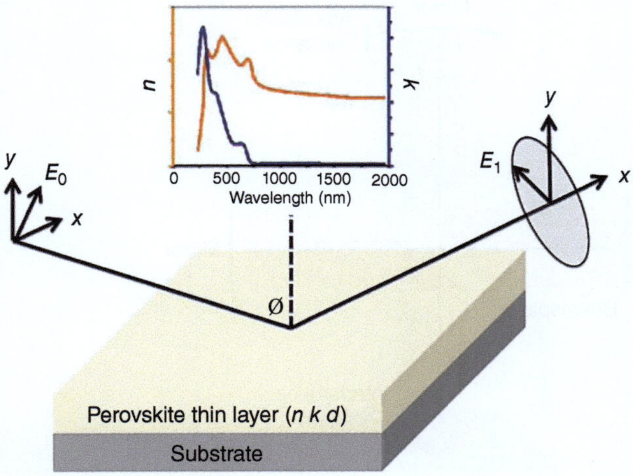

Figure 2.10 Principle of measurement in ellipsometry.

thicker films. For ultrathin films below 1 μm, the technique achieves remarkable sub-0.01 nm sensitivity, while micron-scale thicknesses can typically be resolved with better than 1 nm precision.

ELL achieves exceptional refractive index measurement sensitivity exceeding 5×10^4, representing the highest precision available among noncontact optical techniques, particularly for thin film characterization. The technique maintains nearly equivalent sensitivity for extinction coefficient determination, enabling detection of absorption coefficients as low as 10–100 cm in materials.

ELL's multilayer analysis capability is primarily constrained by the optical contrast between adjacent layers, yet it can routinely characterize stacks of ten or more layers. The technique offers several distinct advantages over alternative metrological approaches: its completely noninvasive and nondestructive nature, minimal sample energy deposition, direct dielectric function determination, and compatibility with any optically transparent measurement environment.

The measurement principle is illustrated in Figure 2.10. When linearly polarized light reflects off a sample surface (requiring adequate flatness and reflectivity), the polarization state transforms from linear to elliptical polarization through the reflection process. The technique quantitatively characterizes two optical properties: (i) $\tan \Psi$, representing the amplitude ratio of in-plane versus out-of-plane electric field components, and (ii) $\cos \Delta$, denoting their phase shift. Full 360° phase resolution is achieved by incorporating a retarder to generate circularly polarized light for secondary measurements, yielding both $\tan \Psi$ and $\sin \Delta$ values.

The sample's properties are encoded in the complex ratio between the Fresnel reflection coefficients for the parallel (r_p) and perpendicular (r_s) components of the incident plane-polarized electric field:

$$\rho = \frac{r_P}{r_S} \tag{2.75}$$

The fundamental equation of ELL

$$\rho = \tan \Psi e^{i} \tag{2.76}$$

describes the connection between the measured quantities Ψ and Δ and the sample properties contained in the coefficients of reflection, and hence in ρ. In principle, this complex equation can be solved analytically for pure substrates only. For complex sample structures, an accurate characterization requires both optical modeling of the sample and optimization of the calculated $\rho(\Psi,\Delta)$ to match experimental measurements. While Ψ and Δ represent directly measured values (assuming proper instrument function), derived parameters like film thickness, refractive index, and extinction coefficient are model-dependent quantities. The validity of these calculated properties fundamentally relies on selecting an appropriate optical model for the sample.

The interference of light waves reflected from the front and back surfaces of a film depends on its thickness, which in turn affects the overall reflectance. For absorbing films, thickness measurements are only feasible when the back surface still contributes significantly to the sample's reflectance. In the case of metallic layers, the typical measurable thickness is usually below 50 nm.

The accuracy of ellipsometric measurements relies on the incident angle, as the reflection coefficients (r_s and r_p) vary with angle. Above 50°, a pronounced difference between r_s and r_p is observed, with r_p reaching its minimum at the Brewster angle. Transparent materials exhibit a Brewster angle between 50° and 60°, while absorbing materials generally have one above 70°. Thin films on substrates are most effectively analyzed when measurements are performed close to the substrate's Brewster angle. Furthermore, optical models can be verified by conducting measurements at multiple angles.

The two ellipsometric angles, Ψ and Δ, measured at a fixed wavelength and angle of incidence, allow the determination of up to two sample properties—such as the thickness and refractive index of a transparent film. By conducting measurements at multiple angles, the number of obtainable parameters increases, though even then, the maximum number of measurable properties typically remains limited to 4–5. Spectroscopic ELL, which measures wavelength-dependent responses, significantly expands the range of measurable properties and allows analysis of more complex sample structures. This technique enables the determination of both refractive index dispersion and extinction coefficients, while also facilitating characterization of multilayer systems, interfaces, material gradients, and composition variations. By employing parameterized dielectric functions, the method effectively reduces the number of unknowns compared to the quantity of measured ellipsometric angles, making complex analyses tractable.

While spectroscopic ELL provides valuable insights into the dielectric properties of materials within a layered structure, it does not directly yield compositional information. To obtain a comprehensive understanding of complex unknown samples, ellipsometric analysis can be effectively combined with compositional characterization techniques such as Auger electron spectroscopy or XPS, along with cross-sectional transmission electron microscopy (XTEM) for structural verification. This multimodal approach enables the development of accurate optical models by correlating dielectric responses with material composition and vertical structure.

Ellipsometers can be implemented in various configurations, but all share fundamental components: a light source, polarization optics, a sample stage, and a detection system. The key distinction among different designs lies in their polarization generation and analysis methods. Typically, one polarizer prepares the incident beam's state, while another analyzes the reflected light, with advanced systems incorporating retarders to manipulate phase. Light sources vary significantly, ranging from monochromatic lasers to broadband options like xenon arc lamps or halogen lamps (often coupled with monochromators), and even infrared-specific emitters such as silicon carbide glowbars.

Broadband detectors used in ELL consist of photomultiplier tubes (for vacuum UV), silicon photodiodes (UV-Vis range), Ge or InGaAs photodiodes (near-infrared), and mercury cadmium telluride (MCT) (HgCdTe), along with deuterated triglycine sulfate (DTGS) detectors (mid-infrared). When paired with a white light source, these detectors require a monochromator for wavelength selection. For rapid measurements, spectroscopic ellipsometers employ photometers or diode arrays, allowing simultaneous multiwavelength detection instead of sequential scanning. In the infrared range (0.8–100 μm), advanced systems utilize modulated white light from an FTIR instrument combined with broadband detectors. The reflected signal undergoes a Fourier transformation to generate intensity spectra. Ellipsometric measurements are typically conducted at a fixed angle of incidence, though configurations may operate at discrete wavelengths or in spectroscopic mode (with variable angles).

The null ellipsometer operates by dynamically adjusting rotating polarizers and a compensator to control incident polarization, while a rotating analyzer evaluates the reflected beam. The key measurement principle involves iteratively aligning these components until the detector signal reaches a null (zero intensity). Despite being one of the most precise ellipsometric techniques, its reliance on mechanical rotation inherently limits measurement speed. Modern ELL commonly employs rotating element configurations where either the polarizer (RPE) or analyzer (RAE) rotates continuously at 10–40 Hz, enabling rapid automated measurements. The time-dependent detector signal undergoes Fourier analysis to extract the ellipsometric parameters Ψ and Δ. In RPE systems, measurement accuracy benefits from immunity to detector polarization sensitivity, accommodating various detectors like photomultiplier tubes or monochromators. This operational advantage explains the widespread adoption of such configurations in contemporary instruments.

2.3.2.4 Other Photon-detecting Techniques

Appearance potential spectroscopy techniques share a common fundamental principle: the detection of ionization thresholds in shallow core levels and their associated near-edge fine structures. The distinguishing feature among these methods lies solely in their detection mechanisms. Experimentally, they employ a consistent approach where (i) primary electron energy is progressively increased from near-zero to sample-specific values and (ii) primary current is maintained at a constant level throughout the measurement. When the incident electron energy exceeds successive core-level ionization thresholds of surface atoms, the resulting excited states decay through either characteristic X-ray emission or Auger electron ejection—the two principal recombination pathways for ionized core holes.

Soft X-ray Appearance Potential Spectroscopy In soft X-ray appearance potential spectroscopy, the detection of emitted X-ray photons can be achieved through two approaches: (i) traditional indirect measurement via photoelectron emission from a photosensitive surface, or (ii) direct detection using modern high-sensitivity X-ray detectors. When the incident electron energy exceeds the threshold for core-level excitation, the X-ray emission probability becomes dependent on the available unoccupied electronic states. Since the excitation process involves both the incident and excited electrons, the resulting spectral features are proportional to the self-convolution of the density of unoccupied states.

Disappearance Potential Spectroscopy The operating principle of disappearance potential spectroscopy (DAPS) relies on monitoring the elastic reflection current of primary electrons as a function of incident energy. As the beam energy surpasses core-level ionization thresholds, the resulting electron loss creates detectable dips in the reflected current profile. This detection approach presents an interesting contrast to Auger electron appearance potential spectroscopy (AEAPS); spectral comparisons between the two techniques for identical surfaces demonstrate complementary, mirror-image relationships. Despite this elegant symmetry, both methods share practical limitations, including significant background contributions that must be accounted for in DAPS measurements.

Compared with X-ray appearance potential spectroscopy (XAPS), both AEAPS and DAPS exhibit a significant advantage: their operation requires substantially lower primary electron currents. This characteristic minimizes potential disturbances to adsorbed species, making them more suitable for delicate surface analysis.

In this chapter, we have systematically investigated the characterization techniques for chalcogenide compounds, encompassing both well-established methodologies and emerging approaches. Our comprehensive exploration has been conducted from multiple perspectives, including structural analysis, compositional characterization, optical property evaluation, and other critical aspects that are essential for understanding these complex materials.

3

Growth of Chalcogenide Films: Mechanisms and Strategies

With the rapid development in the field of low-dimensional materials, thin film science and technology have garnered widespread attention, not only from the perspective of fundamental scientific research but also in terms of applied technology. Understanding and mastering the mechanisms and key factors involved in the thin film growth process are of crucial significance for selecting optimal growth conditions and producing high-quality thin films more effectively, ultimately achieving twice the result with half the effort.

Thin film growth is a complex nonequilibrium process that integrates kinetic factors (such as growth rate and growth temperature) and thermodynamic factors (such as lattice mismatch degree, surface energy, and interfacial energy). Its basic core content or typical picture is as follows: The thin film is formed by the condensation of gaseous atoms of the material. In the initial stage of thin film formation, the condensation of atoms occurs in the form of three-dimensional nucleation, followed by a diffusion process, during which the nuclei grow to form a continuous thin film (Koma 1992).

Figure 3.1a illustrates the atomic motion and the thin film formation process during film deposition. Atoms or molecules directed toward the substrate or film surface collide with the substrate surface, with some being reflected and others remaining on the substrate surface. The portion that stays on the surface undergoes surface migration and diffusion under the combined influence of its own energy and the energy of the substrate. Some of these atoms/molecules re-evaporate and detach from the surface, while others settle in regions of lower potential energy, becoming trapped within potential barriers and adsorbed onto the surface. This process is known as condensation, which involves nucleation and growth, island formation, coalescence, and growth, ultimately leading to the formation of a continuous film layer (Ratsch and Venables 2003).

Figure 3.1b is the Terrace–Step–Kink (TSK) model of thin film growth, i.e., terrace, step, and kink (The model also writes the steps as ledge, so that the TSK model is also written as the TLK model, which refers to the same concept.). In the actual growth process, there is a very important barrier on the terrace: the Ehrlich–Schwoebel (ES) barrier (Lagally and Zhang 2002) (at present, the ES barrier of this two-dimensional (2D) case has been extended to other dimensions (Zhang and Lagally 1997, Lagally and Zhang 2002)). In simple terms, the barrier refers to the obstacles that need to

Chalcogenide Nanophotonics, First Edition. Tun Cao, Ying Su, Yinghan Li, and Hongze Gao.

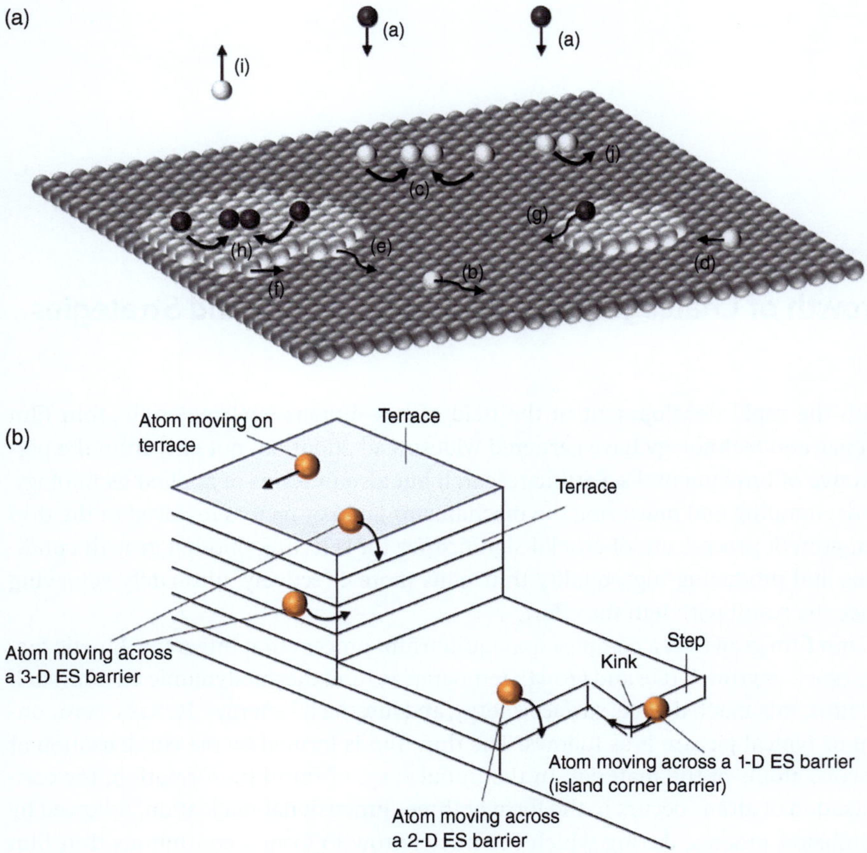

Figure 3.1 (a) Some basic atomic processes in the film growth process. (b) TSK model and ES barrier during film growth.

be overcome to deposit atoms on one terrace to cross other steps. The larger the potential barrier of ES is, the more difficult it is for atoms to move between layers and tend to be stacked on the terrace to form islands. On the contrary, the atoms tend to cover each terrace of the substrate horizontally. In other words, the ES barrier is an important parameter affecting the film growth mode. The following will analyze several common growth modes of thin films.

From the thermodynamic point of view, the different modes of growth mainly depend on the following three factors: the surface energy of the deposited layer γ_A, the surface energy of the substrate γ_B, and the size of the interface energy between them γ_{AB}. Figure 3.2 describes three basic modes of thin film growth (Vook 1984). Specifically:

1) **Layered Growth Mode:** Also known as Frank–van der Merwe (FM) growth. When $\gamma_B \geq \gamma_A + \gamma_{AB}$, and the lattice mismatch is small, the extra atoms tend to bond to the dangling bonds of the substrate and grow into a flat film layer by layer.

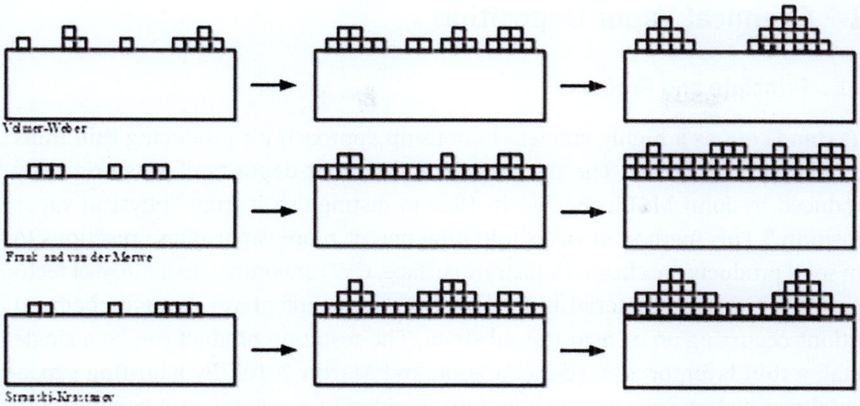

Figure 3.2 Three growth modes of thin films.

2) **Island Growth Mode:** Also known as Volmer–Weber (VM) growth. When $\gamma_B \leq \gamma_A + \gamma_{AB}$, and the lattice mismatch is large, the foreign atoms tend to preferentially combine with other atoms to form a three-dimensional island.
3) **Layered and Island Growth Mode:** Also known as Stranski–Krastanov (SK) growth. When $\gamma_B \geq \gamma_A + \gamma_{AB}$, and the lattice mismatch is large, there is a critical thickness. When the film is less than this critical thickness, it tends to grow in layers with stress accumulation; when the film exceeds this critical thickness, it will grow in an island mode to release stress.

In addition, there exists a special case known as van der Waals epitaxy, which is commonly observed in heteroepitaxial growth of layered compounds. Due to the weak interaction between the substrate and the epitaxial layer, there is no need to have too high requirements for lattice adaptation during growth, and the mismatch-induced stress can be almost completely released (Zhang and Lagally 1997). A variety of materials grown on graphene substrates belong to this category (Song et al. 2011, Peng et al. 2015).

It should be pointed out that the actual film growth process will be affected by various objective factors. Although there are many models and theoretical calculations to explain (Fischer et al. 1992, Ahlawat et al. 2013), it is not easy to obtain high-quality single-crystal films. Film growth is more like cumulative knowledge that depends on the continuous attempts and experience of researchers.

The specific preparation method of the film is mainly based on vapor deposition, which is divided into physical vapor deposition (PVD) and chemical vapor deposition (CVD) according to whether the chemical reaction is included in the preparation process. PVD is subdivided into thermal evaporation and sputtering. CVD includes conventional CVD and metal organic CVD. Other film preparation methods include nonvapor deposition methods such as liquid phase epitaxy, solid phase epitaxy, and Langmuir–Blodgett method. This section provides a detailed overview of four widely used thin-film fabrication methods: CVD, thermal evaporation, pulsed laser deposition (PLD), and sputtering.

3.1 Chemical Vapor Deposition

3.1.1 Principle and Process

CVD stands out as a highly efficient bottom-up approach for producing thin films and nanoparticles (NPs). The phrase "chemical vapor deposition" was originally introduced by John M. Blocher, Jr. in 1960 to distinguish it from "physical vapor deposition." This method involves initiating one or more vapor-phase reactions to form solid products on a heated substrate surface. CVD encompasses a range of techniques where a solid material is deposited from a vapor phase through chemical reactions occurring on or near the substrate. The resulting product can be a single crystal, a thin layer, or a powder (Carlsson and Martin 2010). By adjusting experimental parameters such as substrate type, temperature, gas mixture composition, pressure, and flow rates, materials with tailored physical, tribological, and chemical properties can be developed (Dobkin and Zuraw 2003). A key advantage of CVD is its exceptional throwing power, enabling the creation of uniform, low-porosity coatings on complex-shaped surfaces. Additionally, CVD allows for selective or localized deposition on patterned substrates, making it a versatile and precise fabrication technique (Zhang et al. 2013).

CVD is utilized in a variety of contexts, with the specific methods for initiating chemical reactions differing based on the operational conditions. These methods can be categorized according to the pressure environment:

1) **Atmospheric Pressure CVD (APCVD):** This process occurs at standard atmospheric pressure.
2) **Low-Pressure CVD (LPCVD):** Conducted below atmospheric pressure, LPCVD reduces unwanted gas-phase reactions and improves coating uniformity across the substrate.
3) **Ultrahigh Vacuum CVD (UHVCVD):** Operated at extremely low pressures, typically below 10^{-6} Pa (around 10^{-8} torr), UHVCVD is sometimes defined by a threshold as low as 10^{-7} Pa in certain applications.
4) **Sub-Atmospheric CVD (SACVD):** Performed at pressures below atmospheric levels, SACVD is often employed to deposit silicon dioxide (SiO_2) using tetraethyl orthosilicate (TEOS) and ozone, particularly for filling high-aspect-ratio silicon structures.

CVD is a method of preparing thin films by heating at high temperature, evaporating raw materials into a gaseous state, then transporting them to the substrate through carrier gas, generating target products through chemical reaction, and then depositing them on the substrate. It is the most widely used technology for preparing thin films in the semiconductor industry, and it is also a better method for preparing chalcogenide thin films. Figure 3.3 illustrates the basic process of CVD (Li et al. 2021).

The primary objective of CVD is to produce thin films with consistent and controllable properties, achieve cost-effective growth rates, and minimize interference with prior processing steps. Key film attributes include thickness, composition, purity,

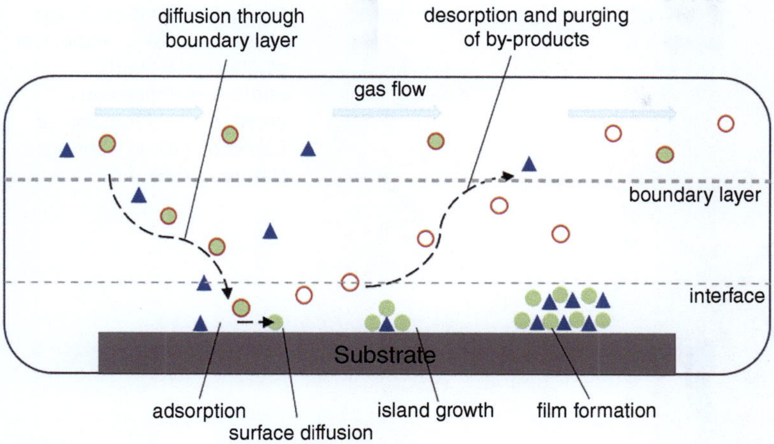

Figure 3.3 Basic process of CVD.

crystalline quality, and surface morphology (Morgan et al. 2018). The acceptable range for these properties varies by application, with electronic materials requiring significantly stricter tolerances compared to those in the chemical industry. Process parameters such as molar flow rates, energy supplied, system pressure, and substrate parameters can be modified according to the deposition chemistry and the design of the reactor. Ideally, predicting process outcomes based on these variables would enable optimization of CVD for specific applications. However, understanding the relationship between process variables and results is often complex. Consequently, much of the research has relied on empirical approaches to correlate film properties with process conditions, with limited emphasis on understanding the underlying chemistry and reactor dynamics. Furthermore, differences in impurity levels and reactor configurations complicate the comparison of results across studies (Matsui et al. 2016).

3.1.2 Applications in Chalcogenide Films

CVD methods for preparing chalcogenide thin-film materials are mainly divided into one-step and two-step methods. The one-step method is to use a gas containing chalcogenide to deposit on the substrate under high-temperature and low-pressure conditions, and finally obtain the required film on the substrate. Gao et al. used WO_3 and sulfur powder as precursors to grow WS_2 on gold foil by one-step CVD (Gao et al. 2015), as shown in Figure 3.4.

The two-step method refers to depositing a layer of other material film on the substrate first, and then sulfurizing or selenizing or tellurizing the precursor film by the CVD method to obtain the target material film. The first step to deposit other material films is generally PVD, such as evaporation coating, magnetron sputtering and molecular beam epitaxy, atomic layer deposition, and so on. Song et al. used a two-step method (Song et al. 2013), as shown in Figure 3.5. The first step was to deposit a layer of WO_3 film on the substrate by the atomic layer deposition method.

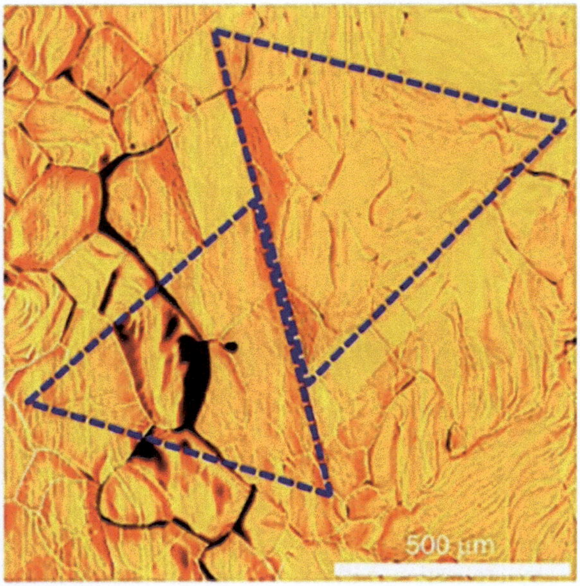

Figure 3.4 Optical image of triangular WS_2 monolayer single crystals with millimeter dimensions grown on gold foil. *Source:* Gao et al. (2015) / Springer Nature / CC BY 4.0.

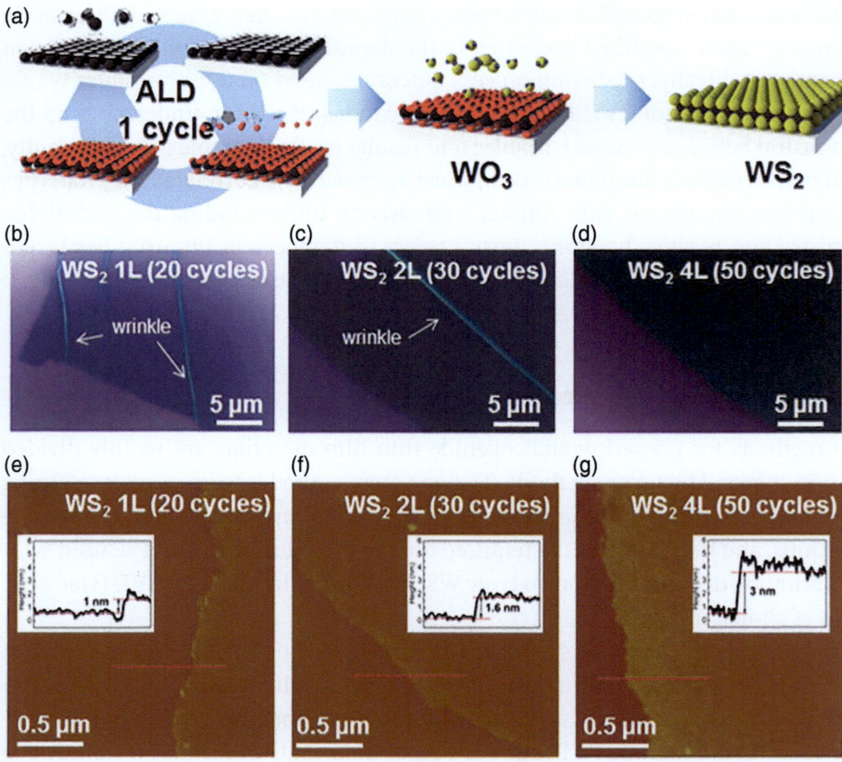

Figure 3.5 (a) ALD synthesis protocol for WS_2 nanosheets. (b–d) Optical microscopy characterization of mono-, bi-, and tetra-layer WS_2 transferred onto SiO_2 substrates. (e–g) Corresponding AFM topographic analysis with layer thickness profiles (insets). *Source:* Song et al. (2013) / American Chemical Society.

The second step was to obtain a large area of WS_2 film by sulfurization in the CVD system.

Thin films play a crucial role in the development of solid-state devices and materials. Many devices are constructed by stacking multiple layers of precisely patterned films. These films, ranging in thickness from nanometers to micrometers, can serve as protective coatings, alter device characteristics, or function as active elements within the device itself. CVD and associated methods are extensively employed for depositing various functional thin films, including dielectric layers, conductive coatings, protective passivation films, oxidation barriers, transparent conductive oxides, high-temperature resistant coatings, as well as wear- and corrosion-resistant surfaces, along with epitaxial layers for microelectronic applications (Dapkus 1982). CVD serves as an effective method for synthesizing heat-resistant materials, including tungsten and ceramic compounds. Additionally, it facilitates the production of solar energy conversion devices, high-performance fiber-reinforced composites, and precisely sized particles. Contemporary applications extend to the growth of carbon nanotubes and high-Tc superconducting materials (Pierson 1999). The precise control of oxygen partial pressure during deposition eliminates the need for post-growth oxygen annealing to attain superconductivity (Pierson 1999).

3.1.3 Advantages and Disadvantages

The following are some advantages of the CVD technique:

1) **Large Area Deposition and Good Film Uniformity:** The film with uniform thickness and performance can be obtained on a large area substrate, which can meet the requirements of industrial production for product consistency.
2) **Suitable for Depositing Diverse Materials such as Metals, Semiconductors, and Ceramics:** It is also cost-effective and easy to use. Appropriate precursors can be selected according to the requirements, and thin films of different materials can be deposited in the same process, which provides the possibility for the composite and multi-functionalization of materials.
3) **Strong Step Coverage Ability and High Rates of Deposition:** Good film coverage can also be achieved on the surface of the substrate with complex morphology, which can be uniformly coated on the surface of the micro-nano structure to meet the strict requirements for film coverage in the field of micro-nano processing.

The following are some disadvantages of the CVD technique:

1) **Limited Control Over Properties and Quality of Film:** Although the CVD method is widely used, the reaction characteristics of the precursor and the complexity of the reaction environment make it difficult to accurately control the microstructure and composition uniformity of the film, thus affecting the control of film performance and quality.
2) **Substrate Limitations:** The CVD method has limitations in substrate selection because its reaction conditions (such as high temperature and specific gas environment) may be incompatible with some substrates, and different substrates will affect the film growth process and final performance, resulting in not all substrates being suitable for this method.

3.2 Thermal Evaporation

3.2.1 Principle and Process

Evaporation is a fundamental process that was first utilized by Faraday in 1857 during the discovery of PVD. In this early experiment, Faraday melted a metal wire via the evaporation process. He employed a substrate and placed the metal on a heating source. When an electric voltage was applied, the metal source melted and evaporated, producing fumes that were deposited onto the substrate, thus forming a thin film with thicknesses ranging from nanometers to microns. However, the desired properties of the thin film were not fully realized at that time. Subsequently, with the advancement of technology, the evaporation process found extensive applications. It adheres to the principles of the typical PVD technique, involving the use of a substrate, anode, cathode, and an applied voltage. The process essentially entails heating a substance by applying an appropriate voltage, which is determined based on the substance's melting point. As the material is heated, its atoms start to vibrate and eventually dislodge (evaporate), dispersing into the air. Due to the presence of a uniform gas environment, the evaporated material travels directly toward the substrate and adheres to it. The amount of coating is accurately measured using a digital device known as a quartz crystal. After deposition, the thin film is subjected to further analysis for various applications. This evaporation process has been employed in numerous commercial processes (Popov 2005). It is cost-effective and can produce thin films with relatively good quality compared to some other methods. The deposition rate in evaporation is quite high, approximately 1000 atoms per second, meaning that 1000 atoms are incorporated into the film within one second, effectively contributing to the film's growth. During the evaporation process, the selection of the source material can be made based on its melting temperature (Deshpandey and Bunshah 1991). The melting point is a crucial factor; if not properly considered, the process may fail, necessitating a shift to alternative methods. Initially, this process was widely used, but over time, its usage decreased due to significant challenges associated with thin-film deposition.

Typically, the substrate is initially affixed to the anode, while the source material is positioned adjacent to the heating element on the cathode. Subsequently, the chamber is sealed, and a vacuum is established within it. Once the targeted vacuum level is achieved, the relevant operations related to vacuum creation are concluded (Westwood 1989). Thereafter, the voltage applied is determined in accordance with the melting point of the source material. If the melting point is relatively high, a higher voltage is required. However, applying a high voltage may generate a large quantity of dense fumes. These fumes can impact the substrate, resulting in surface roughness. Thus, it is necessary to precisely adjust the voltage. When the voltage is applied, the filament begins to heat up, causing the source material to heat as well and produce fumes. These fumes are then deposited onto the substrate. The amount of deposition is accurately measured by an electrical device known as a quartz crystal. This device is capable of detecting fumes from any location within

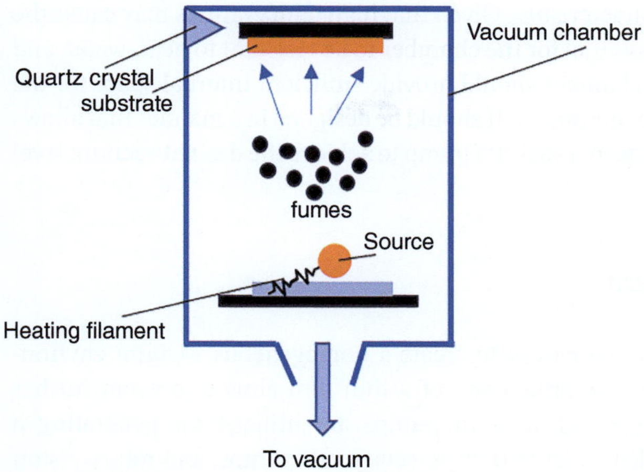

Figure 3.6 Schematic diagram of the thermal evaporation technique.

the chamber. After the deposition process is complete, the film is removed from the chamber for subsequent testing and characterization of its properties. This entire procedure is referred to as the thermal evaporation process (Chopra and Das 1983). Figure 3.6 depicts the schematic diagram of the thermal evaporation technique (Chopra and Das 1983).

The working principle of the thermal evaporation process is centered around the creation of a vacuum environment. A pumping system is employed to establish the vacuum. Once the target vacuum level is reached, the evaporation process proceeds uniformly, and the gas flux stabilizes. The equipment used for thermal evaporation includes several key components necessary for thin film formation. In this method, the heated material vaporizes and then deposits onto the substrate. This process follows a principle similar to the three-step PVD technique. First, the solid material to be deposited (e.g., aluminum, gold, or silver) is heated—usually by applying an electric voltage—causing it to vaporize. Next, the vapor condenses back into a solid film upon contacting the substrate. Through these stages, a thin film is formed and deposited by evaporation (Deshpandey and Bunshah 1991). The construction consists of the following components:

3.3 Vacuum Chamber

The assembly can be housed in a chamber made of either metal or glass. This chamber is a crucial component in the evaporation process, as the thin-film deposition cannot be accomplished without it. The deposition of the film onto the substrate occurs due to gas collisions, high pressure, or the presence of a mixture of different gases and air. For greater accuracy, a glass bell jar is commonly used as the enclosure, with multiple devices attached to it. This chamber needs to tolerate high voltages and have heat-resistant characteristics. Given that high temperatures may cause the chamber to explode, it is essential for the chamber to be resistant to heat, water, and voltage. Furthermore, the chamber should provide sufficient internal space for the installation of all necessary equipment. It should be designed in a manner that allows for easy connection to an external vacuum pump to achieve the desired vacuum level (Safarian and Engh 2013).

3.4 Pumping System

Various vacuum pumps are employed to create a homogeneous vacuum environment within the chamber. The production of a thin film aims to prevent further gas collisions. Multiple types of vacuum pumps are utilized for generating a vacuum, including oil-sealed vacuum pumps, rotary van pumps, and rotary piston pumps. Additionally, diffusion pumps, cryogenic pumps, and gathering pumps are frequently employed. Each type of pump exhibits distinct vacuum-generating speeds and is associated with specific equipment configurations. The selection of a pump depends on the required vacuum level and range. Generally, to achieve high purity, a low deposition rate is preferred. A low temperature is also maintained to prevent surface roughness. Moreover, a uniform vacuum and consistent flux density are essential for the successful formation of a high-quality thin film.

3.5 Substrate

A substrate serves as the material upon which deposition takes place to form a thin film. It can be fabricated from any material as per specific requirements. Typically, the substrate is positioned on the anode or the electrode and remains in place until the deposition process is completed. Subsequently, it is removed from the chamber. The substrate must possess the ability to endure the pressure exerted during deposition and the impact of ions. In the event that it fails to withstand the maximum pressure and fractures, the thin film cannot be successfully formed, and the desired properties of the film cannot be achieved.

3.6 Source

A source refers to the material that gets deposited onto the substrate. It can encompass various types, including gold, silver, chromium, magnesium, and nickel. Nevertheless, the selection of the source material ought to be determined by its melting point. When the material has a low melting point, it can be conveniently utilized, and the resulting film is smooth. Conversely, if the material has a high melting point, a high voltage is applied. Under high-voltage conditions, the filament heats up more rapidly, producing a large number of fumes. These fumes are then deposited on the substrate, leading to a rough film surface. Therefore, the melting point is a key factor influencing whether the deposited film is rough or smooth.

3.7 Filament

The filament or heater is typically positioned on the cathode. One end of it is connected to the voltage supply, and the other end is linked to the source material. Once the voltage is applied, the filament begins to heat up and emit light. A relatively rapid increase in its temperature indicates that it can quickly transfer heat to the source material. As a result, the source material melts and transforms into a vapor state. This process is inherently straightforward. Low-voltage, high-power transformers are employed to supply the necessary voltage to the filament. For applications involving high temperatures, graphite filaments or heaters are commonly utilized (Deshpandey and Bunshah 1991).

3.8 Voltage Supply

A voltage source is necessary to start the reaction. Low-voltage, high-power transformers typically provide the required voltage to the filament, though a high-voltage supply is generally employed during the process.

3.9 Quartz Crystal

A quartz crystal monitor is a digital device installed inside the chamber, displaying readings externally. It measures the vapor deposited on the substrate and provides real-time thickness data (Couturier et al. 2009). By utilizing this tool, precise control over the thin film's thickness can be achieved, allowing users to set and maintain the desired thickness with ease.

3.9.1 Applications in Chalcogenide Films

The electronics field continues pursuing cost-efficient methods to produce large-scale, stable semiconductors exhibiting high mobility. Liu et al. successfully fabricated wafer-sized ultrathin metal chalcogenide semiconductors through room-temperature thermal evaporation, showcasing their potential for high-performance complementary electronics (Liu et al. 2022), as shown in Figure 3.7. N-type bismuth sulfide films transition from conductive to high-performance semiconducting behavior via mild thermal annealing accompanied by spontaneous phase reorganization, enabling transistors with electron mobilities exceeding 10 cm^2 V^{-1} s^{-1} and current switching ratios above 10^8. Combined with p-channel tellurium transistors (showing hole mobilities >50 cm^2 V^{-1} s^{-1}), these materials form complementary logic circuits displaying sharp voltage transfer curves with gains reaching 200. The developed deposition methodology offers a straightforward, cost-efficient route to scalable electronics manufacturing that can be monolithically integrated with existing commercial devices, including organic light-emitting diodes. By rationally designing the device structure and

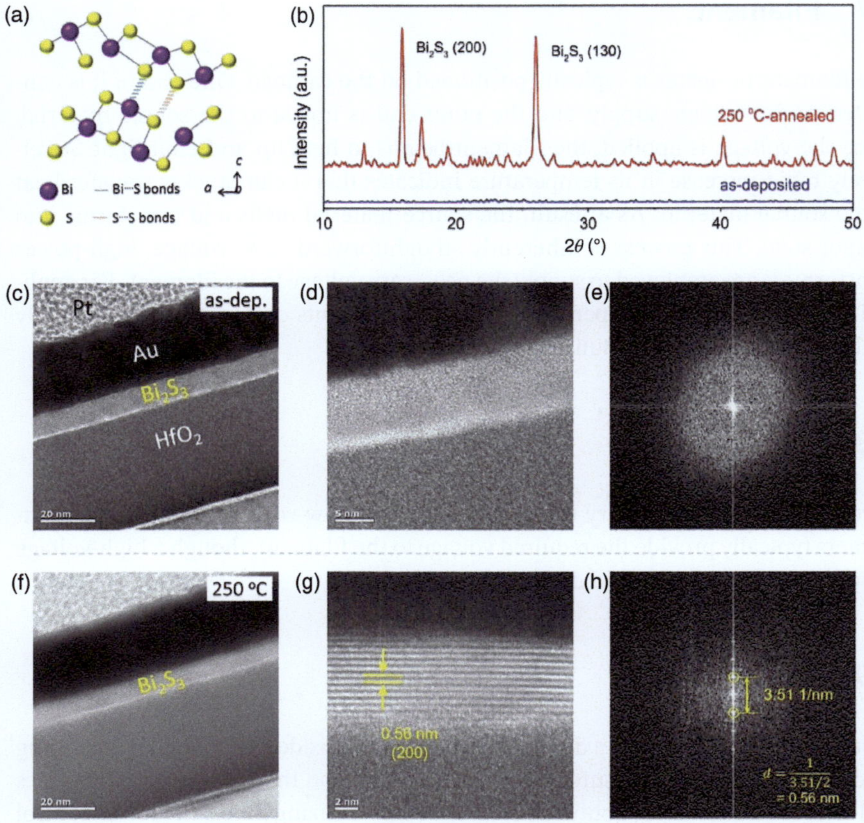

Figure 3.7 Microstructure analysis of Bi_2S_3 thin films. (a) Crystal structure of Bi_2S_3. (b) X-ray diffraction patterns comparing pristine Bi_2S_3 films with those thermally treated at 250 °C. (c–e) High-resolution transmission electron micrographs and selected-area electron diffraction of the as-evaporated films. (f–h) Corresponding HRTEM and SAED data for the annealed samples, with fast Fourier transform analysis (from panel g) revealing lattice spacings of 0.56 nm (3.51 nm^{-1} reciprocal spacing). *Source:* Liu et al. (2022) / Springer Nature / CC BY 4.0.

utilizing the unique electrical and optical properties of chalcogenide films, excellent performances such as low power consumption, high switching ratio, and fast response speed are achieved, which provide a new material system and technical ideas for the development of next-generation high-performance complementary electronic devices. The application of the prepared nano-chalcogenide films in the field of complementary electronic devices, such as integrated circuits, sensors, and display technologies, plays a crucial role.

Evaporative deposition plays a significant role in photovoltaic applications, particularly for fabricating thin-film solar cells utilizing semiconductor materials such as cadmium telluride (CdTe) and copper indium gallium selenide (CIGS). This vacuum-based technique enables precise deposition of photoactive layers onto substrates, yielding films with optimized optoelectronic properties, including enhanced light absorption and charge transport characteristics. The process typically employs rapid thermal evaporation (RTE) to achieve high-quality, uniform coatings essential

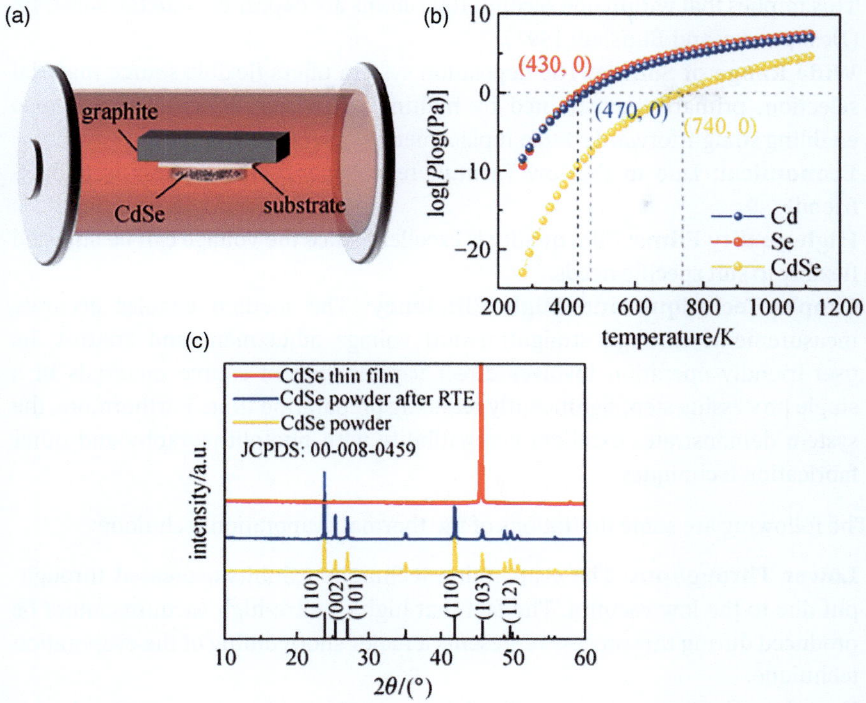

Figure 3.8 (a) Illustration of the rapid thermal evaporation setup. (b) Temperature-dependent vapor pressure plots for Cd, Se, and CdSe. (c) X-ray diffraction patterns comparing CdSe in powder and thin-film forms. *Source:* Li et al. (2021) / Springer Nature.

for efficient solar cell operation. Taking advantage of the high deposition rate (\~2 Å/min) and excellent uniformity of CdSe source materials, Li et al. employed RTE technology to enhance the utilization efficiency of CdSe. They successfully fabricated hexagonal CdSe thin films characterized by large grain size and stable purity (Li et al. 2021), as illustrated in Figure 3.8. The resulting CdSe films exhibited a bandgap of 1.72 eV, a sharp photoluminescence peak, and rapid photo-response. Through optimization of device architecture and layer thickness, the researchers obtained an initial power conversion efficiency of 1.88% in CdSe-based photovoltaic devices, demonstrating the potential of this thin-film solar cell technology. These films serve as active layers in thin-film photovoltaic devices, offering enhanced flexibility, reduced weight, and improved cost efficiency compared to conventional silicon solar cells. CdSe is now deposited via evaporation techniques, enabling the development of lightweight, adhesive solar cells with superior light absorption and economic viability.

3.9.2 Advantages and Disadvantages

The following are some advantages of the thermal evaporation technique:

1) **High Deposition Rate:** The evaporation method offers a significant benefit due to its high deposition rate, reaching approximately 1000 atoms per second.

This implies that within one second, 1000 atoms are deposited onto the substrate (Deshpandey and Bunshah 1991).

2) **Wide Range of Source:** The deposition system offers flexible source material selection, primarily constrained by melting point considerations, while also enabling straightforward source replacement when needed.
3) **Economical:** Due to the low vacuum requirements, the process is budget-friendly.
4) **High-quality Films:** Film quality is excellent since the voltage can be adjusted flexibly to suit specific needs.
5) **Simple Technique and High Efficiency:** The method enables accurate measurements through straightforward voltage adjustment and control. Its user-friendly operation involves direct vaporization of source materials in a single processing step, significantly reducing preparation time. Furthermore, the system demonstrates excellent compatibility with photolithography and other fabrication techniques.

The following are some limitations of the thermal evaporation technique:

1) **Lower Throughput:** The evaporation technique exhibits decreased throughput due to the low vacuum. The fact that high or ultra-high vacuum cannot be produced during this process represents a major shortcoming of the evaporation technique.
2) **Poor Step Coverage:** Additionally, it demonstrates subpar step coverage. In this technique, due to the low vacuum and insufficient voltage supply, the source material fails to deposit completely. It is also challenging to attain uniform deposition across the substrate; some areas receive excessive deposition while others remain relatively unchanged.
3) **Small Area Deposition:** This technique is not applied in large-scale production.
4) **Poor Universality of Materials:** Generally, this technique is inappropriate for a wide range of materials. Instead, it is restricted to just a handful of substances, all of which tend to be costly.
5) **Poor Purity:** It is hard to maintain the material's purity because it has the potential to react with other substances, giving rise to roughness and inaccurate deposition (Jamkhande et al. 2019).

3.10 Pulsed Laser Deposition

3.10.1 Principle and Process

Shortly after the first ruby lasers became available in the early 1960s, the initial reports emerged regarding the use of laser radiation pulses to "ablate," or eliminate, particles from a solid (or liquid) target. Given the clear efficiency of the material ablation process, it did not take long for the pulsed laser technique to be employed as a means for thin-film deposition. PLD, also known as pulsed laser ablation (PLA), is a multifunctional and relatively simple experimental method, which is widely

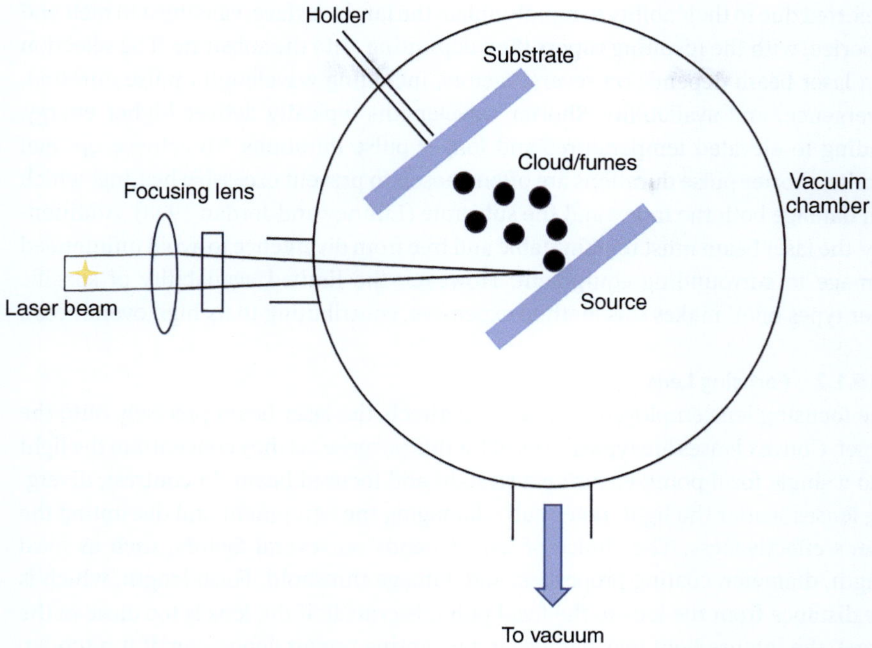

Figure 3.9 Schematic diagram of PLD.

used in thin film deposition and multilayer research to process various materials. The underlying principle relies on a physical deposition process. In this process, condensed substances are transformed into the vapor phase and subsequently revert to the condensed phase. While the technique appears straightforward in theory, its apparent simplicity masks a complex array of fascinating and still poorly understood molecular physics.

The process involves several stages, starting with the initial interaction between the laser and the target, which transforms solid material into a gaseous state. This is followed by the dynamics of the ejected material cloud, including its propagation and processing, and ultimately concludes with the deposition of gas-phase species onto the substrate to form the desired coating (Ashfold et al. 2004). This process utilizes a high-energy laser beam that is focused on the target material. When the laser strikes the target, its high energy causes the material to heat and turn into vapor. The vapor then settles on the substrate, creating a thin film. A schematic representation of PLA can be seen in Figure 3.9 (Chopra and Das 1983).

The structure consists of various elements, each possessing distinct characteristics:

3.10.1.1 Laser Beam

The laser beam's wavelength is chosen based on the desired properties of the thin film, spanning from the UV to the IR spectrum. Since its discovery in 1960, lasers have been utilized across a wide range of wavelengths, from nanometers to micrometers, with varying energy and temperature levels. For PLD, high-energy beams are preferred due to their ability to rapidly ablate the target surface, causing it to melt and vaporize, with the resulting vapors then depositing onto the substrate. The selection of a laser beam depends on several factors, including wavelength, pulse duration, divergence, and availability. Shorter wavelengths typically deliver higher energy, leading to elevated temperatures and longer pulse durations. To achieve optimal results, shorter pulse durations are often chosen to prevent excessive heating, which can damage both the target and the substrate (Lunney and Jordan 1998). Additionally, the laser beam must remain stable and free from divergence to avoid unintended damage to surrounding equipment. However, the limited availability of specific laser types often makes this method expensive, contributing to its high overall cost.

3.10.1.2 Focusing Lens

The focusing lens employed in this setup directs the laser beam precisely onto the target. Convex lenses are typically used for this purpose, as they concentrate the light into a single focal point, ensuring a straight and focused beam. In contrast, diverging lenses scatter the light, potentially damaging the equipment and disrupting the laser's effectiveness. The choice of lens depends on several factors, such as focal length, diameter, coating properties, and damage threshold. Focal length, which is the distance from the lens to the focal point, is crucial: if the lens is too close to the target, the intense light may damage it, preventing proper deposition; if it is too far, only a portion of the target will be ablated. Coating properties are also important, as the lens should minimize reflection, maximize transmission, and possess high optical quality to avoid damaging the target. Additionally, the lens must have a damage threshold that matches the intensity of the laser to prevent burning or harming the target material. Safety is another consideration, as direct exposure to the laser beam can cause severe eye and tissue damage (Stafe et al. 2014). The lens selection is further influenced by the desired spot size and contamination concerns, with specific choices tailored to meet the experimental requirements.

3.10.1.3 Rotor

In this process, a rotor or vacuum chamber is utilized. The rotor is essential because continuously directing the laser beam onto the same spot on the target could damage the vacuum chamber. To prevent this, the target is mounted on the rotor, which rotates continuously, ensuring the beam strikes different areas of the target and enabling uniform ablation. Similarly, the substrate is also rotated to ensure even deposition across its entire surface, as stationary placement would result in deposition only on a single spot. Both the rotational speed and temperature must be carefully regulated. If the rotation is too slow or too fast, it can harm the target and substrate, hindering proper deposition. Additionally, excessive rotational speed can lead to increased temperatures, which, combined with the high temperature of the laser beam, may damage the setup.

3.10.1.4 Source and Substrate

The source and substrate are influenced by various factors, particularly temperature, as they must withstand the high temperatures and intense light from the laser beam. Their optical properties, such as absorption and transmission, are also critical, as the source needs to absorb the light effectively, reaching its melting point to vaporize and deposit onto the substrate. Chemical stability is essential to prevent reactions with the intense light, and the substrate size must match the source, neither too large nor too small, to ensure uniform coverage. Additionally, the substrate must endure continuous rotation and exposure to high-temperature vapors during deposition. Selection criteria for both the source and substrate include electrical conductivity, resistivity, optical characteristics, and thermal stability to ensure optimal performance.

3.10.1.5 Pumping System

To achieve ultra-high vacuum levels, specialized vacuum pumps with high pumping speeds are employed. These pumps must be efficient, reliable, and capable of maintaining consistent performance. Key features include compact size, high energy efficiency, and the ability to produce ultra-high vacuum conditions, which are essential for depositing high-quality thin films. Additionally, the pumps should be cost-effective, with low maintenance requirements, and incorporate robust safety features to ensure secure operation (Hablanian 1984).

3.10.2 Applications in Chalcogenide Films

WTe_2 is a 2D-layered material classified as a transition metal dichalcogenide (TMD) semimetal with a distinctive 1T (Td) crystal structure. It holds significant promise for applications in spintronics, electronics, and optics. The WTe_2 film can be used to generate femtosecond mode-locked laser pulses, control the spin-orbit torque, and serve as a metal contact material for 2D semiconductors. In addition, WTe_2 is also predicted to play an important role in tunneling field-effect transistors. However, the transport behavior of WTe_2 is significantly different with the change of thickness, especially electrostatic gating can control the interplay between weak localization and weak antilocalization. In order to meet the needs of future device applications, it is very important to develop ultra-thin material preparation methods that can provide a large area and high yield. Gao et al. prepared large-area, uniform, and high-grade WTe_2 ultrathin films via the PLD technique, followed by an annealing process (Gao et al. 2017), as shown in Figure 3.10. Through the manipulation of the Te vapor pressure during annealing, they succeeded in boosting the transport properties. This was because a significant number of Te vacancies were removed. MR tests demonstrated that in WTe_2 ultrathin films, weak localization and weak antilocalization competed with each other, and this competitive relationship could be controlled by the level of Te vacancies.

3.10.3 Advantages and Disadvantages

The following are some advantages of the PLD technique:

1) **Highly Controllable:** The high controllability of this process represents its greatest benefit. It allows for the adjustment of every component in line with

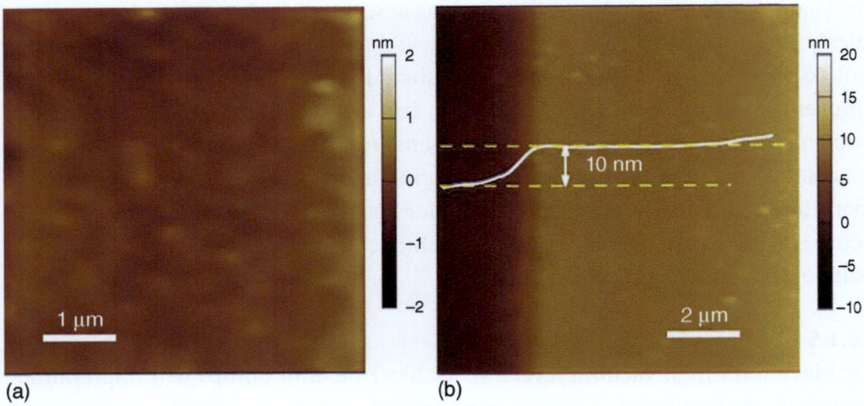

Figure 3.10 Surface morphology and thickness of the WTe_2 ultrathin film. (a) AFM analysis of the surface roughness. (b) Thickness of the WTe_2 ultrathin film. *Source:* Gao et al. (2017) / AIP Publishing.

our requirements, ensuring a highly accurate coating over the entire surface. Through the manipulation of factors including temperature, rotational speed, intensity, voltage, and material selection, a fine thin film coating can be produced. Moreover, PLD can be executed using a computer-controlled setup, making it a remotely operable system. After placing the material and substrate, the process can be managed through computer-based systems, thus enabling remote processing.

2) **Environment-Friendly:** Since this process does not utilize any lubricants or other fluids, it is considered a clean and dry method. As a result, it is also regarded as environmentally friendly, as it does not cause any environmental pollution.
3) **High Universality of Materials:** This method is highly adaptable, as it can be utilized with various materials depending on the wavelength and intensity of the laser light employed. Lasers are capable of operating across a broad spectrum, from nanometers to micrometers, and offer a range of intensities with distinct energy levels and temperatures, making them exceptionally versatile.
4) **High Purity and Quality:** This process ensures a high level of purity, and the resulting films are both highly pure and of superior quality. Such films have various uses across different domains, including medical and engineering.

The following are some disadvantages of the PLD technique:

1) **Temperature Sensitive:** If the temperature of a single component does not meet the specified criteria, it might inflict significant harm on the remaining components, causing the film to become rough.
2) **Highly Specialized:** Nonprofessional handling of this can cause harm to the eyes, nerves, and biological tissues, and there is also a risk of explosion if not used correctly. The process is highly challenging to execute because of its intricacy. Only professional physicists or engineers are capable of using it, but such specialists are scarce. Furthermore, a large number of them lack the knowledge to operate it and manage the entire setup.

3) **Costly:** This process incurs substantial expenses. Each component is imported from abroad, and the laser light, which is also costly, is often in short supply. Consequently, should any damage occur to the equipment, the maintenance expenses will be extremely high.

3.11 Sputtering

3.11.1 Principle and Process

Sputtering is a commonly used method for depositing thin films onto surfaces. It involves bombarding a target material, which acts as the source, with ions. When these ions strike the target, they cause sputtering—a physical process that produces vapor from the material. This places sputtering within the category of PVD techniques. In this process, gas is ionized by applying a voltage, and the resulting ions erode the target's surface. Additionally, the ionization generates secondary electrons, which may interact with the substrate and potentially cause surface damage (Depla et al. 2010). The reactive gas initiates the sputtering and deposition process. An important aspect is the formation of compound layers, where the interaction between the reactive gas and target material causes nonlinear effects on deposition parameters due to variations in gas flow. To accurately describe this behavior, it is essential to determine the fluxes of different species reaching the target. Similarly, the fluxes arriving at the substrate are critical because they affect the reactivity of the sputtering process and influence the structure of the resulting

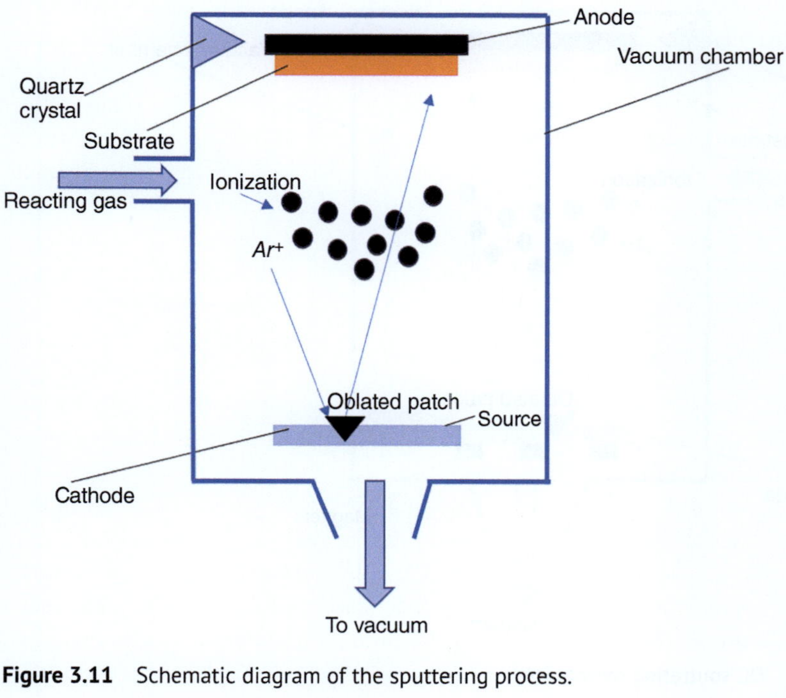

Figure 3.11 Schematic diagram of the sputtering process.

film. The characteristics of the particles arriving at the substrate significantly impact the microstructure of films produced by magnetron sputtering. A schematic illustration of the sputtering process is shown in Figure 3.11 (Chopra and Das 1983).

According to the different source types, sputtering is divided into radio frequency (RF) sputtering and direct current (DC) sputtering.

DC sputtering is a widely used sputtering method that employs a DC to deposit material layers onto substrates. It is particularly suitable for metals because it allows for large-area vaporization and produces a stable, uniform plasma. In this technique, the cathode acting as the source must be electrically conductive to ensure effective ion bombardment and material deposition on the substrate. Although DC sputtering is relatively simple to operate, it tends to be time-intensive and expensive, mainly because it requires an extra electrode for the source. Secondary electrons are generated at the cathode, but magnets placed around the cathode deflect these electrons, causing them to circulate around the cathode rather than reaching the substrate. This technique is also referred to as magnetron sputtering. A limitation of DC sputtering is that the plasma is confined to the cathode and does not interact effectively with the reactive gas. Adjusting the magnet arrangement can address this issue, but it often results in reduced efficiency and nonuniform plasma (Safi 2000). Figure 3.12 depicts the DC sputtering mechanism. By using magnets around the cathode, the generation of secondary electrons is controlled, ensuring they circulate around the cathode and do not interfere with the substrate, thereby aiding in the formation of smooth thin films.

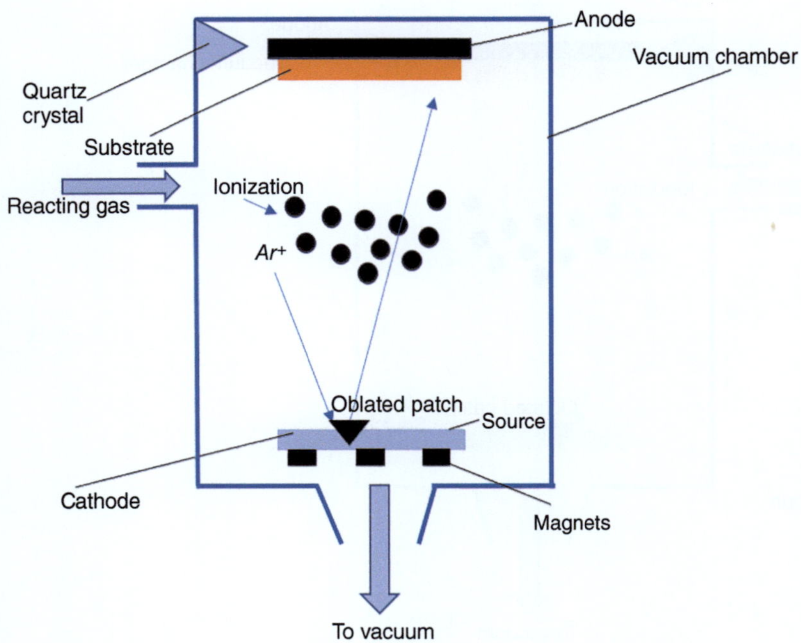

Figure 3.12 DC sputtering mechanism.

RF sputtering, also known as RF or AC sputtering, involves the use of magnets during the sputtering process. In this method, a specific RF—usually below 50 kHz—is applied to regulate the behavior of secondary electrons. This alternating potential causes the electrons to oscillate near the cathode by periodically reversing the voltage. This ensures that the plasma is not confined solely to the cathode, allowing for more uniform plasma distribution and effective reactive gas sputtering. At frequencies above 50 kHz, electrons do not discharge from the cathode, and the applied potential flows between the two electrodes, enabling sufficient ionization. In this alternating voltage setup, the process operates in cycles. During one half-cycle, electrons ablate the surface and move toward the substrate for sputtering, while in the next half-cycle, they prevent charge buildup on the surface. This method is particularly advantageous for brittle insulating materials, though it has a significant drawback: insulating materials often have low thermal conductivity, and applying high voltages can cause the target material to break, halting further deposition (Depla et al. 2010). Figure 3.13 illustrates the RF sputtering mechanism (Kiani et al. 2013). By using RFs below 50 kHz, alternating voltages are generated, which prevent secondary electron deposition and create a glow around the cathode, resulting in high-quality thin films compared to reactive gas sputtering (Orjuela et al. 2012).

The underlying principle relies on physical deposition, as this method falls under the category of PVD. It involves the transformation of condensed matter, but instead of generating vapors, a portion of the source material is ablated through ion bombardment. Ions are produced via the ionization of gas when voltage is applied. Upon

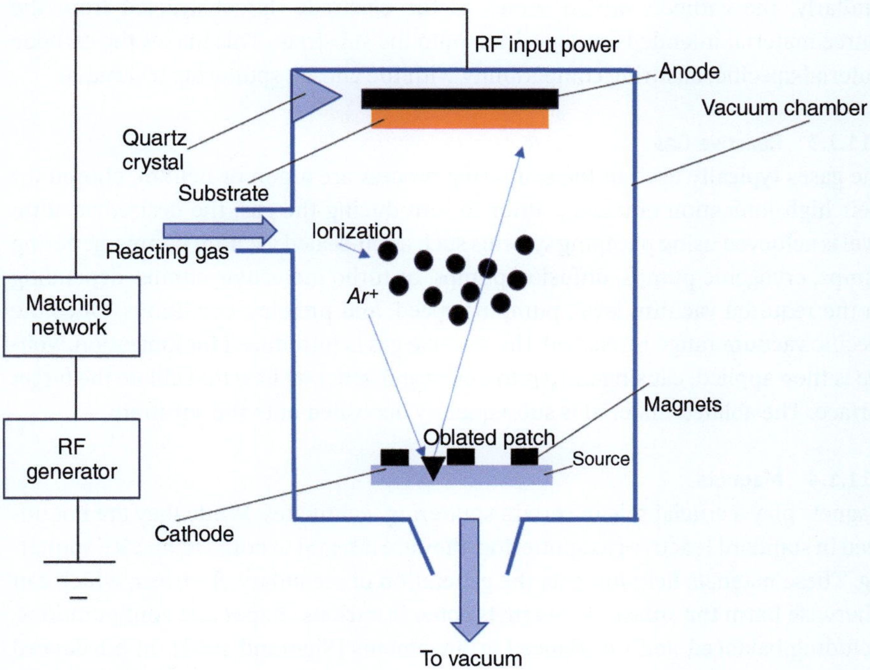

Figure 3.13 RF sputtering mechanism.

supplying voltage, the gas ionizes and ablates a section of the source material, which is then deposited onto the substrate to form a thin film. Both RF and DC sputtering operate on this principle, though they incorporate magnets to deflect secondary electrons. The setup includes various components as follows:

3.11.1.1 Vacuum Chamber

A vacuum chamber is typically constructed from materials like glass, steel, or metal, designed to withstand extreme conditions such as high temperatures, pressure, water exposure, light, and heat. It must be durable, unbreakable, and capable of accommodating all necessary equipment. The chamber should be leak-proof, resistant to heat and water, and equipped with pressure regulation and high pumping speed to achieve an ultimate vacuum level. Seals, valves, and feed-throughs must be thoroughly inspected and environmentally safe (Stuart 2012). These criteria guide the selection and construction of a vacuum chamber, which is essential for physical deposition processes. The requirements are similar to those for other physical deposition techniques.

3.11.1.2 Substrate and Cathode

The substrate used for thin-film deposition can be composed of various materials, with the choice depending on the required temperature range. For metals, conductors, and insulators, reactive gas sputtering is employed. DC sputtering equipment is utilized for conductive materials, while RF sputtering is preferred for insulating materials, as each method is tailored to the specific properties of the material. Similarly, the cathode, which serves as the electrode, is constructed from the source material intended for deposition onto the substrate. This makes the cathode material-specific, ensuring compatibility with the chosen sputtering technique.

3.11.1.3 Reactive Gas

The gases typically used in the sputtering process are argon or helium, chosen for their high ionization efficiency. Prior to introducing the gas, the desired vacuum level is achieved using pumping systems such as oil-sealed vacuum pumps, gettering pumps, cryogenic pumps, diffusion pumps, or turbo molecular pumps, depending on the required vacuum level, pumping speed, and pressure conditions. Once the specific vacuum range is reached, the reactive gas is introduced for ionization. Voltage is then applied, causing the gas to ionize and generate ions that ablate the target surface. The ablated material is subsequently deposited onto the substrate.

3.11.1.4 Magnets

Magnets play a crucial role in certain sputtering techniques. While they are not utilized in standard reactive gas sputtering, they are integral to both DC and RF sputtering. These magnets help mitigate the generation of secondary electrons, which can otherwise harm the substrate. Magnets come in various shapes and configurations, including balanced and unbalanced arrangements (Sigmund 1993). In a balanced setup, magnets are organized in a specific pattern to deflect electrons, causing them to circulate around the cathode and create a glow. However, in DC sputtering, this arrangement confines the plasma to the cathode, limiting its interaction with the gas. To address this, an unbalanced arrangement is employed, allowing some electrons to reach the substrate, while others continue to orbit the cathode (Depla and Mahieu 2008). This dual functionality enhances the sputtering process.

3.11.1.5 RF Generator

In RF sputtering, an RF generator is employed, operating within a specific frequency range below 50 kHz. Frequencies above 50 kHz prevent electrons from discharging from the cathode, causing the voltage to remain confined between the electrodes, which can lead to film roughness. The frequency range is monitored using a matching network, a digital circuit designed to detect and calculate the RF, displaying the results accordingly (Rossnagel 2001). The matching network is connected to the anode, with RF input power supplied to it, while the entire chamber is grounded.

3.11.2 Applications in Chalcogenide Films

For many years, the development of sensors has been an important research direction in the field of optics, one of the important aspects of which involves the creation and integration of optical elements for micro-optical sensors. Central to any sensing platform is the optical waveguide, which facilitates the propagation of light or laser beams. Planar or ridge waveguides are particularly effective due to their ability to guide and propagate light within a confined space. Materials with high refractive indices, such as glassy substances, enable precise light control at microscopic scales, making them ideal for high-index-contrast photonic devices. Chalcogenide glasses have emerged as promising materials for infrared sensors. These glassy materials are mainly composed of chalcogen elements such as sulfur, selenium, and tellurium, often alloyed with elements like arsenic, germanium, gallium, or antimony. They can be processed into optical fibers or manufactured as thin-film waveguides. Due to their distinctive features—such as a wide optical transmission window spanning from the visible to the far-infrared region (extending to around 20 μm), high refractive indices (generally ranging from 2 to 3), and notable nonlinear optical properties—they are gaining significant interest for use in advanced photonic applications.

Verger et al. successfully prepared amorphous chalcogenide films and gold NPs by RF sputtering and DC sputtering, demonstrating their application in surface-enhanced infrared absorption spectroscopy (Verger et al. 2013), as shown in Figure 3.14. Due to its anti-crystallization and excellent optical properties in the infrared band, $Ge_{25}Sb_{10}Se_{65}$ thin films were selected as the ideal substrate for gold NP deposition. The prepared Au/Ge-Sb-Se heterostructure exhibits an enhancement factor of about 24 times in the NO_2 symmetric stretching vibration band at 1336 cm^{-1}, demonstrating its potential for detecting biochemical molecules in the infrared spectral range. This paper demonstrates the application of sputtering technology in the preparation of high-quality chalcogenide thin films and gold NPs, and provides new ideas for the development of infrared spectral sensors.

(a) (b) (c)

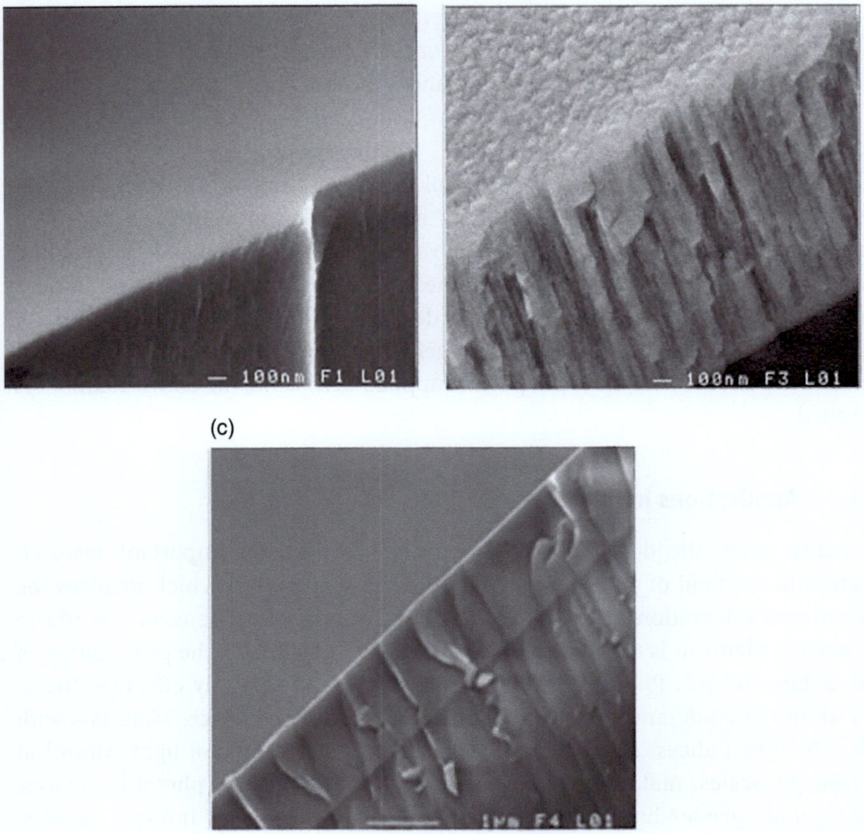

Figure 3.14 SEM cross-sectional pictures of $Ge_{25}Sb_{10}Se_{65}$ thin films: (a) 5.10^{-3} mbar, 40 W; (b) 5.10^{-} mbar, 40 W; and (c) 5.10^{-3} mbar, 60 W. *Source:* Verger et al. (2013) / Optica Publishing Group.

Chalcogenide thin films used in phase-change random access memory (PRAM) are of considerable technological value due to their ability to rapidly and reversibly switch between amorphous and crystalline states when heated. These materials—typically composed of sulfur, selenium, and tellurium—are key components in PRAM devices. A widely used example is the $Ge_2Sb_2Te_5$ (GST) alloy, which has been adopted in commercial mobile devices and Intel Optane Memory. GST's suitability stems from its excellent switching endurance (capable of millions of cycles), nanoscale scalability down to 1.4 nm while maintaining phase-change functionality, and full compatibility with CMOS processes. To induce crystallization (Set operation), GST must be heated above its crystallization point (T(C) = 483 K) but below its melting point (T(M) = 873 K). Conversely, transforming it into the amorphous phase (Reset operation) requires a brief but intense thermal pulse, delivered optically or electrically. However, this high energy requirement poses challenges for applying phase-change materials in reconfigurable and tunable photonic systems.

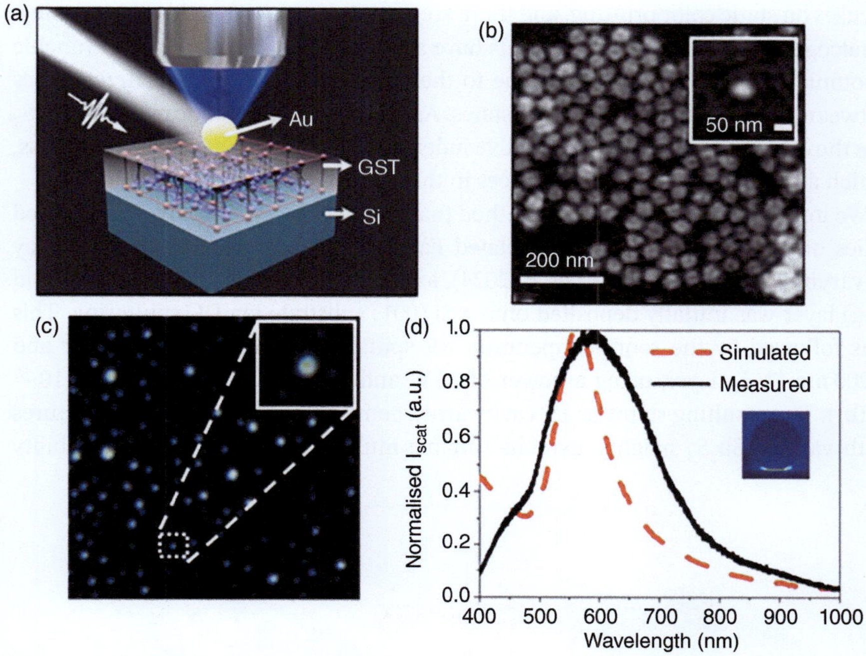

Figure 3.15 (a) The optical setup applied for analyzing the scattering spectrum and performing dark-field measurements of isolated Au NPs deposited onto a GST thin film. (b) An image obtained via electron beam lithography (EBL) that shows the Au NPs. (c) A dark-field microscopy image presenting 25 nm radius Au NPs on the GST film. (d) The measured and calculated normalized Iscat spectra of the Au NPs in combination with the GST film. *Source:* Cao et al. (2020) / John Wiley & Sons.

To reduce power, we deposited GST films on silicon substrates by RF sputtering and deposited gold NPs on GST films by drop casting (Cao et al. 2020), as shown in Figure 3.15. Raman spectroscopy results indicate that a GST film can crystallize with an optical power as low as 2 mW. This power level is approximately one-tenth of that required by materials without embedded NPs. The reduced power consumption is attributed to the increased light absorption enabled by the strong localized surface plasmon resonance (LSPR) of the Au NPs. The Au NPs embedded in the GST film scatter light at a wavelength of $\lambda = 587$ nm, which is near the LSPR wavelength of around 535 nm for Au NPs. A 532 nm laser is employed to measure the Raman scattering of the Au–GST system. Compared to a pure GST film, the Raman scattering of the Au–GST system is enhanced by a factor of 12. These findings suggest that the combination of GST film and Au NPs is well-suited for high-speed, low-power phase-change memory applications and can serve as a novel type of tunable surface-enhanced Raman scattering substrate.

Traditional color filters rely on pigments, which have problems such as low resolution, high cost, and poor color durability. Structural colors produce colors through optical resonance or interference effects of nanostructures, which have the advantages of high resolution, low cost, and high stability. However, most of the research focuses on static color printing, and there are few studies on dynamic color printing. Chalcogenide phase-change materials have attracted extensive attention in tunable photonic devices in recent years due to their fast phase transition characteristics between amorphous and crystalline states. As a typical chalcogenide material, Sb_2S_3 has the characteristics of high refractive index contrast, wide band gap, and low loss, which are suitable for photonic devices in the visible light band.

We introduced a color printing method that can switch dynamically. This method relies on Sb_2S_3-based stepwise pixelated Fabry–Pérot (FP) cavities with an array of varying cavity lengths (Liu et al. 2024), as shown in Figure 3.16. A 100 nm gold (Au) layer was initially deposited onto a Si (001) substrate via DC sputtering. This was followed by the room-temperature RF sputtering of a 5 nm Si_3N_4 layer and a 200 nm Sb_2S_3 layer, using a power of 30 W and a chamber pressure of $5 \times 10{-}^{6}$ mTorr. The resulting stepwise FP cavity array, composed of air/Sb_2S_3/Au structures with varying Sb_2S_3 heights, exhibits tunable multicolor behavior. This tunability

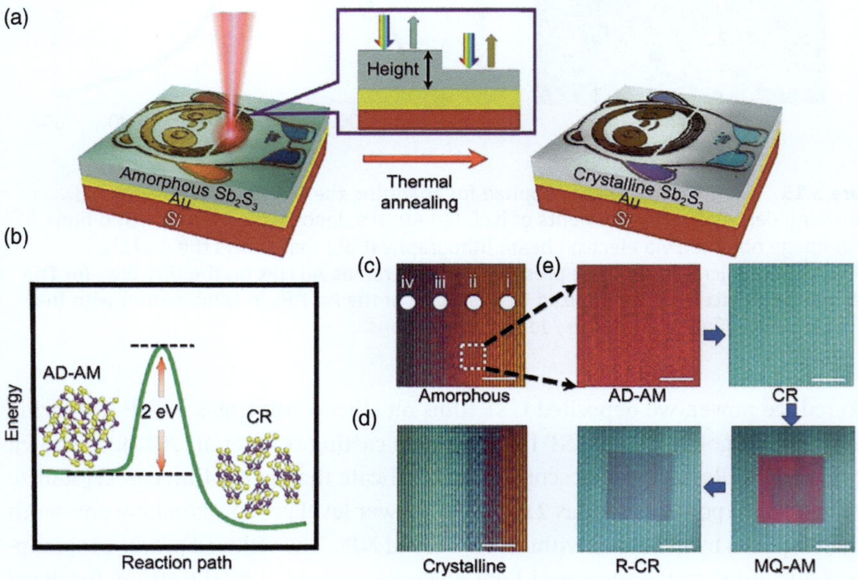

Figure 3.16 (a) A diagram illustrates a dynamic color display system that employs a stepwise FP cavity array based on Sb_2S_3. The system shows the material in its amorphous state (left column) and crystalline state (right column). The stepwise FP cavity array is composed of an air/Sb_2S_3/Au structure, where the Sb_2S_3 strips vary in height. The colors reflected by this structure depend on the phase and thickness of the Sb_2S_3. An inset provides a detailed schematic of the stepwise FP cavity array. Pixelated Sb_2S_3 strips of different heights, fabricated through a lithography-free femtosecond (fs) laser patterning technique, are placed between an air-capping layer and an Au mirror reflector. In the FP cavity, resonant modes are formed, enabling the selective reflection of light of specific colors. (b) Atomic structures of Sb_2S_3 in its amorphous and crystalline phases are presented on an enthalpy-order schematic plot. Photographs of the stepwise FP resonator array are shown (c) prior to the crystallization of the Sb_2S_3 strips within the cavity and (d) after the crystallization process. (e) Reversibly switchable photographs of strip (ii) from figure (c) are also included. The scale bar indicates a length of 100 μm. *Source:* Liu et al. (2024) / Institute of Optics and Electronics, Chinese Academy of Sciences / CC BY 4.0.

arises from the significant refractive index change in Sb_2S_3 across the visible spectrum as it transitions between amorphous and crystalline phases. Additionally, sub-50-nm nano-gratings were directly patterned on an ultrathin Sb_2S_3 film using microsphere-assisted femtosecond laser irradiation (800 nm) in the far field. By adjusting the Sb_2S_3 film thickness, the grating feature size was further reduced to approximately 45 nm ($\lambda/17$). These rapidly reconfigurable Sb_2S_3-based photonic structures show strong potential for applications such as rewritable paper, dynamic displays, information encryption, adaptive camouflage, and anti-counterfeiting. Notably, our work explores the possibilities of quickly fabricating and rewriting periodic structures with nanoscale resolution, and it can provide guidance for the future development of chalcogenide-based photonics components.

3.11.3 Advantages and Disadvantages

The following are some specific advantages of sputtering:

1) **Highly Controllable:** By adjusting the sputtering parameters (such as sputtering power, gas pressure, and target composition), the thickness, composition, and structure of the film can be accurately controlled.
2) **Extensive Material Applicability:** Sputtering can be used to deposit a variety of materials, including metals, alloys, oxides, nitrides, sulfides, etc. Almost any solid material can be deposited into a thin film by sputtering. And the sputtering technique is not sensitive to temperature.
3) **Large-area Deposition, High Step Coverage, and High Quality:** The sputtering technique can be used for thin film deposition on large-area substrates, which is suitable for industrial production. The sputtering film has good step coverage on complex surfaces (such as steps and holes), which is suitable for multi-layer structures in microelectronic devices. During the sputtering process, the purity of the target directly affects the purity of the film, so high-purity films can be obtained by using high-purity targets. High-quality films with good uniformity and density can be prepared by sputtering. The film has strong adhesion and a smooth surface.

The following are some disadvantages of sputtering:

1) **Costly and Time-consuming:** Changing the source proves to be difficult. During evaporation, the source can be readily replaced for deposition purposes. In contrast, sputtering requires the creation of a new cathode, which takes a significant amount of time and comes with a hefty price tag.
2) **Low Target Utilization:** During the sputtering process, the utilization rate of the target material is low. Usually, only 30%–40% of the target material is effectively utilized, and the rest may be deposited on the chamber wall.

4

Optical Properties

Chalcogenide materials exhibit unique and technologically significant optical characteristics that render them indispensable for a wide range of applications in optoelectronic devices, including but not limited to optical sensors, microelectronic components, photodetectors, and laser systems. Both bulk chalcogenide semiconductors and their nanoscale counterparts have attracted considerable research attention due to their versatile functional properties. This chapter systematically reviews fundamental theoretical frameworks and spectroscopic characterization techniques pertinent to chalcogenide materials.

In Section 4.1, we presented the foundational principles of electromagnetic theory and established its application framework for the electromagnetic analysis of chalcogenide materials. Subsequently, Section 4.2 systematically investigated the methodological framework for dielectric function characterization, with a particular focus on elucidating the optoelectronic properties of representative chalcogenide semiconductor systems. The discussion then progressed to Raman spectroscopic analysis in Section 4.3, where we conducted a detailed examination of both the theoretical underpinnings of Raman spectroscopy and the distinctive spectral signatures exhibited by various chalcogenide compounds. Furthermore, Section 4.4 provided a comprehensive analysis of emission spectrum characteristics, specifically addressing the unique photoluminescent behaviors observed in nanoscale chalcogenide structures. Finally, Section 4.5 encompassed an in-depth exploration of light scattering mechanisms and their characterization methodologies as applied to chalcogenide material systems.

4.1 Macroscopic Electrodynamics

This segment initiates with an exposition of the foundational principles inherent to density functional theory (DFT). Commencing with the seminal Hohenberg–Kohn theorem, it subsequently encapsulates the pivotal concepts and approximations that are cornerstones of DFT computations, incorporating the Kohn–Sham framework and the local density approximation (LDA). These constructs facilitate the reduction of complex, many-body interactions to a simplified, single-particle scenario, thereby facilitating the computation of a solid's electronic ground state. Additionally,

Chalcogenide Nanophotonics, First Edition. Tun Cao, Ying Su, Yinghan Li, and Hongze Gao.

the discourse delves into the utility of DFT within the realm of chalcogenide semiconductors, highlighting its efficacy in discerning the structural nuances across crystalline, liquid, and amorphous forms, while elucidating both the fundamental characteristics of the electronic ground state and the intricacies of bonding dynamics. DFT emerges as an indispensable instrument in this context, demonstrating its profound capacity for structural analysis and electronic structure understanding.

4.1.1 Overview of Electrodynamics in Chalcogenides

DFT constitutes a highly efficacious computational methodology, particularly renowned for its capacity to model the electronic ground state of multiatomic assemblies accurately. It is increasingly pivotal in chalcogenide semiconductor research. To achieve computational tractability, a suite of simplifications is employed, grounded in the premise that the rapid dynamics of electronic states permit treating them as static relative to ionic displacements. This principle, known as the Born–Oppenheimer approximation (Born and Heisenberg 1985), facilitates a distinct decoupling of electronic and ionic degrees of freedom. The inherently complex many-electron scenario is thereby ingeniously transformed into a more manageable single-particle scenario.

While the Hartree–Fock approach represents an early endeavor in this simplification, approximating the many-body wave function through a single Slater determinant, it inherently disregards electron correlation beyond the level of Pauli's exclusion principle, encompassing only the Hartree and Fock terms. This limitation led to the ascendency of DFT, which, in theory, offers a more comprehensive and exact treatment of the electronic ground state. Despite aiming for exactitude, practical DFT implementations necessitate the use of a hierarchy of approximations, collectively refining the description of electron interactions beyond what is achievable with the Hartree–Fock framework. These advancements underscore DFT's pivotal role in contemporary materials science, where it balances accuracy with computational feasibility.

In DFT, the comprehensive depiction of ground-state characteristics, including the total energy

$$E = \int v(r)\rho(r)dr + F[\rho(r)] \tag{4.1}$$

is fundamentally rooted in the charge density distribution, a principle initially elucidated by Hohenberg and Kohn (1964). Their groundbreaking theoretical contribution formulated a fundamental theorem stating that (i) there exists a universal functional $F[\rho(r)]$ that completely determines the system's total energy through its dependence on the electron density distribution $\rho(r)$. This functional encapsulates the complexity of electron–electron interactions within a single variable, thereby simplifying the many-body problem. (ii) The theorem further establishes that the total energy E reaches its global minimum E_0 if and only if the electron density $\rho(r)$ equals the true ground-state density $\rho_0(r)$ for a specified external potential $v(r)$. This profound insight underscores that, in the absence of ground-state degeneracy, the unique mapping between the energy and the true charge density is established. Consequently, the Hohenberg–Kohn theorem not only rigorously establishes the existence of this functional relationship but also offers a computational methodology for ground-state determination through variational minimization of the energy functional with respect to $\rho(r)$. This methodology provides the theoretical foundation for DFT-based computational approaches, enabling the determination of both electronic energies and structural properties while circumventing the need for explicit solution of the many-body Schrödinger equation. Levy's (Levy 1979) contribution to the field refined and expanded upon the Hohenberg–Kohn theorem by providing an alternative, rigorous formulation is given by

$$E[\rho(r)] = \min_{[\rho(r)]} \langle \psi \mid H \mid \psi \rangle \tag{4.2}$$

In this condition, the relation between wave functions Ψ and charge density distribution $\rho(r)$ is

$$\rho(r) = \left\langle \varphi \left| \sum_i \delta(r - r_i) \right| \psi \right\rangle \tag{4.3}$$

where H denotes Hamiltonian and is the index of spatial coordinate. So, we have the following expression:

$$E[\rho(r)] \geq E[\rho_0(r)] = E_0 \tag{4.4}$$

As of yet, the precise formulation of the total energy functional remains elusive; nonetheless, the methodology for ascertaining electronic ground-state characteristics can be conceptually delineated. This entails initially identifying an appropriate approximation for $E[\rho_0(r)]$, subsequently employing the variational principle to interrogate the total energy functional

$$\delta E[\rho(r)] = 0 \tag{4.5}$$

so that the number of particles N is given by

$$\int \rho(r)\mathrm{d}r = N \tag{4.6}$$

We also can use the Lagrange parameter μ to denote this constraint

$$\delta \left\{ E[\rho(r)] - \mu \left[\int \mathrm{d}r\rho(r) - N \right] \right\} = 0 \tag{4.7}$$

The methodology for determining the ground-state electronic configurations of intricate systems, as initially outlined by Kohn and Sham within the framework of DFT (Kohn and Sham 1965), constitutes the focus of the ensuing section. This pioneering approach systematically unfolds the computational strategy for such systems. The charge density is

$$\rho(r) = \sum_{i=1}^{N} |\Phi_i(r)|^2 \tag{4.8}$$

In this equation, T can be denoted as

$$T[\rho(r)] = -\sum_{i=1}^{N} \int dr\Phi_i^*(r)\frac{1}{2}\nabla^2\Phi_i(r) \tag{4.9}$$

In their seminal work, Kohn and Sham reformulated the total energy functional presented in Eq. (4.1) through an alternative perspective as follows:

$$E[\rho(r)] = T[\rho(r)] + E_{\text{tot}}[\rho(r)] \tag{4.10}$$

$$= T[\rho(r)] + \frac{1}{2}\int drdr'\frac{\rho(r)\rho(r')}{|\, r - r'\,|} + \int dr v(r)\rho(r) + E_{\text{xc}}[\rho(r)] \tag{4.11}$$

Beyond the kinetic energy T represented by the initial component, Eq. (4.10) encompasses an entity E_{tot}, which is comprised of diverse elements. These include the Hartree energy, encapsulating the electrostatic interactions within the electronic system (as denoted by the second term in Eq. (4.11)), the energy due to the interaction with an external field (the third term therein), and the exchange-correlation energy E_{xc}. The latter term, accounting for the intricate electron behaviors of exchange and correlation, remains enigmatic. Unlike the readily computable terms on the equation's right-hand side, E_{xc} defies precise determination for complex, non-uniform systems. Consequently, formulating an accurate depiction of E_{xc} constitutes the fundamental hurdle in DFT computations.

Under this framework, the paradigm shifts from a system of mutually interacting particles subjected to an external potential v to one comprising non-interacting particles that are governed by an effective potential V_{tot}. In this revised context, the individual particle wave functions, as detailed in Eq. (4.8), adhere to orthogonality principles and satisfy the norm condition

$$\langle \Phi_i(r) \mid \Phi_i(r) \rangle = 1 \tag{4.12}$$

When the variation with respect to Φ_i is invoked within the variational principle for the total energy E, the scenario evolves such that the norm assumes the role of a constraint, superseding the preservation of particle number. This manipulation yields

$$\delta\left\{ E\left[\Phi_i, \Phi_i^*\right] - \sum_{i=1}^{N}\varepsilon_i \left(|\, \Phi_i(r) \mid \Phi_i(r)\right) - 1 \right\} = 0 \tag{4.13}$$

The resolution of Eq. (4.13) entails the systematic tackling of the effective single-particle Schrödinger equations, individually for each Φ_i, to accommodate a system comprising N particles:

$$\left\{ -\frac{1}{2}\nabla^2 + v(r) + \frac{1}{2}\int dr'\frac{\rho(r')}{|\,\text{r} - r'\,|} + \frac{\delta E_{\text{sr}}\rho(r)}{\delta\rho(r)} \right\}\Phi_i(r) = \varepsilon_i\Phi_i(r) \tag{4.14}$$

The equation referred to is the renowned Kohn–Sham (KS) equation, with the effective one-electron Hamiltonian, intimately linked to Eq. (4.14), designated as the KS Hamiltonian. It is imperative to note that, despite its efficacy in modeling electronic band structures and charge distributions within solids, the eigenvalues εi and the associated single-particle wave functions *Φi* do not bear direct physical significance. Specifically, Φi functions as an intermediary entity, intricately linked to the actual charge density ρ, and are henceforth designated as KS wave functions. Consequently, the Slater determinant constructed from these Φ_i does not embody the genuine many-electron wave function. Instead, Eq. (4.14) embodies a single-electron approximation, wherein the complexities of many-body interactions are encapsulated within the effective potential V_{tot}

$$V_{\text{tot}}(r) = V_{\text{ext}}(r) + V_H(r) + V_{\text{xc}}(r) \tag{4.15}$$

The effective potential comprises V_{ext} for the external influence, V_H denoting the Hartree potential and V_{xc} embodying the exchange-correlation potential. Notably, this potential is functionally dependent on the charge density, which itself is a consequence of the wave functions, thereby establishing a reciprocal relationship. This intricate interdependence necessitates the self-consistent resolution of an ensemble of equations.

DFT inherently relies on approximations, most notably in its treatment of exchange-correlation effects within inhomogeneous electron systems.

The LDA, first proposed by Kohn and Sham (1965), provides a fundamental simplification by employing the homogeneous electron gas as a reference system to address this theoretical challenge. This approach rests on the premise of treating $\rho = \rho(r)$ as a position-independent density, thereby disregarding its spatial variability. Thus, the exchange-correlation energy at any spatial coordinate r is evaluated using its counterpart in a uniform electron gas of identical density ρ, a concept that can be mathematically expressed as follows:

$$E_{\text{xc}}^{\text{hom}} = \rho \varepsilon_{\text{xc}}^{\text{hom}} \tag{4.16}$$

where in εhom xc signifies the spatially uniform exchange-correlation energy density within the homogeneous electron gas. When transitioning to an inhomogeneous system, one derives

$$\varepsilon_{\text{xc}}^{\text{LDA}}(r) = \varepsilon_{\text{xc}}^{\text{hom}}[\rho(r)] \tag{4.17}$$

So

$$E_{\text{xc}}^{\text{LDA}}[\rho(r)] = \int \mathrm{d}r \rho(r) \varepsilon_{\text{xc}}^{\text{LDA}}(r) \tag{4.18}$$

and

$$V_{\text{xc}}^{\text{LDA}}([\rho(r)]) = \frac{\delta E_{\text{sc}}^{\text{LDM}}[\rho(r)]}{\delta \rho(r)} = \frac{\delta \left[\rho(r) \delta_{\text{xc}}^{\text{LDA}}(r)\right]}{\delta \rho(r)} \tag{4.19}$$

Slater and colleagues (Slater et al. 1969) introduced the most rudimentary variant of the LDA for the exchange-correlation potential, which conceptualizes V_{xc} as

$$V_{\text{xc}}([\rho]) = -\left(\frac{3\alpha}{2}\right)\left(\frac{3\rho}{\pi}\right)^{1/3} \tag{4.20}$$

The methodology employed for the exchange-correlation functional, colloquially referred to as the $X\alpha$ approach, typically anchors its α coefficient on established outcomes from electron gas studies. A notable congruence exists in the form of the exchange term between this gas and the method, leading to α being precisely determined as 2/3 when solely the electron gas's exchange term is under scrutiny. In their original derivation, Slater and coworkers obtained $\alpha = 1$ by performing an average over the exchange interaction term in the Hartree–Fock formalism.

Building upon the foundational studies of Ceperley and Adler (Ceperley and Alder 1980), alternative parameterizations have been advanced by researchers such as Perdew and Zunger (1981) and Goedecker et al. (Goedecker et al. 1996). These functional forms have demonstrated considerable efficacy in predicting electronic ground states across a broad spectrum of materials. The insight provided by Gunnarsson et al. (1979) underscores that the LDA's success partly lies in its ability to accurately capture the spherically-averaged exchange-correlation hole, a pivotal quantity in determining the potential, despite the complexity arising from the electronic density distribution.

The generalized gradient approximation (GGA) constitutes a significant improvement upon LDA by extending the exchange-correlation energy functional to include both the local electron density and its gradient field, thus enabling a more sophisticated physical description:

$$E_{\mathrm{xc}}^{\mathrm{GGA}}[\rho(r)] = \int \mathrm{d}r\rho(r)\varepsilon_{\mathrm{xc}}^{\mathrm{GGA}}(\rho(r), \nu\rho(r)) \tag{4.21}$$

In typical parameterizations, the exchange-correlation energy density, εxc, is crafted as an analytical form that conforms to specific criteria, such as ensuring that the exchange hole exhibits a negative value uniformly, indicative of a missing electron. Notable among these GGA formulations are those devised by Perdew and Yue (1986) and the one introduced by Perdew et al. (1996), which have become cornerstones in the field.

4.1.2 Application of DFT in the Field of Chalcogenides

Early numerical investigations into materials now recognized as chalcogenide semiconductors were pioneered by O'Reilly et al. (1981) and Rabe and Joannopoulos 1987). While O'Reilly et al. (Kolobov et al. 2004) employed empirical potentials in their simulations, deviating from a first-principles approach, their work, alongside Rabe's, exemplifies the profound potential of electronic structure calculations. Notably, O'Reilly and coworkers systematically investigated the electronic state distribution across multiple structural configurations of non-crystalline GeTe, a prototypical chalcogenide phase-transition compound. By correlating their computational results with experimental X-ray photoelectron spectroscopy (XPS)/UPS spectroscopy data, they were able to identify a local structural framework for the amorphous phase. This framework underscores that the amorphous phase is predominantly chemically ordered. Specifically, Ge atoms are usually coordinated by four neighboring atoms, while Te atoms are typically surrounded by two neighboring atoms. The experimental verification of such local structures, historically challenging, has only recently been achieved through EXAFS studies, which confirm the consistency with O'Reilly et al.'s early predictions. These findings illustrate the groundbreaking role of computational simulations in elucidating material properties that elude direct experimental determination, marking a seminal instance in the integration of theory and experiment.

Rabe and Joannopoulos (1987) investigated the crystallographic transition between trigonal and rock salt polymorphs in GeTe, a process inherently linked to a ferroelectric phase change. Their computations anticipated a phase-transition temperature of 657 ± 100 K, aligning well with empirical observations, exemplifying the efficacy of ab initio methods in capturing finite-temperature phenomena. In contemporary practice, DFT-based numerical simulations for chalcogenide semiconductors have crystallized into two main avenues: First, they are utilized to forecast structural characteristics, achieved through comparing the total energy of competing structures or via ab initio molecular dynamics (MD) for liquids and amorphous states. Second, calculations are harnessed to determine electronic properties, elucidating bond characteristics and establishing correlations between the electronic and structural facets of these materials. The subsequent section will offer a panoramic view of the ongoing research endeavors in this domain.

The utilization of chalcogenide semiconductors hinges on phase transitions between crystalline and amorphous forms, necessitating a profound understanding of structures across these phases, including the liquid phase due to its role in amorphization. Specifically, characterizing the liquid and amorphous systems presents a significant challenge owing to their lack of long-range order and periodicity, precluding the definition of a unit cell or precise atomic coordinates. Despite this, researchers rely on statistical properties, such as structure factors and correlation functions, to analyze medium- and short-range order, thereby elucidating the structure of amorphous materials. Techniques spanning neutron diffraction, EXAFS spectroscopy, and computational methods like MD, with recent advancements in ab initio MD, have been pivotal in this pursuit. In ab initio MD simulations, an initial random structure is configured within a computational cell, with atom counts large enough for statistical validity (often between 100 and 1000 atoms). Employing a Nosé thermostat for thermal coupling creates a simulated environment (Nosé 1984), where the system is heated, randomized, cooled, and thermalized to achieve a distribution of configurations reflective of the target temperature. Atom trajectories are then calculated via Langevin dynamics (Chelikowsky et al. 2001), capturing their dynamic behavior within this simulated framework:

$$M_i \frac{dv_i}{dt} = -M_i \gamma v_i + G_i(\gamma, T) + F_i \tag{4.22}$$

In this context, the dynamics involve quantities such as the mass, M_i, of each particle i^{th}, its velocity, v_i, subject to interatomic forces, F_i. Additionally, the system incorporates a viscosity, γ, and a stochastic force, G_i, which is contingent on both temperature and viscosity, mediating the interaction with the thermal reservoir.

In alignment with this approach, the microstructural characteristics and dynamical behaviors of various materials, such as Si, Ge, GaAs, and GeTe, in their liquid phase have undergone extensive investigation. These studies yield structure factors and pair correlation functions at diverse temperature points, subsequently benchmarked against experimental findings from neutron diffraction. The observed concurrence is striking (as illustrated in Figure 4.1), facilitating deeper analyses that include the derivation of partial structure factors and pair correlation functions—parameters that are inherently challenging to obtain experimentally.

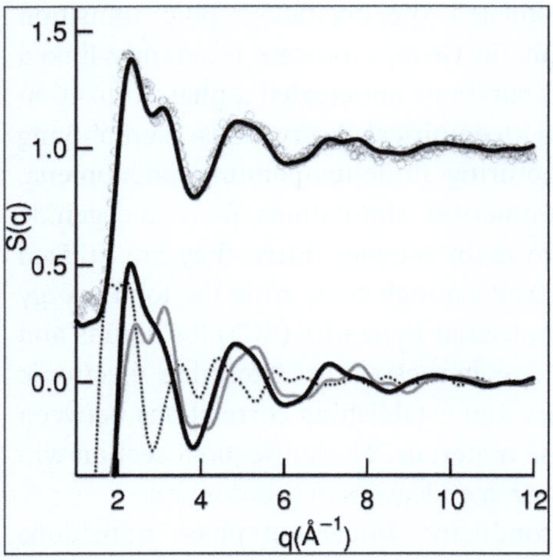

Figure 4.1 Top: A comparison is depicted between the computed structure factor for GeTe at a temperature of 1000K, represented by a continuous line, and neutron diffraction experimental data denoted by symbols, illustrating congruence in their trends. Bottom: Partial structure factors are presented, each offset by −1 for clarity: The S_{GeTe} contribution is indicated by a bold line, the S_{GeGe} component by a shaded line, and the S_{TeTe} partial structure factor by dotted points, collectively offering a detailed breakdown of the system's structural intricacies. *Source:* Raty et al. (2000) and Blaineau et al. (2003) / with permission of Elsevier.

Regarding the local structural dynamics, a notable temperature-sensitive phenomenon emerges in the liquid phase of the phase-change material (PCM) GeTe. Below its melting point, the structure exhibits a reentrant Peierls distortion, characterized by alternating short and long bonds, echoing its trigonal crystalline state. This feature gradually fades with increasing temperature, concomitant with a shift from a semiconductor to a metal state. For the alloy $Ge_{15}Te_{85}$, evidence of a Peierls-like transition has been confirmed through neutron scattering, EXAFS spectroscopy, and first-principles MD simulations. However, comparable investigations into the liquid phase of ternary chalcogenide semiconductors, including $Ge_2Sb_2Te_5$ and $Ge_1Sb_2Te_4$, remain unreported in the extant literature.

Studies on the structural characteristics of binary IV–VI glassy alloys have been undertaken (Blaineau et al. 2003), albeit with recent advancements enabling feasible MD simulations of non-crystalline GeTe and Ge–Sb–Te ternary systems. Prior to this, DFT calculations incorporated simplified structural models informed by experimental findings. Given DFT's accuracy in estimating energy differences among structures, these models effectively assess stability. An EXAFS-inspired model for $Ge_1Sb_2Te_4$ was proposed, featuring a structural motif similar to spinel geometry, with tetrahedrally coordinated Ge cations and octahedrally coordinated Sb/Te anions akin to their crystalline arrangement. This model aligns well with experimental parameters, such as bond lengths and density changes relative to the crystalline state, and is confirmed to be energetically stable, with an energy difference of 30 meV per atom compared to the crystalline phase. This is consistent with crystallization energy measurements ranging from 28 to 42 meV per atom. Advancements in computational power have facilitated ab initio MD simulations for amorphous GeTe and $Ge_2Sb_2Te_5$, yielding an intricate model of $Ge_2Sb_2Te_5$

local structure. Agreement between calculated and experimental structure factors validates the model, which then delineates atomic coordination: Ge and Sb atoms typically have four nearest neighbors, while Te atoms have three. Bond angles peak at 90 and 180 °C, indicative of distorted rock salt-like environments. Notably, 38% of Ge atoms are tetrahedrally coordinated, especially when homopolar bonds occur, an observation further analyzed through a local order parameter, q

$$q = 1 - \frac{3}{8}\sum_{i>k}\frac{1}{3} + \cos(\theta_{\text{ijk}}) \tag{4.23}$$

The summation encompasses bonded atom pairs relative to a focal atom j, with θ_{ijk} denoting the angle between these atoms, where j assumes the central position. An in-depth analysis of ring configurations within amorphous GeTe and $Ge_2Sb_2Te_5$ (Akola and Jones 2007) yields additional insights. In $Ge_2Sb_2Te_5$, four-membered rings predominate, often exhibiting an ABAB alternation pattern, where A represents Ge and/or Sb, and B denotes Te. This pattern is also evident in GeTe, albeit without a distinct prevalence for any specific ring topology. MD simulations further uncover the vacancy distribution within these amorphous alloys, revealing that void spaces constitute 11.8% of the total volume in $Ge_2Sb_2Te_5$ and 6.4% in GeTe. These vacancies, typically bordered by Te atoms, exhibit mutual repulsion, leading to a scarcity of di- or polyvacancy configurations. These observations contribute to an understanding of the phase-transition dynamics, proposing it as a reorientation process of ABAB rings, facilitated by vacancies. Given the higher abundance of ABAB rings and reduced incidence of homopolar bonds in $Ge_2Sb_2Te_5$ relative to GeTe, the crystallization pathway is inherently more expedited in this ternary chalcogenide system.

A significant challenge in elucidating the structural characteristics of crystalline chalcogenide semiconductors resides in the intricate nature of their architectures. These materials, predominantly found along the (GeTe)x(Sb_2Te_3)y pseudo-binary phase diagram, typically adopt a stable hexagonal structure comprised of alternating GeTe and Sb_2Te_3 motifs, alongside a metastable rock salt phase. In this phase, Te atoms inhabit one lattice sublattice, while Ge, Sb atoms, and vacancies populate the other, defying perfect randomness. EXAFS studies and DFT computations collectively reveal that the rock salt structure experiences deviations from ideal symmetry. Specifically, DFT analysis uncovers that the pristine rock salt structure of $Ge_1Sb_2Te_4$ is dynamically unstable, undergoing relaxation into a lattice with varied bond lengths, resembling a distorted rock salt configuration. This non-uniform arrangement boasts a lower total energy state and reduced interatomic force magnitudes, thereby constituting a more energetically favored structure (Figure 4.2).

Experimental determination of the exact stacking sequence of structural units in the hexagonal phase, along with deciphering local order within the Ge/Sb sublattice of the rock salt phase, poses significant analytical challenges. Consequently, DFT calculations are employed to aid in unraveling the structures of crystalline PCMs. An investigation into the $Ge_2Sb_2Te_5$ hexagonal phase suggests that the configuration with the minimum total energy—and thus the most stable—exhibits a [0001]-oriented layered sequence of Te–Ge–Te–Sb–Te–Te–Sb–-Te–Ge, corresponding to the ground-state atomic arrangement. The same study explored alternative

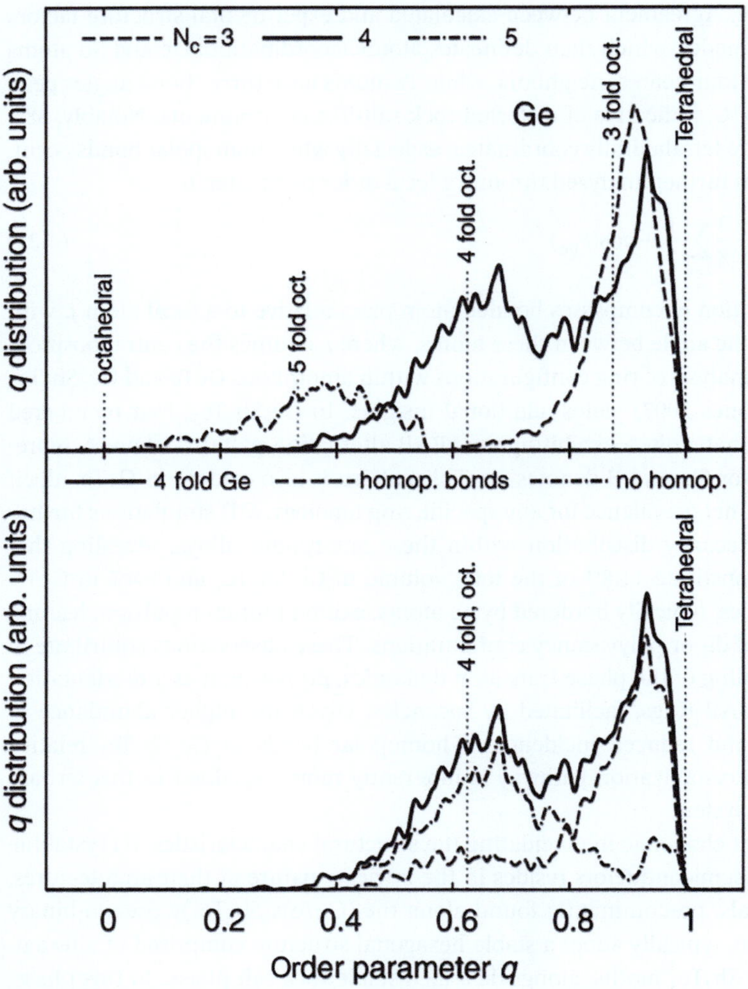

Figure 4.2 The distribution of the local order parameter q for germanium atoms, classified by their distinct coordination numbers Ne, exhibits characteristic trends. In an ideal tetrahedral arrangement, q approaches q–1, while for a perfectly sixfold coordinated octahedral site, q is 0, and a four-coordinated site with an octahedral defect yields q–5/8. The upper segment of the analysis presents the g-distribution, meticulously segmented according to the coordination number of Ge atoms. Conversely, the lower segment delves into the q-distribution, specifically for four-coordinated Ge atoms, further refined to distinguish between those forming at least one homopolar bond with either Ge or Sb and those exclusively bonded to Te without any homopolar bonds, illuminating nuanced structural variations within the amorphous matrix. *Source:* Adapted from Caravati et al. (2007).

structural configurations of the metastable rock salt phase, revealing that the Ge/Sb sublattice occupation is not arbitrary but adheres to a specific periodicity along the [111] crystallographic axis (equivalent to [0001] in the hexagonal phase). Te–Ge–Te–Sb–Te–v–Te–Sb–Te–Ge shows the minimum total energy, where 'v' denotes a vacancy layer. This aligns with the most stable hexagonal structure identified. However, it is crucial to acknowledge that the energetically favored configurations differ marginally (exag2 meV for honal vs. 7 meV for rock salt structures), indicating entropy's potential significance at room temperature. Moreover, the finite exploration of configurations suggests the possibility of even more energetically favorable arrangements. Contrasting findings by (Eom et al. 2006), who observed random vacancy distribution in the metastable crystalline phase, add to the ongoing debate about the crystalline phase's local structure. A potential resolution to this ambiguity may lie in integrating DFT calculations with data mining techniques for predicting crystal structures, offering a sophisticated approach to computational materials science (Sun et al. 2006).

In addition to structural insights, ab initio computations provide valuable electronic structure information. Remarkably, the DFT-derived charge density from single-particle approximations accurately reproduces the many-body system's electron distribution. Given the intrinsic link between structure and electronics, DFT is frequently employed to decipher the character and energetic positioning of defects within materials. PCMs, notably harboring significant vacancy concentrations in their crystalline phases—such as $Ge_1Sb_2Te_4$, where 12.5% of lattice sites are unoccupied—emerge as compelling subjects for electronic structure studies focusing on defect states. For GeTe, such defects have undergone ab initio scrutiny in both its metastable rock salt phase and its trigonal ground state. The methodology entails incorporating a point defect within a supercell, which is a larger structural ensemble consisting of multiple Wigner–Seitz cells, to determine both defect formation energies and their electronic states. Formation energies are computed by

$$\Delta E_{\text{form}}(q) = E_d(q) - E_p(q = 0) - \sum_{i}^{n}(N_i\mu_i) - q(\varepsilon_F) \tag{4.24}$$

In this context, $E_d(q)$ denotes the total energy of the supercell containing a defect in a specific charge state q, whereas $E_p(q = 0)$ represents the energy of an uncharged, pristine supercell. The Fermi energy, ε_F, is referenced from the valence band (VB) maximum. The term incorporating i and N_i accounts for the chemical potential associated with the ith atomic species, where N_i represents the change in the number of atoms added or removed. The electrochemical level referred to by (n−1) /n signifies the Fermi level at which the defect's occupancy transitions from n electrons to $n-1$, marking the addition of one electron. This level is computed through

$$E((n-1)/n) = [E_d(n-1) + E_p(n)] - [E_d(n) + E_p(n-1)] \tag{4.25}$$

The energies attributed to an ideal supercell and one incorporating a defect are denoted as E_p and E_d, respectively.

In the context of DFT analyses conducted for both the trigonal and rock salt phases of GeTe, it is revealed that the Ge-vacancy assumes the role of the energetically favored defect across the entire range of Fermi energies examined. Specifically, at Fermi energy levels exceeding approximately 0.25 eV, the spontaneous generation of Ge vacancies is forecasted (refer to Figure 4.3).

Additionally, the 0/+1 energy level resides beneath the VB's edge. Considering entropy contributions under realistic conditions, the Fermi level at ambient

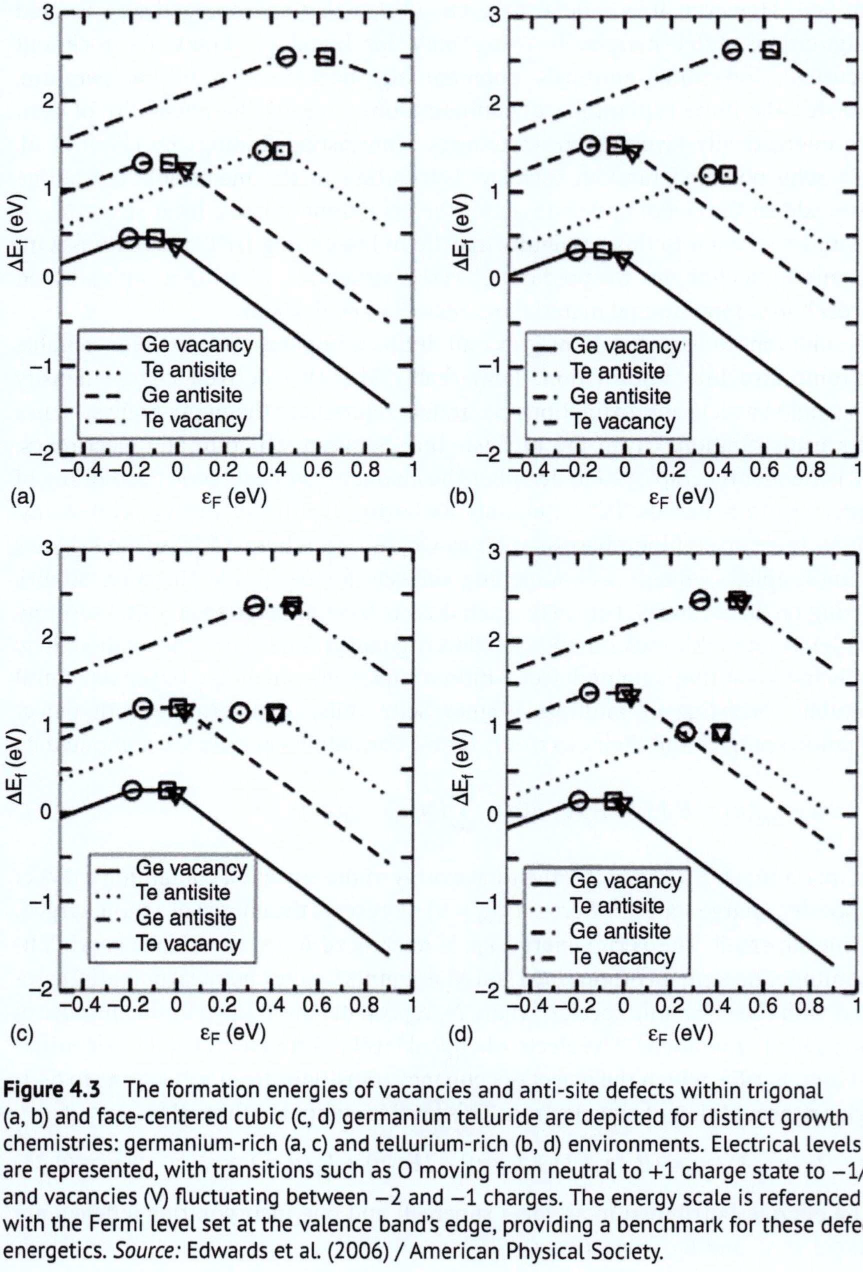

Figure 4.3 The formation energies of vacancies and anti-site defects within trigonal (a, b) and face-centered cubic (c, d) germanium telluride are depicted for distinct growth chemistries: germanium-rich (a, c) and tellurium-rich (b, d) environments. Electrical levels are represented, with transitions such as O moving from neutral to +1 charge state to −1/0, and vacancies (V) fluctuating between −2 and −1 charges. The energy scale is referenced with the Fermi level set at the valence band's edge, providing a benchmark for these defect energetics. *Source:* Edwards et al. (2006) / American Physical Society.

temperatures and above is positioned at or beneath the VB edge. This circumstance holds significant consequences for transport characteristics, leading to the generation of a substantial hole population near the VB maximum, thereby imparting p-type conductivity to the material. Consequently, this insight elucidates the presence of the metallic Drude peak observed in the optical spectra of crystalline GeTe, aligning with the described electronic structure.

Inherently, ternary Ge–Sb–Te alloys within the rock salt structure exhibit a substantial vacancy content. The underlying electronic mechanisms responsible for these elevated vacancy levels can be elucidated through DFT computations. By simulating systems spanning the stoichiometric gradient from $Ge_1Sb_2Te_4$ to $Ge_2Sb_2Te_4$ in the rock salt phase, one can quantitatively assess the impact of varying vacancy densities on electronic characteristics. Specifically, $Ge_1Sb_2Te_4$ comprises a 25% vacancy rate on the Ge/Sb sublattice, whereas $Ge_2Sb_2Te_4$ sees these vacancies supplanted by Ge atoms. To delve into the nature of chemical bonding, the Chemical Orbital Hamilton Population (COHP) analysis serves as an insightful method, enabling the band structure energy to be reformulated in relation to atomic and bonding interactions, as expressed in the following equation:

$$\mathbf{COHP}(\varepsilon) = \sum_{R,L}\sum_{R',L'} \langle x_{R,L} \mid H \mid \chi_{R',L'} \rangle \underbrace{\sum_j u^*_{R,L,j} u_{R',L',j'} \delta(\varepsilon_j - \varepsilon)}_{\text{DOS-matrix}} \tag{4.26}$$

In this framework, $\chi_{R',L'}$ denote local orbitals that, in conjunction with eigenvectors $u^*_{R,L,j}$, form the basis for single-particle wave functions. The indices R and L, along with quantum numbers l and m, respectively, specify spatial and momentum states, while f_j represents the occupation number. The energy ε and eigenvalue ε_j for state j define the context for the COHP analysis. Essentially, COHP represents the multiplicative result of the density of states and the Hamiltonian matrix element associated with the overlap between orbitals $\chi_{R,L}$ and $\chi_{R',L'}$. Figure 4.4 illustrates the COHP analysis for covalent Ge–Te and Sb–Te bonds within $Ge_1Sb_2Te_4$, $Ge_{1.5}Sb_2Te_4$, and $Ge_2Sb_2Te_4$, both in their unrelaxed rock salt configuration (upper panels) and after structural relaxation inducing local lattice distortions (lower panels). Notably, $Ge_1Sb_2Te_4$ displays a significant prevalence of antibonding interactions near the Fermi level in the highest occupied bands, indicative of an excessive valence electron concentration (VEC). To enhance stability, a reduction in VEC is necessitated, achieved through atom removal, with Te, due to its high electronegativity and dominance in lower VBs, being less energetically favorable to eliminate compared to Ge and Sb, whose orbitals contribute significantly near the VB edge. Analysis reveals that increasing vacancy numbers in compositions like $Ge_{1.5}Sb_2Te_4$ and $Ge_1Sb_2Te_4$ mitigates these antibonding interactions, which eventually disappear. However, this also reduces bonding states overall. Among these, $Ge_{1.5}Sb_2Te_4$ emerges as the configuration with optimal stability and minimal total energy, a prediction confirmed by computational studies prior to experimental synthesis (Wuttig et al. 2007), highlighting the pivotal role of COHP analysis in understanding compound stability.

Additionally, upon undergoing structural relaxation (as depicted in the lower portion of Figure 4.4), a substantial enhancement in the energy gap separating the valence and conduction bands (CBs) is observed. This phenomenon, characterized by an electronic energy reduction concurrent with the emergence of a bandgap due to structural deformation, is referred to as Peierls instability, a concept previously mentioned. Consequently, DFT is harnessed not only to illuminate the physicochemical traits of PCMs but also to forecast novel materials that may exhibit enhanced characteristics, thereby advancing materials science.

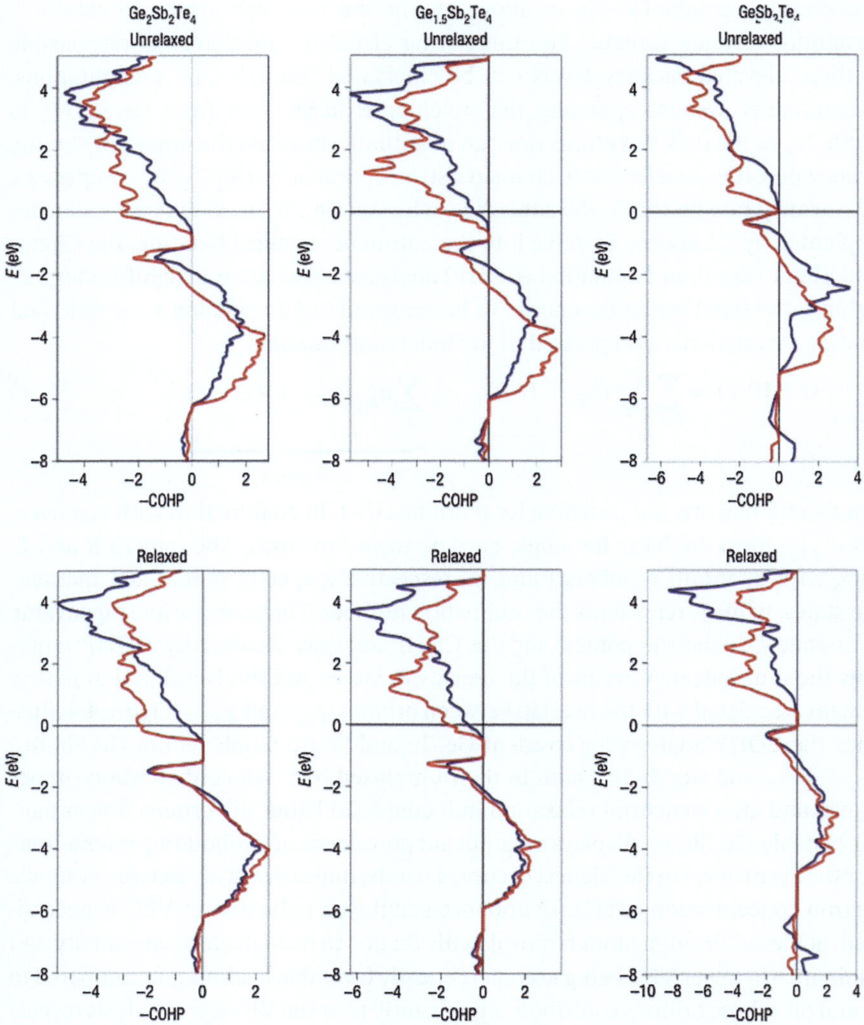

Figure 4.4 In the realm of Crystal-Orbital Hamilton Population (COHP) analyses, the depicted graph illustrates bonding characteristics for Ge–Te (represented by a continuous line) and Sb–Te (indicated by a dotted line) within three compounds: $Ge_2Sb_2Te_4$ (left), $Ge_{1.5}Sb_2Te_4$ (center), and $Ge_1Sb_2Te_4$ (right). The upper and lower panels, respectively, present data for the unrelaxed, NaCl-type lattice structure and the fully atomically relaxed configurations. For standardization, the Fermi level is positioned at zero energy. The graphical representation designates bonding interactions on the right side and antibonding interactions on the left, offering insights into the electronic structure and bonding nature of these phase-change materials. *Source:* Wuttig et al. (2007) / Springer Nature.

The predictive capabilities of DFT extend to a broader stoichiometric spectrum, transcending GeSbTe alloys. An investigation into the structural stability and bonding characteristics of $XInTe_2$ and $XSbTe_2$, where X encompasses Cu, Ag, and Au, reveals that solely those alloys incorporating antimony manifest p-type bonding, a feature characteristic of PCMs that facilitates the formation of a rock salt crystal structure. Conversely, alloys with indium display sp^3 hybridization, conducive to a chalcopyrite structure characterized by tetrahedral bonding. Chalcopyrite structures, however, have been demonstrated to possess inadequate optical contrast and excessively lengthy recrystallization periods, thereby rendering them inappropriate for phase change technology applications. Through stability assessments, it becomes evident that alloys predisposed to adopt the rock salt structure typically possess a count of valence s- and p-electrons surpassing 4.1, whereas those inclined toward the chalcopyrite arrangement fall below this critical threshold.

The investigation of the amorphous phase utilizing ab initio techniques poses a considerably greater challenge due to the indistinct nature of its structure. Despite this, simplified models have been employed with notable efficacy to ascertain the electronic properties of amorphous PCMs. An illustrative example involves modeling amorphous $Ge_1Sb_2Te_4$ with a spinel-like structure, where germanium occupies tetrahedral coordination sites, revealing an expanded bandgap relative to its crystalline counterpart. This finding holds significant implications for electronic data storage applications, as the amorphous state's conductivity is governed by trap states, necessitating a substantial bandgap. Contrary to conventional wisdom associating these trap states with valence alternation pairs (VAPs), typically involving divalent tellurium and their lone pairs, the analysis, even in more sophisticated models informed by MD simulations, reveals that tellurium atoms in the amorphous phase are predominantly trigonal coordinated. A comprehensive examination of electronic structure computations elucidates the correlation between local structural rearrangements and the alteration in electronic properties, leading to the bandgap's emergence. Notably, while germanium atoms undergo a transition from octahedral to tetrahedral coordination, it is the tellurium and antimony atoms that exhibit the most profound changes in electronic characteristics. Specifically, the Te–p states, which dominate the VB edge in the crystalline phase, experience a shift to lower energies, contributing decisively to the bandgap opening. This transformation is visually depicted in Figure 4.5, which compares the calculated electronic density of states between the two phases, crystalline and amorphous, emphasizing the differences that underlie the material's functional behavior.

A positive magnitude signifies a greater contribution to the rock salt phase; conversely, a negative magnitude indicates a stronger affiliation with the amorphous phase. Notably, substantial weighting in the crystalline structure is evident immediately beneath the Fermi level at 0 eV. In contrast, the amorphous phase exhibits heightened weighting in the −4.5 to −3 eV energy interval, a feature attributed to enhanced s–p state hybridization characteristic of the spinel-like local ordering within the amorphous state. The electronic profile of Ge atoms, however, undergoes insignificant alteration. The findings of this rudimentary structural model align with the outcomes from MD simulations for GeTe and $Ge_2Sb_2Te_5$, which yield a more intricate yet precise structural depiction. Despite the enhanced complexity, the alterations in electronic properties remain strikingly analogous. The density of states near the Fermi level is more pronounced in the crystalline phase, mainly due to the contributions from Te–p, Sb–s, and Sb–p, d orbitals. In contrast, the energy band below the Fermi level is narrower in the amorphous phase. This difference is

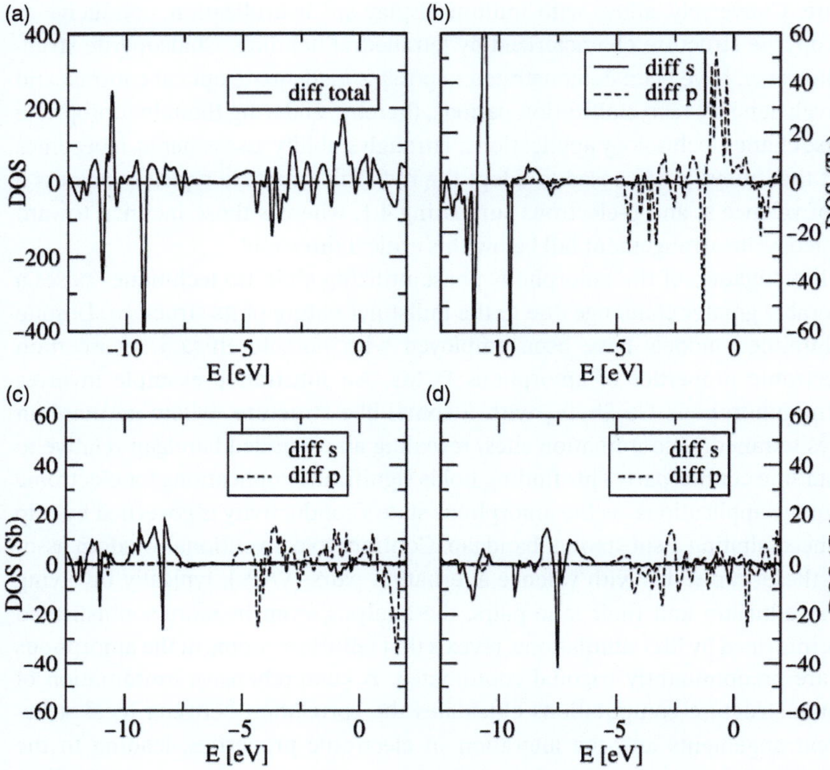

Figure 4.5 The comparative analysis of the density of states (DOS) between the crystalline, rock salt configuration, and its amorphous counterpart encompasses all atomic contributions, with a particular focus on the s- and p-orbital bands for individual elements. Values that are positive denote a greater spectral weight in the crystalline phase; conversely, negative values signify a stronger presence in the amorphous phase. Notably, a discernible positive shift is observed immediately beneath the Fermi level, ranging from 0 eV, equivalent to the Fermi energy (=Er), down to −2 eV. This disparity in DOS is predominantly attributed to the p-orbitals of tellurium (Te), whereas germanium (Ge) exhibits significant variations primarily within its s-band. Antimony (Sb), in this context, exhibits spectral changes in both its s- and p-bands that occupy a middle ground in terms of magnitude, bridging the observed extremities of Ge and Te. *Source:* Wełnic et al. (2006) / Springer Nature.

attributed to a transition from predominantly sp3 bonding to p-type bonding during the crystallization process, as indicated by the computational studies.

In the field of PCMs, the optical and transport characteristics are extremely important. These properties influence the differences in optical reflectivity and electrical conductivity between the amorphous and crystalline states of the material. As a result, they play a crucial role in determining whether a material is suitable for phase-change applications. DFT, albeit foundational for understanding the electronic ground state, inherently falls short in capturing optical and transport phenomena, which involve excited electronic states. Consequently, computational exploration of these properties necessitates an extension beyond the confines of DFT to accurately account for these dynamic aspects of electronic behavior.

Time-dependent DFT (TDDFT) and many-body perturbation theory have been successfully used to determine excited electronic states in various solids. To explore the optical properties of PCMs and delineate the distinctions between their amorphous and crystalline phases, TDDFT has been employed, with specific utilization of the random-phase approximation (RPA). This approach has been instrumental in simulating the "optical contrast," a term encapsulating the experimentally noted shifts in optical behavior upon transitioning to the amorphous state, for GeTe and $Ge_1Sb_2Te_4$, utilizing simplified structural models that incorporate germanium atoms in partially tetrahedral coordination. These computations enable a detailed investigation of such phenomena, including the determination of the dielectric function's imaginary component, ϵM, being executed in accordance with Fermi's golden rule, thereby enhancing our understanding of the underlying electronic transitions:

$$\mathrm{Im}(c_M(\omega)) = 2\pi \lim_{q\to 0} v_0(q) \sum_{v\in} |\langle c \mid e^{-iq\cdot r} | v\rangle|^2 \delta(\omega - (\varepsilon_e - \varepsilon_v)) \tag{4.27}$$

In this context, where the eigenstates are designated as c and v, with corresponding eigenenergies ε_c and ε_v, and the frequency of excitation is denoted by ω, the analysis entails a decomposition. This dissection reveals a component that stems fundamentally from the band structure, referred to as the joint density of states (JDOS):

$$\mathrm{JDOS}/\omega^2 = \frac{1}{N_k \omega^2} \sum_{v,c,k} \delta(\varepsilon_{ck} - \varepsilon_{vk} - \omega) \tag{4.28}$$

The optical transition process is characterized by velocity matrix elements $m_{v,c,k} \equiv \langle c|v|v\rangle$, which, alongside JDOS, plays a pivotal role. Conventionally, in semiconductors, it is postulated that amorphization leads to a broadening of the JDOS, subsequently causing a modest alteration in optical absorption, with matrix elements assumed invariant. Conversely, in the case of GeTe and $Ge_1Sb_2Te_4$, the observed reduction in optical absorption within the amorphous phase, as illustrated in Figure 4.6a,c, defies explanation solely through variations in JDOS, necessitating a more nuanced analysis.

In both alloys, a substantial portion of the investigated spectral range exhibits a higher JDOS in the amorphous phase. Consequently, to comprehend the optical contrast, it is indispensable to consider the impact of altered oscillator strength associated with the optical transitions. This critical contribution is illustrated in Figure 4.6b for $Ge_1Sb_2Te_4$ and Figure 4.6d for GeTe. Specifically, $Ge_1Sb_2Te_4$ displays stronger matrix elements in its crystalline phase up to approximately 2 eV, while for crystalline GeTe, this enhancement spans the entire spectral range, albeit more prominently. These computational results highlight that the change in local ordering during PCM amorphization leads to a notable outcome: a significant decrease in the oscillator strength associated with optical transitions. This, in turn, gives rise to the pronounced optical contrast pivotal to optical data storage applications.

The realm of computational physics encounters a significant hurdle in accurately modeling transport properties through ab initio computations, a task of utmost relevance to PCMs. Given their burgeoning potential in electronic data storage, a meticulous, nanoscopic comprehension of transport characteristics is imperative. While advancements, incorporating techniques such as TDDFT in conjunction with nonequilibrium Green's functions, have significantly contributed to the elucidation

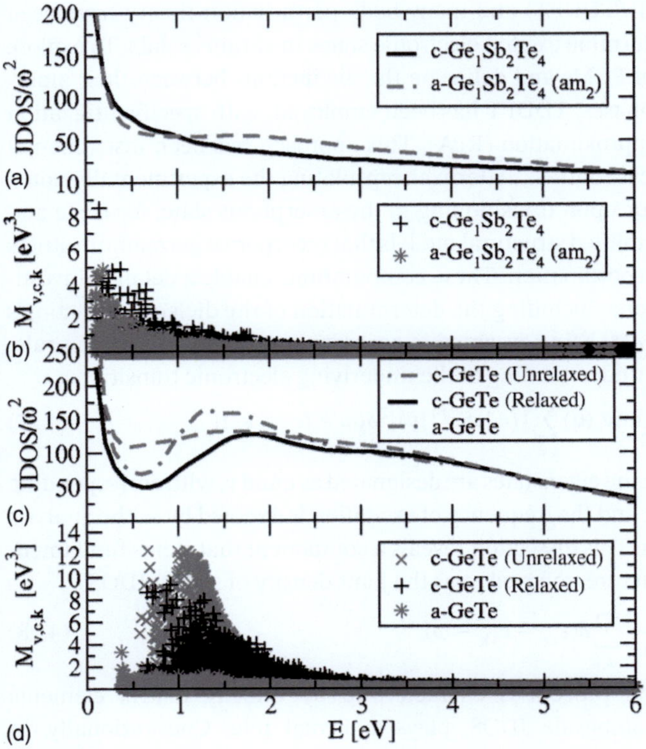

Figure 4.6 The joint density of states (JDOS) per unit cell and k-point, normalized by ω^2, for $Ge_1Sb_2Te_4$ (a) is depicted in both its crystalline and amorphous forms, revealing a distinction in transition density. Notably, up to an energy of 1.4 eV, the JDOS within the crystalline phase exhibits a higher density of transitions. Conversely, for GeTe (c), an intriguing inversion is observed, with the amorphous phase hosting a greater number of transitions up to 1.7 eV relative to its crystalline counterpart. In accounting for the observed diminution in optical absorption upon transitioning to the amorphous state for both substances, consideration of the velocity matrix elements becomes imperative shown in (b) ($Ge_1Sb_2Te_4$) and (d) (GeTe). *Source:* Wełnic et al. (2007) / American Physical Society.

of quantum transport, a universally accepted theoretical framework remains elusive in the current scientific landscape.

The future integration of a method for first-principles calculation of transport properties, coupled with its numerical realization, promises to unleash novel avenues for the application of computational methodologies within the domain of PCMs. This advancement is poised to enrich our understanding and predictive capabilities, thereby enhancing the role of computational approaches in this field.

4.2 The Dielectric Function

The operational efficacy of chalcogenide semiconductor optical devices hinges on the profound alteration of their optical characteristics accompanying phase transitions, a phenomenon that necessitates meticulous regulation. Consequently, a comprehensive grasp of these optical properties is pivotal to the advancement of technological applications. Although initial progress has largely relied on empirical strategies, recent foundational research has illuminated deeper insights into these properties. This segment delves into the methodologies for assessing optical constants, alongside an in-depth exploration of the optical traits of the archetype PCMs, the GeSbTe alloy series, and the emerging low-loss materials, Sb_2S_3 and Sb_2Se_3, addressing both their amorphous and crystalline structural states to showcase recent advancements in the field.

4.2.1 Measurement Techniques: Ellipsometry and Reflectance

Ellipsometry, a time-honored optical methodology dating back a century, hinges on the scrutiny of polarization state alterations in light reflected from coating surfaces or interfaces. This technique serves as an invaluable tool for discerning the thickness, optical characteristics, and compositional makeup of thin films. The foundational principles of ellipsometry were first laid out by Drude in 1887, predating its formal designation by Rothen in 1945. Drude's pioneering work included the inaugural ellipsometric measurement in 1890, aimed at elucidating the optical properties of metallic substances. Despite its early inception, ellipsometry experienced a protracted period of sluggish development, particularly in its nascent stages. The field was largely confined to single-wavelength ellipsometers due to the intricate nature of multi-wavelength systems and the analytical complexities they entailed. Furthermore, these early instruments lacked the speed necessary for real-time analysis of dynamic systems. A significant turning point emerged in the 1970s, with advancements in photometric instrumentation, including enhanced light sources and detectors, alongside the advent of minicomputers. These innovations propelled ellipsometry into a rapid evolutionary phase, characterized by substantial enhancements in wavelength coverage, measurement accuracy, and operational speed, thereby transforming its utility and impact within materials science.

The ellipsometric methodology revolves around the precise manipulation of light polarization through a combination of polarizing components and the sample under study. Figure 4.7 illustrates the fundamental polarization–state–controlled sample–analyzed (PCSA) ellipsometer setup, comprising a light source, a linear polarizer (P), a retarder (also referred to as a compensator, C), the sample (S), another linear polarizer serving as an analyzer (A), and a detection system. The assembly involving the source, polarizer, and retarder is tasked with establishing a controlled polarization state for the incident light. Conversely, the analyzer and detector assembly is employed to quantify the polarization alteration imparted by the sample. An alternative configuration of the PCSA approach can be achieved by relocating the compensator to a position between the sample and the analyzer.

In the depicted scenario of Figure 4.7, a planar sample is considered, where the incidence angle is defined as φ. The incident light and reflected light beams are confined to the plane of incidence, which includes a line perpendicular to the sample's surface. Oblique incidence, where the angle φ is greater than zero, is often used because it significantly affects the polarization state of the reflected light. This effect arises from the different responses of p-polarized and s-polarized light:

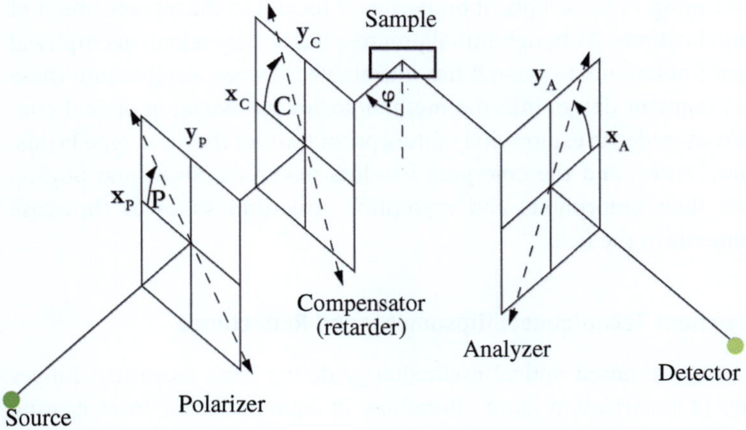

Figure 4.7 The polarizer–compensator–sample–analyzer (PCSA) arrangement in ellipsometry.

p-polarized light is aligned with the plane of incidence, while s-polarized light is orthogonal to it. The distinction between these two types of polarization traces back to the German term "senkrecht" meaning perpendicular. In isotropic materials, the induced currents caused by the light wave align with the direction of the electric field. This results in different behavior for s-polarization, where currents flow along the sample surface or at layer interfaces, and for p-polarization, where currents are partially perpendicular to the surface. Conversely, anisotropic samples exhibit polarization alterations even at normal incidence (φ equals zero), as the induced currents may not align straightforwardly with the electric field due to the material's directional properties.

Within the framework outlined in Figure 4.7, the orientations of linear polarization as it passes through the polarizer and analyzer are designated by angles P and A, respectively. The orientation of the compensator's fast axis is designated as C. These angular measurements are referenced from the local x-axes located within the plane of incidence, with positive angles corresponding to clockwise rotations when viewed along the positive z-direction, following the right-hand rule convention. To understand how the detected field amplitude depends on the relevant parameters, a systematic procedure is employed that includes

i) Postulate the complex amplitude E_P of the linearly polarized radiation that passes through the polarizer. It is pertinent to mention that for an incident wave that is either unpolarized or circularly polarized, E_P remains unaffected by the polarizer's azimuth angle P.
ii) The wave reaching the compensator constitutes a superposition of two components: one parallel to the fast axis with a complex amplitude $E_P\cos(P-C)$ and the other orthogonal, aligned with the slow axis, having an amplitude $E_P\sin(P-C)$. These components are then subjected to modification as they pass through the compensator, each being multiplied by their respective complex transmittance coefficients, t_{fast} and t_{slow}. Consequently, the field amplitudes for the fast and

slow axes of the compensator are determined as follows:

$$E_{\text{fast}} = t_{\text{fast}} E_p \cos(P - C), E_{\text{slow}} = t_{\text{slow}} E_p \sin(P - C). \tag{4.29}$$

iii) Upon reaching the sample, the incident wave comprises p- and s-polarized components, oriented parallel to the plane of incidence and orthogonal to it, respectively. The complex amplitudes for these components are represented as Efastcos$C$ – Eslowsin$C$ for the p-polarization and Efastsin$C$ + Eslowcos$C$ for the s-polarization. Subsequently, the reflected waveforms are altered through multiplication by their respective complex reflectance coefficients, r_p and r_s, leading to the reflected amplitudes of the p- and s-polarized components from the sample being computed as follows:

$$\begin{aligned} E_{rp} &= r_p(E_{\text{fast}} \cos C - E_{\text{slow}} \sin C) \\ E_{rs} &= r_s(E_{\text{fast}} \sin C + E_{\text{slow}} \cos C) \end{aligned} \tag{4.30}$$

iv) Ultimately, the wave that emerges from the analyzer is a consequence of the coherent summation of the incident radiation along with its p- and s-polarized components, all aligned according to the azimuth angle A. By incorporating the intermediate expressions derived from Eqs. (4.29) and (4.30), the final outcome can be formulated as follows:

$$\begin{aligned} E_A &= E_{\text{rp}} \cos A + E_{\text{rp}} \sin A \\ &= E_p[r_p \cos A H_{\text{freq}} \cos(P - C) \cos C - t_{\text{sbou}} \sin(P - C) \sin C] \\ &\quad + r_s \sin A[I_{\text{sb}} \sin(P - C) \cos C + t_{\text{fra}} \cos(P - C) \sin C]]. \end{aligned} \tag{4.31}$$

The intensity of light detected by the sensor bears a proportional relationship to the squared modulus of the complex amplitude $|E_A|^2$.

The polarizer–sample–analyzer (PSA) setup constitutes a more streamlined ellipsometric arrangement, achieved through the exclusion of the compensator from the PSAC design. This is practically realized by orienting the compensator's axes in parallel with the polarizer's azimuth, effectively bypassing its role in the optical path. Consequently, the field amplitude detected is derived from Eq. (4.31), with the simplification that the transmittances for both fast and slow axes, t_{fast} and t_{slow}, are set equal to 1, reflecting the absence of the compensator's influence:

$$\begin{aligned} E_A &= E_P(r_p \cos P \cos A + r_s \sin P \sin A) \\ &= E_p r_s(\rho \cos P \cos A + \sin P \sin A). \end{aligned} \tag{4.32}$$

ρ can be expressed as

$$\rho = \frac{r_p}{r_s} = \tan \psi e^{i\Delta}. \tag{4.33}$$

This is commonly represented in the context of ellipsometry by utilizing the parameters ψ and Δ, which are herein defined.

Field amplitudes are conventionally determined through the employment of Jones vectors and matrices within judiciously selected coordinate frameworks.

Specifically, the analytical process for the PSA configuration may be encapsulated as follows:

$$E_A \begin{bmatrix} 1 \\ 0 \end{bmatrix} = \begin{bmatrix} 1 & 0 \\ 0 & 0 \end{bmatrix} \begin{bmatrix} \cos A & \sin A \\ -\sin A & \cos A \end{bmatrix} \begin{bmatrix} r & 0 \\ 0 & r_s \end{bmatrix} \begin{bmatrix} \cos P \\ \sin P \end{bmatrix} E_p. \tag{4.34}$$

In this context, the normalized Jones vector representing the linearly polarized light emanating from the polarizer, situated within the x–y coordinate system of the source arm, is positioned on the right-hand side. This vector undergoes sequential multiplication from the left by three matrices. Initiated by the sample's diagonal Jones matrix, it encodes the amplitude reflectivities for both p- and s-polarizations. Subsequently, a rotation matrix intervenes, effecting a $-A$ angle rotation of the detection arm's x–y coordinate system, aligning the first coordinate with the analyzer's transmission axis. Consequently, the analyzer's Jones matrix in its eigenpolarization reference frame becomes applicable. It is noteworthy that the relationship between the scalar field amplitudes E_A and E_P, derived from Eq. (4.34), mirrors that outlined in Eq. (4.32). The utility of the matrix-based approach becomes particularly evident when dealing with imperfect, partially polarizing optical elements, where the employment of Stokes vectors and their corresponding matrices offers a streamlined methodology.

4.2.2 Spectroscopic Ellipsometry for Dielectric Function Analysis

Chalcogenide PCMs, especially those in the Ge–Sb–Te (GST) family, have attracted significant attention due to their remarkable ability to undergo rapid and reversible transitions between amorphous and crystalline states. This process leads to substantial changes in their optical properties. These unique characteristics have been utilized in various photonic applications, such as optical switches, reconfigurable meta-optics, and nonvolatile photonic memories. The optical constants of these materials, particularly the refractive index (n) and the extinction coefficient (k), are crucial in determining their performance in these applications.

In Figure 4.8, the refractive index (n) and extinction coefficient (k) are depicted for energies up to 2.5 eV, corresponding to a wavelength of 0.5 μm. A salient feature observed in all presented PCMs is a substantial enhancement in the refractive index upon crystallization. Specifically, for PCMs exhibiting resonant bonding, this increase can surpass 50% below the bandgap energy, where interband transitions are suppressed. Reflectance measurements of thin films deposited on metallic substrates clearly reveal these modifications to both the bandgap and refractive index. Remarkably, no other material class exhibits such a marked refractive index shift upon amorphous-to-crystalline transition, a phenomenon that cannot be solely attributed to density increases typical in conventional semiconductors. Instead, it stems from a fundamental alteration in bonding dynamics, a characteristic limited to a select group of chalcogenides. These materials exhibit resonant bonding, a balance between ionicity, as influenced by the electronegativity differential of their elements, and limited s–p valence electron hybridization, preserving near-ideal octahedral coordination. Resonantly-bonded compounds often have bandgaps

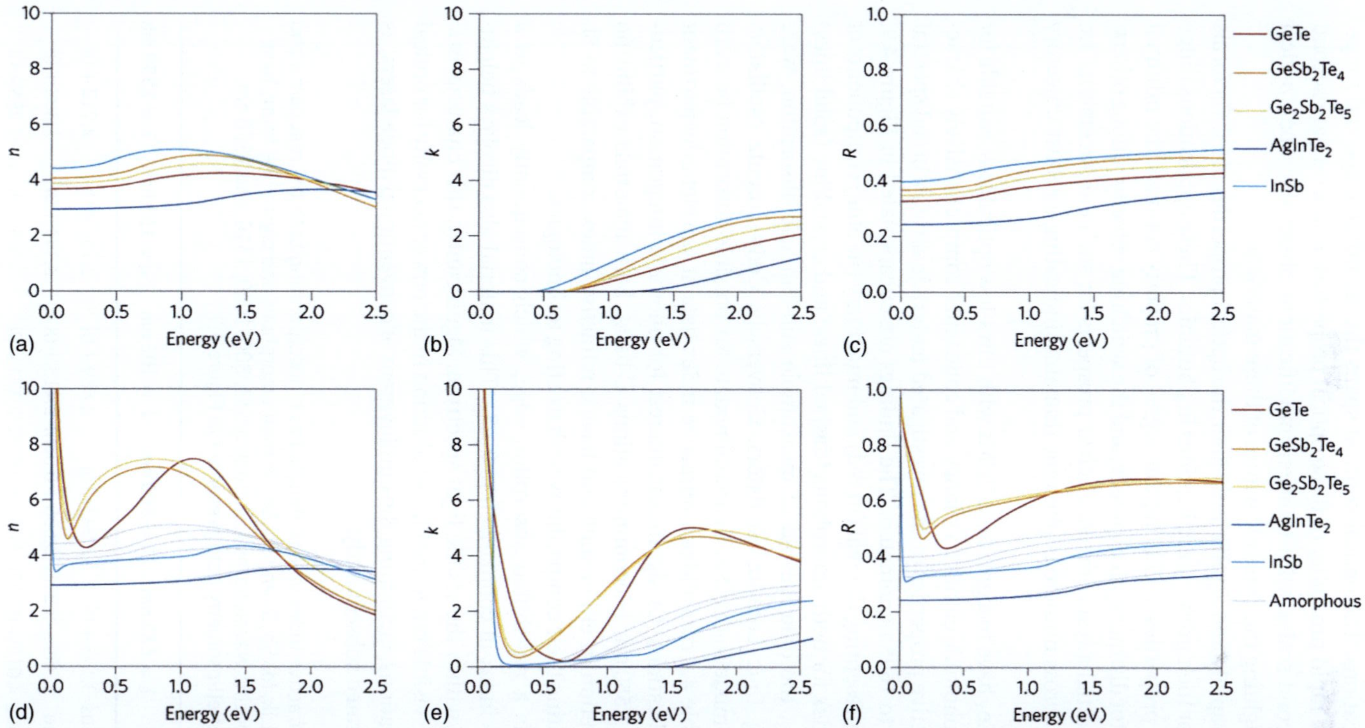

Figure 4.8 The optical characteristics of phase-change materials (PCMs) and conventional covalent semiconductors are meticulously examined, revealing distinct spectral behaviors. For PCMs such as GeTe, GeSbTe, and GeSbTea, alongside the semiconductors InSb and $AgInTe_2$, their refractive indices (n), extinction coefficients (k), and reflectance (R) are plotted against photon energy for both amorphous (panels a–c) and crystalline (panels d–f) phases. A striking transformation in optical properties upon crystallization is evident in the PCMs, a feature not as pronounced in the standard semiconductors, as evidenced by the contrast with their amorphous phase depicted by gray lines. This transition is marked by a rise in refractive index concomitant with a narrowing of the bandgap. Notably, at low-energy regimes, the influence of free carriers becomes discernible, contributing to the dielectric function and subsequently affecting the refractive index, extinction coefficient, and reflectivity in the crystalline state, highlighting a unique aspect of these materials' optical response. *Source:* Adapted from Wuttig et al. (2017).

near 0.5 eV, with significant optical property disparities persisting above the bandgap, accompanied by absorption in both phases. This can pose limitations in specific applications. The optical contrast spans the far-infrared spectral range. Furthermore, the rapid tunability of these optical properties occurs on nanosecond and sub-nanosecond scales, with transient modifications observable upon optical excitation, highlighting the dynamic nature of these materials.

The ability to rapidly and significantly alter the optical properties of PCMs forms the foundation for their growing applications in photonics. These applications hinge on various PCM properties, including the degree of property contrast, the minimal scale achievable for distinct optical phases, and the switching velocity, among others. Currently, substantial advancements or clear prospects for such enhancements are being witnessed across numerous photonic domains, indicating a robust trajectory of development.

Sb_2S_3 and Sb_2Se_3 have recognized as PCMs with ultra-low optical loss, suitable for integration into photonic circuits. Delaney and colleagues comprehensively characterized the refractive properties of the investigated materials across a broad spectral range, from 195 to 1695 nanometers. The findings are encapsulated in Figure 4.9, with (a) and (b) presenting the real (in red) and imaginary (in blue) components of the refractive index for both amorphous (dashed lines) and crystalline (solid lines) states. For Sb_2S_3, the data, devoid of measurable sub-bandgap absorption, were fitted employing Tauc–Lorentz oscillators. Conversely, Cody–Lorentz oscillators were utilized in modeling Sb_2Se_3 to accommodate for possible absorption beneath the bandgap. Table 4.1 consolidates refractive index data at wavelengths pertinent to photonic applications. Of significant interest for photonic integration, particularly within the 1550 nm telecommunications C-band, both materials exhibit no detectable absorption in either state and have a refractive index comparable to Si, thereby highlighting their compatibility with existing technologies.

Sb_2Se_3 exhibits a red-shifted absorption edge, which consequently leads to a higher refractive index in comparison to Sb_2S_3. This material also displays heightened absorption within the visible light spectrum. Significantly, this characteristic absorption in the visible range enables a reduction in the power necessity for optical switching applications, facilitating the employment of conventional diode lasers, as will be demonstrated subsequently.

Table 4.1 The refractive index values, denoted as n′ incorporating both the real part n and the imaginary part ik, for Sb_2S_3 and Sb_2Se_3 in their amorphous and crystalline forms have been derived at specific wavelengths λ, including 633, 800, 1060, 1310, and 1550 nm, through analysis of ellipsometry data presented in Figure 4.9.

		λ = 633 nm	λ = 800 nm	λ = 106 nm	λ = 131 nm	λ = 1550 nm
Sb_2S_3	Amorphous	3.148 + 0i	2.943 + 0i	2.829 + 0i	2.767 + 0i	2.712 + 0i
	Crystalline	3.989 + 0.276i	3.630 + 0.008i	3.413 + 0i	3.343 + 0i	3.308 + 0i
Sb_2Se_3	Amorphous	4.013 + 0.532i	3.660 + 0.028i	3.372 + 0i	3.305 + 0i	3.285 + 0i
	Crystalline	5.132 + 1.403i	4.939 + 0.402i	4.313 + 0.010i	4.121 + 0.001i	4.050 + 0i

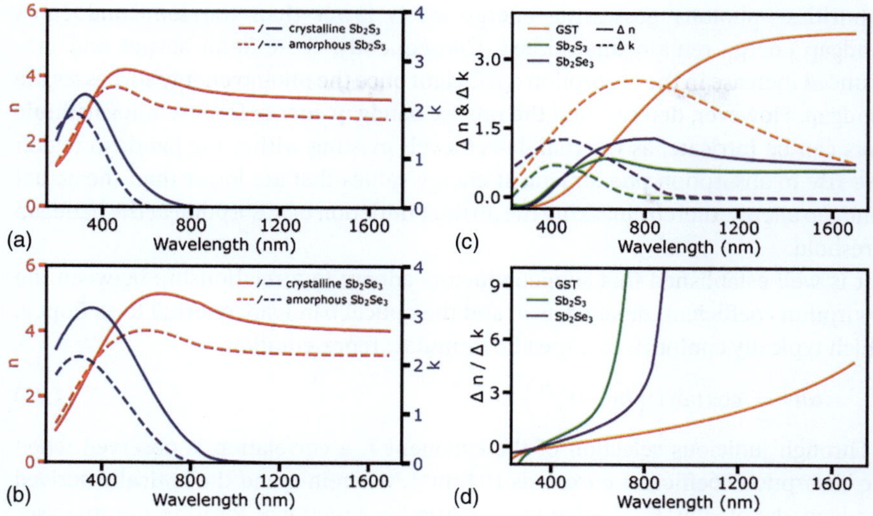

Figure 4.9 The ellipsometry data fittings for (a) S_{b2S3} and (b) S_{b2Se3} are presented, encompassing both their amorphous (represented by dashed lines) and crystalline (depicted with solid lines) phases. Proceeding to (c), we illustrate the variation in the refractive index for GST, alongside S_{b2S3} and S_{b2Se3}, plotted against wavelength, highlighting the disparities among them. In part (d), we delve into the conventional figure of merit (FOM), expressed as $\Delta n/\Delta k$, for GST, S_{b2S3}, and S_{b2Se3}, evaluating their comparative performance in optical properties. *Source:* Delaney et al. (2020) / John Wiley & Sons / CC BY 4.0.

In Figure 4.9c, the switching contrast's real and imaginary parts, denoted as Δn and Δk, are depicted for the materials under examination. At 1550 nm, Sb_2S_3 exhibits a refractive index variation Δn of 0.60, whereas Sb_2Se_3 displays a higher Δn of 0.77. For perspective, GST provides a substantial Δn of 3.56, albeit with a significant imaginary component Δk of 2, indicative of substantial losses in its crystalline phase. This sharply contrasts with the minimal losses reported for both Sb_2S_3 and Sb_2Se_3, where Δk stays below 10^{-5} in both amorphous and crystalline phases across telecommunication wavelengths. The conventional figure of merit, calculated as Δn divided by Δk, is presented in Figure 4.9d for our materials alongside GST. It is noteworthy that as Δk tends toward zero in both material phases, the figure of merit loses its significance due to the near-zero denominator, which underscores the need for alternative metrics in low-loss regimes.

4.2.3 Optical Bandgap of Chalcogenide Semiconductors

The optical bandgap in a semiconductor is characterized by the initial point of optical absorption, which arises when incident photon energy ($h\nu$) matches the energy gap between the topmost filled VB states and the bottommost unfilled CB states. Photons possessing energies exceeding the threshold facilitate electron excitation from the VB to the CB, leading to a steep rise in interband optical absorption due to the enhanced energy levels in both the VB and CB. Conversely, within a perfect semiconductor—one that is entirely free from lattice defects or impurities—photons possessing energy levels lower than the semiconductor's bandgap energy remain unabsorbed. Consequently, there is an abrupt and pronounced increase in the absorption coefficient once the photon energy surpasses the bandgap. However, determining the precise bandgap energy from semilogarithmic plots can be intricate, as internal defect levels existing within the bandgap region give rise to absorption phenomena at energy values that are lower than the actual bandgap energy, thereby masking the distinct initiation of absorption at the bandgap threshold.

It is well-established that semiconductors adhere to a relationship between the absorption coefficient, denoted as α, and the optical bandgap, referred to as Eopt g, which typically conforms to a specific formulaic representation

$$\alpha h\nu = (\cos\tan t)\left(h\nu - E_g^{\mathrm{opt}}\right)^r \tag{4.35}$$

Through judicious selection of the exponent r, a correlation is observed when the absorption coefficient α exceeds 104 cm^{-1}, wherein α and the optically derived bandgap, denoted as E_g^{opt}, adhere to a particular functional form. In our analysis, we employ E_g^{opt} to represent this optically determined energy gap. Compliance with this relationship enables the determination of E_g^{opt} by linear extrapolation; specifically, plotting $(\alpha\ h\nu)^{1/r}$ against $h\nu$ and identifying the abscissa intercept. In the case of crystalline semiconductors characterized by parabolic energy bands, r assumes values of 1/2 for transitions involving direct bandgaps and 2 for those associated with indirect bandgaps, about allowed transitions. Notably, in the context of amorphous semiconductors, Tauc et al. have demonstrated that $r = 2$ can be applied under the premise of parabolic bands (Tauc et al. 1966). It bears clarification that Tauc's method can evoke parallels with indirect gap transitions in crystalline materials. Despite the term "direct" being used for transitions in amorphous semiconductors, which do not require phonon participation, these transitions do not adhere to the momentum conservation rule characteristic of indirect transitions, thereby introducing a nuance in classification.

Figure 4.10 illustrates the optical bandgap characteristics of $Ge_2Sb_2Te_5$ across its various phases. It shows that the relationship with an exponent ($r = 2$), indicative of indirect transitions, is applicable to both the face-centered cubic (fcc) and hexagonal phases, resulting in an estimated optical bandgap of approximately 0.5 eV. For the amorphous phase, Tauc's correlation—also characterized by ($r = 2$)—is relevant here as well, yielding a bandgap energy of about 0.7 eV. This value remains consistent, regardless of the methodology used to derive the optical constants. This underscores the utility of the Tauc–Lorentz dispersion model in analyses where the bandgap serves as a parameter to be fitted. Notably, the substantial disparity in bandgap between the amorphous and crystalline states is remarkable, especially given the facile and rapid nature of the reversible phase changes achievable in this material.

A commonly adopted experimental approach in assessing the optical bandgap of semiconductors, notably amorphous varieties, involves identifying the specific photon energy level where the absorption coefficient attains a value of 104 cm^{-1}, a threshold frequently denoted as E04. Regrettably, past investigations into the optical

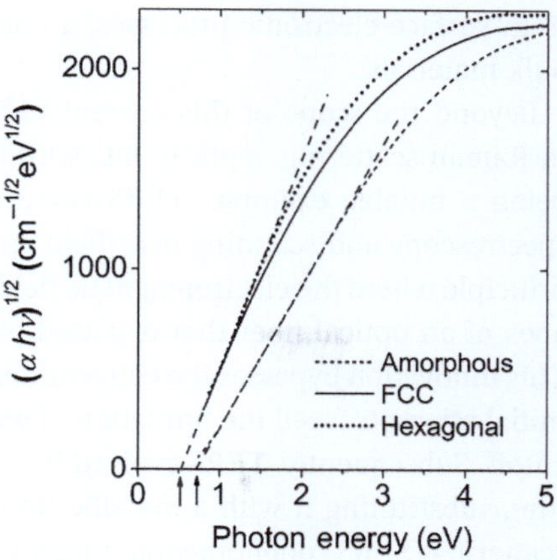

Figure 4.10 Reconstructed diagrams illustrating $(\alpha\ h\nu)^{1/r}$ against photon energy for $Ge_2Sb_2Te_5$ amorphous, face-centered cubic (fcc), and hexagonal phases, reveal a consistent applicability of the $r = 2$ exponent to all phases. Through dashed line extrapolations to the origin, guided by arrows, an optical bandgap of 0.7 eV is discerned for the amorphous phase, referred to as the Tauc gap, whereas the crystalline phases exhibit an indirect bandgap of 0.5 eV. *Source:* Lee et al. (2005) / AIP Publishing LLC.

bandgap of Ge–Sb–Te alloys have employed an ill-suited exponent $r = 1/2$ uniformly across all phases, leading to reported bandgap estimates ranging from 1.5 to 2 eV. These values significantly exceed typical E04 indicators and are at odds with other fundamental properties of $Ge_2Sb_2Te_5$, including the electrical conductivity's temperature dependence and Hall coefficient measurements. Additionally, it has been established that $Ge_2Sb_2Te_5$ alloys incorporate a substantial density of vacancies, contributing to the observation of elevated free carrier concentrations (~1020 cm^{-3}) within their crystalline forms. In such semiconductors, where degeneracy prevails, the accurate assessment of the bandgap can be intricate due to phenomena like band state tails extending into the gap and the Burstein–Moss shift, which can broaden the optical bandgap, complicating the direct correlation with simple models.

4.3 Raman Spectroscopies

For several decades, Raman spectroscopy has been employed as an indispensable technique in the exploration of diverse material properties. This methodology is grounded in the phenomenon of inelastic light scattering by molecular vibrations, a process that yields scattered light at altered frequencies, a testament to the superposition of the combination of the incident photon's frequency with the inherent vibrational patterns or oscillatory states of the material. In contemporary materials science, Raman spectroscopy has cemented its status as a benchmark spectroscopic technique for comprehensive material characterization.

4.3.1 Principles of Raman Scattering

In recent times, Raman scattering has been harnessed as an analytical instrument within the realm of nanoscience and nanotechnology. This application reveals distinct spectral characteristics, which are a consequence of vibration confinement within nanostructures. Additionally, it highlights the impact of local fields, particularly those arising from plasmonic resonances, along with augmented contributions from surface electronic processes, a phenomenon more pronounced compared to bulk materials.

Beyond the scope of this current publication, there are further breakthroughs in Raman scattering applications, with tip-enhanced Raman spectroscopy (TERS) being a notable example. TERS came into being through the fusion of Raman spectroscopy and scanning near-field optical microscopy (SNOM). It operates on a principle where the electromagnetic field is highly concentrated at the metal-coated apex of an optical fiber that is placed in extremely close proximity to the sample. This innovation bypasses the conventional diffraction-limited resolution barrier, yet initial attempts faced the limitation of weak signal collection due to fiber's inherent cutoff. Subsequently, TERS evolved by adopting a design sans an optical fiber aperture, substituting it with a metallic tip that exploits the surface-enhanced Raman scattering (SERS) phenomenon, thereby amplifying the Raman signal intensity.

We can grasp the core principles underlying Raman spectroscopy via the following analysis: When a material is exposed to an externally oscillating optical field denoted as $E(r, t)$, the induced polarization $P(r, t)$ of the material is intricately tied to the material's susceptibility function $\chi(\omega)$, which governs this relationship as per

$$\begin{aligned} P(r,t) &= P(k,\omega)\cos(kr - \omega t) \\ E(r,t) &= E_0(k,\omega)\cos(kr - \omega t) \\ P(k,\omega) &= \chi(\omega)E_0(k,\omega) \end{aligned} \tag{4.36}$$

In this context, the squared modulus of the excitation electric field, $|E_0|^2$, characterizes the intensity of the optical stimulus. The quantization of atomic vibrational modes leads to the formation of phonons, which are elemental quanta of lattice vibrations. These phonons are mathematically represented by atomic displacements that conform to the wave nature, expressed as plane waves in the description of these vibrational dynamics:

$$Q(r,t) = Q(q,\Omega)\cos(qr - \Omega t) \tag{4.37}$$

where q represents the wave vector and the frequency of the phonon, within the confines of the adiabatic approximation, and the function $\chi(\omega)$ may be subjected to a Taylor series expansion about the quantity $Q(r, t)$, which encapsulates the atomic displacements. This expansion allows for a detailed analysis of the system's response under the premise that changes in the electronic states lag behind the nuclear motion:

$$\chi(k,\omega,\Omega) = \chi_0(k,\omega) + \left(\frac{\partial \chi}{\partial Q}\right)_0 Q(r,t) + \cdots \tag{4.38}$$

Upon plugging Eq. (4.38) into Eq. (4.37), the polarization function can be simplified as follows:

$$\begin{aligned} P(r,t,\Omega) &= P_0(r,t) + P_{\text{ind}}(r,t,\Omega) \\ P_0(r,t) &= \chi_0(k,\omega)E_0(k,\omega)\cos(kr - \omega t) \\ P_{\text{ind}}(r,t,\Omega) &= \frac{1}{2}\left(\frac{\partial \chi}{\partial Q}\right)_0 Q(q,\omega)E_0(k,\omega)F(k,\omega) \end{aligned} \tag{4.39}$$

The spectral function can be denoted as follows:

$$F(k,\omega) = \cos[(k+q)r - (\omega+\Omega)t] + \cos[(k-q)r - (\omega-\Omega)t] \tag{4.40}$$

The induced polarization function, P_{ind}, comprises dual supplementary wave components: a Stokes-shifted wave, characterized by a wavevector $kS = k - q$ and frequency $\omega_S = \omega - \Omega$, alongside an anti-Stokes-shifted wave with a wavevector $kAS = k + q$ and frequency $\omega_{AS} = s\omega + \Omega$. These Raman frequencies, ω_S and ω_{AS}, respectively, denote Stokes and anti-Stokes shifts. It is pertinent to mention that Eq. (4.39) encapsulate the scenario of single-phonon Raman scattering. Analogously, more complex arrangements that incorporate combinations of two phonons (which relate to two-phonon or second-order Raman-related scattering phenomena) and Raman processes of higher orders can be deduced based on a comparable line of reasoning.

The intensity of the scattered light is obtained by calculating the time-averaged magnitude of the induced polarization, denoted as Pind, as specified in Eq. (4.39). This gives rise to the notion of the complex second-order Raman tensor (R), which can be precisely defined according to the research conducted by Yu and Cardona (2010):

$$R = \left(\frac{\partial \chi}{\partial Q}\right)_0 \frac{Q}{|Q|} \tag{4.41}$$

Here focus on how the intensity of the scattered light (I_s) varies with respect to the polarizations of both the incident and scattered light, represented by unit vectors e_i and e_s respectively, can be elucidated through Eq. (4.41). Consequently, the magnitude of I_s is determined as follows:

$$I_s \propto |e_i R e_s|^2 \tag{4.42}$$

4.3.2 Raman Spectroscopy in Chalcogenide Semiconductors

Prazakova (2022) delves into the structural transformation dynamics of Germanium-abundant Ge–Sb–Te (GGST) composites, with a focal point on Ge45Ge45GST (45% Ge), juxtaposing it against the benchmark Ge2Sb2Te5 (GST225), through the lens of Raman spectroscopic analysis, systematically examining how the annealing temperature influences these structures.

The Raman spectral profile of the freshly-deposited amorphous GST225 sample displays a notably broadened spectral characteristic that stretches across the range of 100–190 cm^{-1}, which incorporates overlapping peaks at 124 and 154 cm^{-1}, as illustrated in Figure 4.11. This extensive band is complemented by a less intense feature centered around 220 cm^{-1}. The significant peak at 154 cm^{-1} is potentially attributed to vibrational modes reflective of diverse structural elements, with a compelling assignment linking it to the stretching vibrations within Sb–Te bonds in pyramidal $SbTe_3$ units, providing a nuanced understanding of its spectral origin.

The vibrational contributions at 124 and 220 cm^{-1} are attributed to the diverse patterns associated with Ge–Te bonds in tetrahedral arrangements and in octahedral configurations that have structural defects, reminiscent of those found in amorphous GeTe. Typically, vibrational signatures of defective octahedral structures

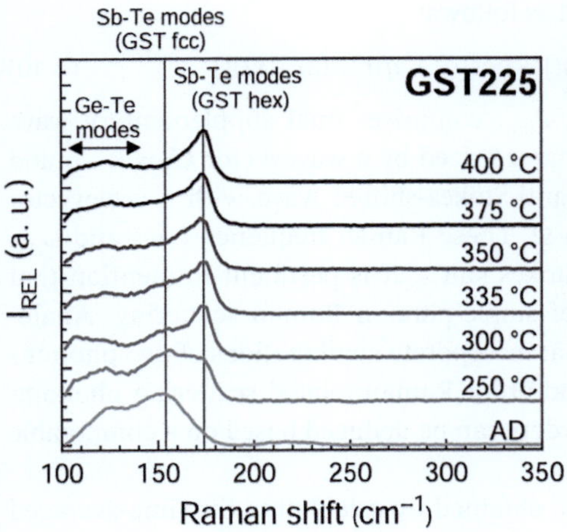

Figure 4.11 The Raman spectroscopic profiles of GST225 in its as-deposited amorphous state and following annealing treatments ranging from 250 to 400 °C illustrate distinct vibrational features. Noteworthy are the vertical markers at frequencies of 154 and 174 cm^{-1}, which emphasize the primary vibrational modes characteristic of SbTe structural components. *Source:* Prazakova et al. (2022) / HAL Science / CC BY 4.0.

are predominantly situated below 190 cm^{-1}, whereas less intense modes associated with Ge atoms in tetrahedral sites emerge at higher frequencies beyond 190 cm^{-1}. By drawing comparisons with the analysis of GeTe, the peak at approximately 124 cm^{-1} aligns with the A1 vibrational mode that is typical of corner-sharing tetrahedra having the chemical formula $GeTe_4$–nGen (where n is 1 or 2). In contrast, the peak at around 220 cm^{-1} is associated with the vibrational characteristics of intact $GeTe_4$ tetrahedral units.

The spectral peak at 154 cm^{-1}, associated with Sb–Te vibrational modes, retains its prominence even after annealing at 250 °C, highlighting the persistence of these structural motifs within the cubic (GST225-fcc) phase of the alloy. Concomitantly, a novel band emerges at 174 cm^{-1}, accompanied by a subtle attenuation in the intensity of modes within the 100 to 140 cm^{-1} range, attributed to Ge–Te interactions. As the temperature during the annealing process rises, there is a gradual transition in the dominance of peaks. The 174 cm^{-1} peak becomes more prominent, while the peak at around 154 cm^{-1} gradually diminishes, which fades progressively. The emerging peak at 174 cm^{-1}, linked to Sb–Te modes in a configuration specific to the hexagonal (GST225-hex) phase, becomes the new spectral hallmark. Following annealing at 350 °C and above, where the hexagonal phase is fully established, the Raman spectra exhibit minimal variation, reflecting a stable phase transition. This systematic evolution of spectral features with increasing temperature underscores the progressive structural transformation from the cubic to the hexagonal crystalline structure in GST225, evidencing phase reconfiguration.

The Raman spectroscopic profile of the as-deposited GGST (Figure 4.12) is characterized by three principal spectral regions [313], which can be delineated as follows: 1) a dominant feature at 154 cm^{-1}, attributed to the Sb–Te stretching vibration within $SbTe_3$ pyramidal structures, a feature consistent with observations in GST225; 2) lower frequency components (<150 cm^{-1}) indicative of vibrational patterns typical to GeTe structural elements; 3) a wide band spanning from 190 to 300 cm^{-1}, encompassing two superimposed peaks at approximately 272 cm^{-1} and 230 cm^{-1}, respectively, associated with transverse optical (TO)- and longitudinal

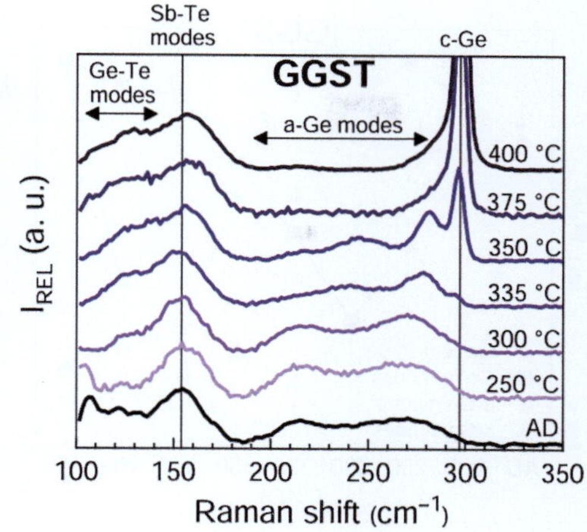

Figure 4.12 In the Raman spectroscopy analysis of both as-deposited amorphous (AD) and annealed GGST, the spectra illustrate distinct peaks, with vertical annotations marking the frequencies at 154 cm^{-1} for Sb–Te vibrational modes and 297 cm^{-1} for crystalline Ge-related modes, highlighting the spectral signatures of structural transitions. *Source:* Prazakova et al. (2022) / HAL Science / CC BY 4.0.

optical (LO)-like modes of amorphous Germanium (a–Ge), revealing the complex vibrational landscape of the material.

Contrary to observations in GST225, the spectral feature at 154 cm^{-1}, associated with SbTe units, remains steadfast in both its position and full width at half maximum (FWHM) up to an annealing temperature of 400 °C. This stability stands in contrast to the noted changes in GST225. The low-frequency range, reflective of GeTe units, undergoes a discernible evolution until 335 °C, beyond which it aligns with the spectral profiles of both as-deposited amorphous and cubic crystalline GST225-fcc (as depicted in Figure 4.11), implicating the robustness of SbTe structures and the lack of a cubic-to-hexagonal phase transition even at elevated temperatures. Within the a-Ge spectral region, a progressive emergence of TO-like and LO-like vibrational modes is noted with rising annealing temperatures. Notably, a distinct peak at 297 cm^{-1}, indicative of the crystalline Ge phase (c–Ge), emerges at 335 °C, marking the onset of Ge crystallization, an event that occurs at a surprisingly lower temperature compared to the crystallization of a standalone amorphous Ge layer.

A comprehensive Raman spectral analysis was conducted to trace the evolution of the structural constituents in the GGST alloy as the temperature changes. This process entailed computing the total integrated intensities of specific characteristic peaks. These peaks include those at 110 and 124 cm^{-1} corresponding to Ge–Te vibrational modes, at 220 and 270 cm^{-1} for a-Ge modes, and at 297 cm^{-1} for c-Ge modes. Each computed sum was then standardized in relation to the unchanging intensity of the Sb–Te mode peak at 154 cm^{-1}, which acts as a consistent reference throughout the examined temperature range. The normalized intensities, plotted against annealing temperature (Figure 4.13), reveal several key trends: (i) Initially, within the as-deposited amorphous layer, the Ge–Te mode intensities decline, hitting a nadir at 300 °C. Conversely, a-Ge mode intensities exhibit a slight rise, peaking at 250 °C. (ii) Following this, the Ge–Te modes linked to the layer's ordering process show a resurgence in intensity post-300 °C, achieving a quasi-steady state after 375 °C. (iii) In tandem with the re-emergence of Ge–Te vibrational modes, the intensity of a-Ge

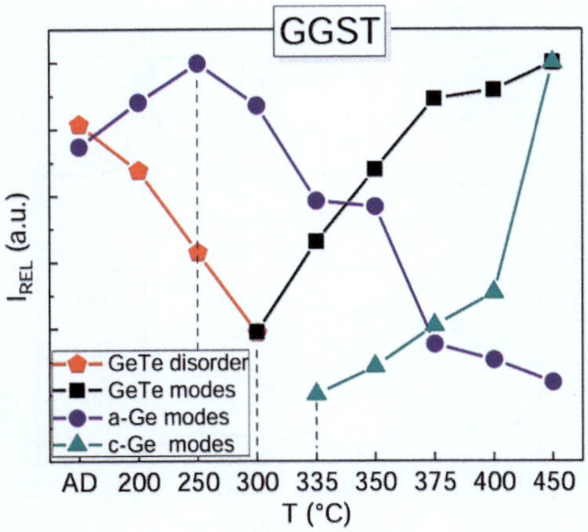

Figure 4.13 The spectral intensity dynamics of Ge–Te, amorphous Ge (a–Ge), and crystalline Ge (c–Ge) modes within GGST's Raman spectra were analyzed as a function of annealing temperature. These intensities were calculated through integration and subsequently normalized against the invariant intensity of the Sb–Te mode contributions. Notably, the diagram demarcates key temperature milestones: the peak intensity for a–Ge modes at 250 °C, a nadir for Ge–Te modes at 300 °C, and the onset of c–Ge modes at 335 °C, each highlighted by dashed lines to underscore these critical transitions. *Source:* Prazakova et al. (2022) / HAL Science / CC BY 4.0.

modes diminishes, gradually giving way to the emergence of a distinct c-Ge peak. The c-Ge peak makes its appearance at 335 °C, and its intensity shows a continuous rise until reaching the maximum annealing temperature used in the experiment.

Raman spectroscopic investigations reveal that the initial temperature escalation to 300 °C initiates a reconfiguration phase, marked by a decline in Ge–Te vibrational modes concomitant with an amplification of a-Ge modes. This phase involves the disruption of suboptimal bonding within mixed GeSbTe complexes (such as GeSbTe3, Ge2SbTe2, and GeSb2Te2, as per references), alongside a reconfiguration of GeTe4–nGen tetrahedral structures (where n spans from 0 to 4). These transformations within the initially disordered amorphous matrix lead to a decrease in Ge–Te mode intensity. Consequently, Ge atoms are released from these mixed units, contributing to a moderate rise in the a-Ge mode intensities. This observation aligns with prior studies on amorphous GeTe alloys, which, grounded in the random bonding paradigm, emphasize that the assembly of diverse tetrahedral complexes is influenced by the layer's stoichiometric balance. At elevated annealing temperatures, the structural transformation escalates in prominence. The Ge–Te mode intensities attain a nadir at 300 °C, a juncture that signals the completion of an initial reconfiguration phase and the liberation of Ge atoms from their mixed complexes. As the temperature rises, there is a gradual uptick in the relative intensity of Ge–Te vibrational modes. This implies an ongoing rearrangement of Ge–Te bonds near Sb–Te clusters, signaling the crystallization process taking place within the GeSbTe phase. Subsequently, the crystalline GeSbTe structure begins to form nuclei and then expands. The diminishment of a-Ge modes above 300 °C is correlated with a methodical reorganization of the Ge network. This process involves the orderly

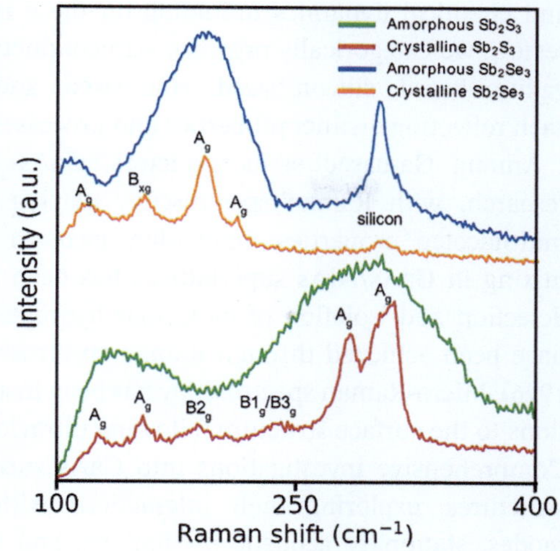

Figure 4.14 Raman spectroscopic profiles for both amorphous and crystalline layers of Sb_2S_3 and Sb_2Se_3, examining their spectral characteristics pre- and post-laser annealing, reveals insightful transformations in their structural phases. *Source:* Delaney et al. (2020) / John Wiley & Sons.

arrangement of Ge atoms within the amorphous framework, which sets the stage for the nucleation and development of the cubic crystalline Ge phase.

Figure 4.14 presents the Raman spectroscopic analysis of both amorphous and crystalline thin film forms of Sb_2S_3 and Sb_2Se_3, capturing their spectral evolution pre- and post-laser annealing. In their pristine, amorphous state, these films exhibit characteristic, broad spectral peaks. Specifically, amorphous Sb2S3 displays signatures at 130 and 300 cm^{-1}, attributed to Sb–S and S–S vibrational modes, respectively. Conversely, amorphous Sb_2Se_3 displays a distinct and broad spectral peak at the 190 cm^{-1} position, indicative of its amorphous structure. A peak corresponding to the silicon substrate, located at 302 cm^{-1}, is discernible in the Raman spectrum of amorphous Sb_2Se_3, albeit with reduced intensity in the crystalline phase due to the film's increased transparency. Notably, in the case of Sb_2S_3, the silicon substrate's spectral feature is less evident, overshadowed by the film's own spectral characteristics.

Upon annealing, the Raman spectra exhibit a marked increase in peak sharpness, indicative of the enhanced crystalline structure. For crystalline Sb_2S_3, the Raman spectrum displays well-defined peaks at 125, 154, 284, and 312 cm^{-1}. These peaks are ascribed to Ag symmetry vibrational modes. Additionally, there are B_1g/B_3g modes at 240 cm^{-1} and a B_2g mode at 188 cm^{-1}. The B_2g mode is linked to the surface oxidation of the film. In the case of crystalline Sb_2Se_3, the structural change gives rise to Ag symmetry peaks at 117, 190, and 211 cm^{-1}. These are accompanied by B_xg peaks at 150 and 153 cm^{-1}, all of which are typical features of its structural makeup. Both Sb_2S_3 and Sb_2Se_3 adopt an orthorhombic crystal structure, a conclusion supported by the observed Raman vibrational modes.

4.3.3 Applications of Raman Spectroscopy in Nanophotonics

Extensive utilization of Raman spectroscopy has been pivotal in elucidating the characteristics of semiconductor surfaces, alongside the intricate physical and chemical dynamics unfolding on these interfaces. Within the scope of this review, we categorically organize semiconductor systems into four pivotal groups: gallium-based, silicon-based, zinc-based, and cadmium-based semiconductors, each reflecting distinct properties and processes.

Among Ga-based semiconductors, GaAs remains the subject of extensive research, with Raman spectroscopy playing a pivotal role in investigating its multifaceted properties. Intervalley electron scattering-induced electronic-state mixing in GaAs/AlAs superlattices has been revealed (Mlayah et al. 1996). The detection and isolation of molecular hydrogen and deuterium on GaAs surfaces have been achieved through Raman spectroscopic techniques (Vetterhöffer et al. 1996). Micro-Raman spectroscopy has been instrumental in documenting modifications to the surface structure following ultraviolet laser ablation (Lyon et al. 1998). Comprehensive investigations into GaAs surfaces have covered quantum—well structures, exploring their interactions with longitudinal—optical vibrational modes, stationary acoustic oscillations, and the characteristic Raman response within the context of a planar microcavity setup (Vina et al. 1996). Photoluminescence (PL) investigations at the interface of GaN and wurtzite structures contribute further insights (Siegle et al. 1995). Regarding GaN thin films deposited on sapphire substrates, which serve as a crucial material in the realm of blue optoelectronics, a characterization process was carried out. This was achieved through the strategic incidence of a laser beam or microscopic focusing, thereby successfully avoiding the interference caused by sapphire's Raman signals (Feng et al. 1997). Additionally, structural characteristics of these GaN films have been detailed (Lyon et al. 1998), highlighting advancements in this field.

Silicon semiconductors continue to hold significant industrial relevance, necessitating thorough characterization studies. The investigation of stress profiles, particularly regarding P-line defects and gradients, has been undertaken at the critical Si/SiO_2 interface (Uozumi and Katoda 1997). Porous silicon, under the irradiation of an Ar^+ laser beam with a wavelength of 514.5 nm, exhibited a time-dependent frequency shift. This phenomenon accompanied by narrowing of spectral lines and an upward energy shift of the peak in the 500–520 cm^{-1} range upon prolonged illumination (Wu et al. 1995). Furthermore, Raman spectroscopy has been instrumental in probing nanostructures on porous silicon surfaces, elucidating quantum size effects within these systems (Lyon et al. 1998).

Studies employing high-pressure conditions, reaching up to 36.0 GPa, have been instrumental in tracking the evolution of longitudinal-optical vibrational excitations and transverse-optical vibrational excitations within ZnSe semiconductors (Lin et al. 1997). The spin-flip Raman scattering technique has been harnessed to explore the impact of nitrogen doping on the compensation phenomena within p-type ZnSe (Townsley et al. 1996). Resonance Raman spectroscopy serves as a highly effective means for examining the characteristics of Cd-based semiconductor surfaces, with nanocrystalline CdSe being classified under the C3v symmetry group (Shiang et al. 1995). The formation of CdS films on Au substrates has led to resonance Raman studies that shed light on the intricate optical characteristics of these films (Boone and Shannon 1996), complemented by descriptions of surface reconstructions occurring in CdS monolayers on Au surfaces (Demir and Shannon 1996). Furthermore, the fabrication of semiconductor molecular wire arrays through electrodeposition has demonstrated that their optical characteristics are contingent upon the diameter of the wires (Routkevitch et al. 1996), highlighting the nanostructured complexity in these systems.

4.4 Emission Spectroscopies

4.4.1 Optical Emission Mechanisms in Chalcogenides

Nanoparticles' luminescence constitutes a keen indicator of their electronic characteristics and serves as the cornerstone for applications across laser technology, optical sensing, and light-emitting diodes (LEDs). This emission phenomenon can be elicited either through photoexcitation, known as PL, or by electrical stimulation, referred to as EL. Within the context of PL, the phenomenon involves the material being excited by photons, with the subsequently emitted light typically possessing lower energy, a characteristic designated as Stokes emission. Conversely, EL emerges from the recombination of electron-hole pairs, a process initiated by the electrical injection of these charge carriers.

PL is categorically divisible into two primary regimes: band edge emission, encompassing excitonic and trap state emissions, with the latter typically exhibiting a redshift relative to the former. Illustratively, CdS nanoparticle PL spectra demonstrate distinct trap state and band edge emissions, the latter occurring at shorter wavelengths (Wu et al. 2000). Two salient observations emerge from this context. Firstly, surface passivation, which mitigates trap states, significantly amplifies the overall PL intensity compared to its unpassivated counterpart. Secondly, the passivated specimen demonstrates an elevated intensity ratio between the band edge emission, which reaches its maximum intensity at approximately 450 nm, and the trap-state emission, attaining its peak around 550 nm. This ratio functions as an indicator of the density and distribution of trap states. An increased prominence of the trap-state emission in comparison to the band-edge emission implies a high concentration of trap states and efficient carrier-trapping processes.

Achieving predominantly band-edge emission in samples is feasible through meticulous surface passivation, exemplified by the use of TOPO capping on CdSe nanoparticles. These samples exhibit a marked dominance of band-edge emission over trap-state emission, thereby serving as a testament to their superior quality and reduced defect density (Figure 4.15).

Enhancement of luminescence is achievable through surface engineering or the employment of core-shell architectures, with notable examples of strong PL demonstrated by CdSe, CdS, and ZnS nanoparticles. Conversely, many nanoparticles exhibit limited or no room-temperature luminescence, a phenomenon often attributed to indirect bandgap characteristics or an abundance of internal and surface trap states, which mitigate emission. Elimination of surface trap states, via surface control, significantly amplifies luminescence and elevates the ratio of

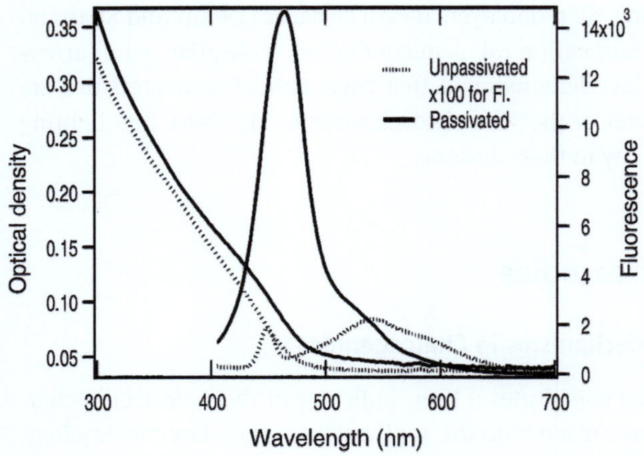

Figure 4.15 Spectroscopic analysis of electronic absorption and fluorescence (excitation wavelength, λ_{ex} = 390 nm) for CdS nanoparticles reveals distinct profiles between unpassivated (UP) and surface-passivated (SP) samples. The representation includes the UP spectrum as a dotted line, contrasted with the SP spectrum depicted by a solid line. To facilitate a nuanced comparison, the fluorescence signal from the unpassivated sample has been amplified by a factor of 100. *Source:* Wu et al. (2000) / with permission of Elsevier.

band-edge to trap-state emissions. This is accomplished through surface capping with organic, inorganic, or biomolecular ligands and ions, which mitigates trap states within the bandgap, thereby inhibiting luminescence quenching. Such strategies are pivotal for applications demanding high luminescent nanoparticles, such as lasers, light-emitting diodes, bioimaging, and optical sensing. PL tends to escalate at lower temperatures, for both band-edge and trap-state emissions, due to reduced electron–phonon scattering, leading to prolonged excited-state lifetimes.

Typically, quantum wells, characterized by one-dimensional (1D) confinement, along with nanorods or nanowires, which are confined in two dimensions (2D), display less significant fluctuations in their optical properties compared to three-dimensionally (3D) confined quantum dots (QDs). This lesser degree of spectral shifts is attributed to a reduced quantum confinement effect. The extent of these changes, however, is intricately linked to the material's chemical composition and the precise geometric parameters employed.

CL denotes light emission resulting from chemical reactions, whereas electrochemiluminescence (ECL) specifically concerns the luminescence arising from electronically excited species formed at electrode interfaces during electron transfer processes. Both phenomena are of keen interest for analytical applications and beyond, owing to their heightened sensitivity and specificity. Nanoparticles have been deployed in diverse roles within CL systems as catalysts expediting chemical reactions critical to CL; as labels, with gold nanoparticles exemplifying tags for specific biomolecules in bioassays enhanced by chemiluminescence; and as luminophores themselves, generating bandgap CL through direct chemical interactions, akin to materials like CdSe/CdS core/shell or CdTe. This CL mechanism, resembling PL, shares a comparable electron-hole recombination process, leading

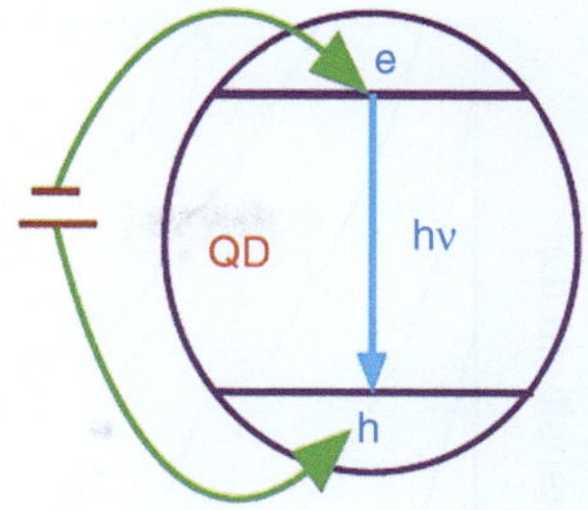

Figure 4.16 A conceptual diagram elucidates the fundamental principle of electroluminescence, which entails the orchestrated injection of electrons and holes into a quantum dot semiconductor, catalyzing the EL phenomenon.

to spectrally similar emissions, despite the distinct origins of excited electrons and holes—chemical reactions for CL versus photoexcitation for PL.

EL relies on the electrical stimulation for the production of light, wherein the fundamental principle involves the recombination of electrons and holes after their injection into a semiconductor matrix due to an applied electric field. Figure 4.16 encapsulates the underlying concept of EL within a semiconductor QD, highlighting this process.

4.4.2 Photoluminescence and Electroluminescence

PL denotes the emission of photons triggered by incident light spanning the ultraviolet to near-infrared spectrum. Owing to their exceptionally high quantum yields (QYs), potentially reaching 100% as previously noted, along with facile synthesis and processability, semiconductor nanocrystals serve as versatile color-tunable nanoscale light sources. Calculations suggest that a monolayer of these nanocrystals, densely packed at a rate of 1012 per square centimeter, with an average radiative lifetime of 20 nanoseconds and a moderate efficiency (η) of 20%, could yield luminous power densities exceeding several watts per square centimeter. Our subsequent discourse will delve into the intricacies of PL in semiconductor nanocrystals.

In their work, Talapin and colleagues (Talapin et al. 2001) synthesized well-dispersed CdSe nanocrystals employing a ternary blend comprising hexadecylamine, trioctylphosphine oxide (TOPO), and trioctylphosphine. Figure 4.17a presents the PL spectra of CdSe nanocrystals synthesized at 300 °C, with a mean diameter of approximately 4.05 ± 0.16 nm, alongside their absorption spectra. Notably, the absorption profiles revealed five distinct electronic transitions, a characteristic that underscores the homogeneity in particle size distribution (PSD) within the CdSe nanocrystal ensemble.

In Figure 4.17b, the PL and absorption spectra of Se/ZnS nanocrystals are depicted, with annotations specifying the ZnS shell thickness and the PL QY percentages. Notably, the absence of a ZnS covering layer results in a PL η of merely 12% for the CdSe nanocrystals. Conversely, the integration of a ZnS cover layer leads to a substantial enhancement, achieving a peak η of 66% at a shell thickness of 1.6 monolayers, a finding consonant with a previously outlined case, thereby precluding redundant discourse.

Dabbousi et al. 1997 examined the dependence of PL intensity on grain size for both uncoated CdSe nanocrystals and those encapsulated with ZnS, forming

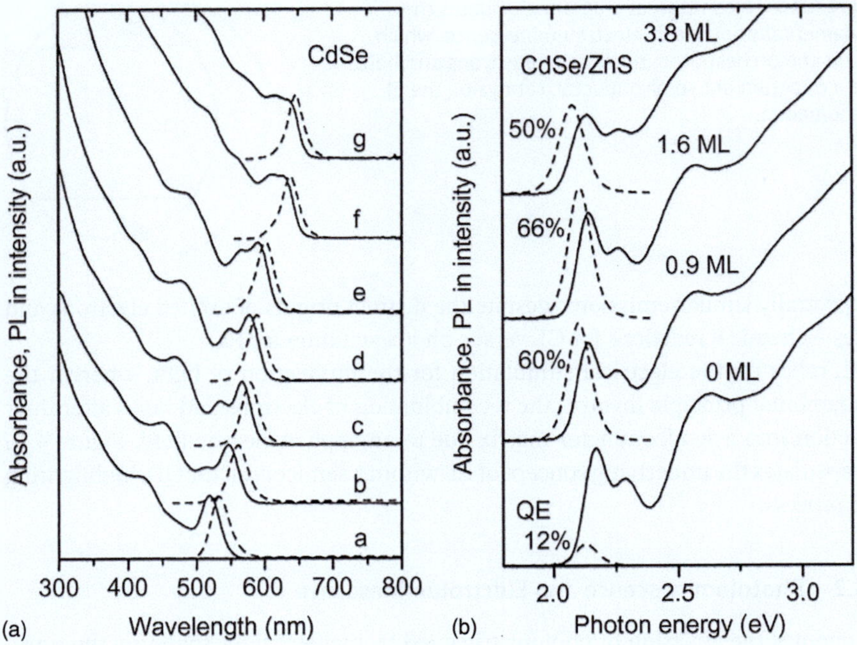

Figure 4.17 At ambient temperature, spectral monitoring of CdSe nanocrystals' growth at 300 °C reveals their absorption and emission profiles (a), while (b) presents the spectra of these nanocrystals both pre- and post-deposition of ZnS shells with varying monolayer thicknesses. *Source:* Talapin et al. (2001) / American Chemical Society.

(CdSe)ZnS composite QDs, at room temperature. Figure 8.13a illustrates these findings, where the dotted line represents the PL of bare CdSe, contrasted by the solid line depicting the (CdSe)ZnS composites. Particle sizes of 2.3, 4.2, 4.8, and 5.5 nm are denoted as a, b, c, and d, respectively. Notably, the application of a ZnS shell significantly enhances PL intensities, achieving QYs of 40%, 50%, 35%, and 30% for these sizes, compared to a substantially lower range of 5–15% for uncoated CdSe nanoparticles.

Additionally, it is demonstrated that an augmentation in particle size leads to a redshift in PL peaks, a phenomenon observed consistently in both uncapped CdSe nanoparticles and their (CdSe)ZnS composite QD counterparts, as shown in Figure 4.18a. At diminutive diameters of CdSe particles, the high surface-to-volume ratio dictates a PL profile heavily influenced by intense emissions from deep trap sites within the TOPA-encapsulated QDs. Introduction of a ZnS shell effectively passivates surface defects and minimizes non-radiative recombination at these sites, thereby mitigating deep-level emissions. Consequently, the PL of (CdSe)ZnS QDs become predominantly governed by band edge radiative recombinations, resulting in a pronounced enhancement in PL intensity across different particle sizes compared to bare CdSe.

Figure 4.18b illustrates the wavelength-dependent PL intensity for (CdSe)ZnS quantum dots, wherein the core of CdSe measures 4.2 nm and is enveloped by ZnS shells of varying thicknesses: (i) 0, (ii) 0.65, (iii) 1.3, (iv) 2.6, and (v) 5.3 monolayers.

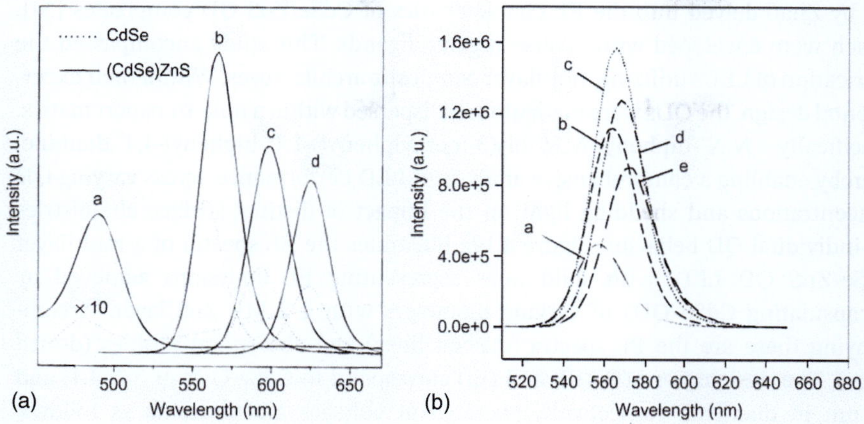

Figure 4.18 (a) PL spectra are depicted for both uncoated (dashed lines) and ZnS-coated (solid lines) quantum dots, with core diameters as follows: (i) 23 Å, (ii) 42 Å, (iii) 48 Å, and (iv) 55 Å. The coated dots exhibit considerably heightened PL intensities, correlating with increased quantum yields of (i) 40%, (ii) 50%, (iii) 35%, and (iv) 30%, respectively. (b) A sequence of the PL spectra is presented for ZnS-coated dots based on 42 Å diameter CdSe cores, with varying ZnS coverage levels: (i) no ZnS, (ii) 0.65, (iii) 1.3, (iv) 2.6, and (v) 5.3 monolayers. These demonstrate the progressive influence of ZnS thickness on PL characteristics. *Source:* Destriau (1939) / Royal Society of Chemistry.

This depiction highlights a consistent trend of PL peak redshift with increasing ZnS coverage. Additionally, it has been noted that the spectral width of the PL peaks broadens concomitantly with the enhancement in the ZnS layer thickness. This behavior, characterized by an initial rise in PL intensity at thinner coatings followed by a decline as the ZnS layer thickens, is a characteristic feature of semiconductor quantum dot architectures.

EL in nanomaterials signifies the transformation of electrical energy into light via the imposition of an electric field, initiating a luminescent process. These EL devices have garnered substantial interest for their potential in digital communication technologies, where they promise to deliver high luminance and contrast, enhance color gamut and resolution, while simultaneously reducing power consumption. Their lightweight, slim profile, and cost-effectiveness further contribute to their appeal. Consequently, significant industrial resources have been directed toward this research area. The scientific literature on EL is extensive, with its origins dating back to Round's observation of light emission from SiC under electric current in 1907. However, it was not until Destriau's work in 1939 (Destriau 1939), reporting ZnS EL under an electrical field, that the term EL came into use. Advancements continued, with Vlasenko and Popkov fabricating an experimental thin-film EL device in 1960 (Vlasenko and Popkov 1960). The commercialization of EL nanocrystalline materials gained momentum in the mid-1990s. Building on the insights from the previous section on PL, our focus now shifts to semiconductor materials and devices in the context of EL.

Previously highlighted, the substantial PL quantum efficiency of organic semiconductors is a subject warranting further elaboration. In 2004, a research team led by Zhao delved into the EL characteristics of CdSe/ZnS QD composites [54], which were enveloped with diverse organic ligands. This study encompassed the fabrication of LEDs utilizing multilayer composite architectures. Within their experimental design, the QDs were meticulously dispersed within a hole-transport matrix, specifically *N*,*N*′-diphenyl-*N*,*N*′-bis(3-methylphenyl)-1,1′-biphenyl-4,4′-diamine, thereby enabling a comprehensive analysis of LED performance across varying QD concentrations and shedding light on the impact of distinct surface chemistries on individual QD behavior. Figure 4.19a illustrates the EL spectra of a multilayer CdSe/ZnS QD LED, with solid lines representing EL emissions achieved by encapsulating CdSe QDs of varying diameters with a 2 ML ZnS layer. Accompanying these are the PL spectra (dashed lines) and absorption profiles (dotted lines). The designations (i), (ii), and (iii) correspond to CdSe QDs of 3.2, 4.1, and 5.4 nm in diameter, respectively, operated at voltages and currents as follows: (i) 5.5 V, 0.11 mA; (ii) 5.0 V, 0.08 mA; and (iii) 4.0 V, 0.06 mA. An inset in the upper right corner provides a schematic of the LED structure, highlighting the use of PEDOT:PSS as a hole-transport layer and PBD (2-(4-biphenyl)-5(4-tert-butyl phenyl)-1,3,4-oxadiazole) in the emission layer, which has an approximate thickness of 20 nm. Notably, the EL spectrum from the CdSe QDs exhibits high intensity, with a full width at half maximum (FWHM) centered around 30 nm. A shift toward green is observed with decreasing QD size, transitioning into a blueshift, although these shifts are relatively minor (a few nanometers) compared to those seen in PL. Furthermore, the systematic correlation between QD size and the extent of redshift or blueshift in EL is not evident, displaying a nuanced behavior distinct from PL trends.

Principally, the quantum confinement effect, synonymous with the Stark effect, can induce a redshift in emission wavelengths. Observations revealed that within a poly(*N*-vinyl carbazole) (PVK) film, 100 nm in thickness, incorporating CdSe/ZnS QDs, both EL and PL exhibit redshifts ranging from 0.1 to 0.2 nm per volt under forward bias voltages spanning 15–25 V. Specifically, at a forward bias of 5 V, the Stark-induced redshift for multilayer devices amounts to 0.5–1.0 nm. This implies that under low forward bias voltages, the EL spectra of devices utilizing *N*,*N*′-diphenyl-*N*,*N*′-bis(3-methylphenyl)-1,1′-biphenyl-4,4′-diamine (TPD) as a host for CdSe/ZnS QDs show negligible redshift due to the Stark effect. A second LED configuration, featuring a 30-nm thick layer of CdSe/ZnS QDs coated with dodecyl amine ligands in lieu of TPD, demonstrated an EL peak energy akin to the TPD-containing LED, albeit with a forward bias current reduced by approximately two orders of magnitude. This suggests a substantial decrease in point-to-point charge transfer in devices without TPD, highlighting the pivotal role of the hole conductor TPD in facilitating charge injection into QDs through intimate contact with embedded QDs, emphasizing its crucial function in enhancing charge injection efficiency.

In Figure 4.19b, the electroluminescence spectra are depicted for CdSe QD LEDs with varying QD concentrations. The thicknesses of the QD layers for samples a through d are 0.3, 0.2, 0.1, and 0.03 monolayers, respectively, while sample e notably lacks CdSe/ZnS QDs within the TPD matrix. EL measurements were conducted across a voltage range of 5.0–6.0 V. Remarkably, all samples, with the exception of

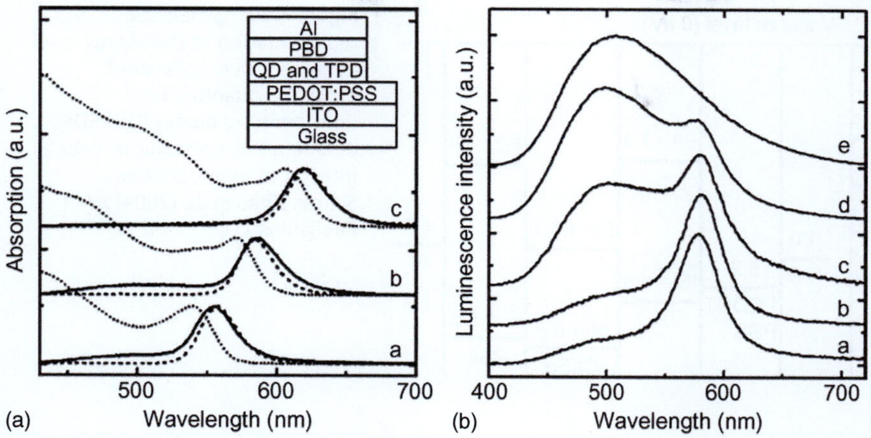

Figure 4.19 (a) Spectral analysis reveals absorption (dotted), photoluminescence (dashed), and electroluminescence (solid) profiles for CdSe QDs of 3.2 nm (i), 4.1 nm (ii), and 5.4 nm (iii) in diameter, each enveloped with a 2 ML ZnS shell. Operated at voltages and currents such as 5.5 V, 0.11 mA (i), 5.0 V, 0.08 mA (ii), and 4.0 V, 0.06 mA (iii), these data are complemented by an inset detailing the multilayered CdSe/ZnS QD LED architecture. (b) Demonstrating the effect of QD concentration, the electroluminescence spectra for CdSe QD LEDs exhibit variations with estimated QD interface coverages of 0.3, 0.2, 0.1, and 0.03 ML for samples a through d. Sample e serves as a control, devoid of CdSe/ZnS QDs within the TPD matrix. These spectra were acquired under a voltage range of approximately 5.0–6.0 V. The abbreviations denote: LED for light-emitting diode, ML for monolayer, QD for quantum dot, and TPD for *N,N′*-diphenyl-*N,N′*-bis(3-methylphenyl)-1,1′-biphenyl-4,4′-diamine, a common hole-transport material. *Source:* Zhao et al. (2004) / AIP Publishing LLC.

sample e, display an electroluminescence (EL) peak at 580 nm against a broad spectral background. This background, which is centered around 505 nm, is likely due to recombination processes involving the excited states of triphenyl-diamine (TPD) and the electron-conducting layer poly (9,9-dioctylfluorene-co-benzothiadiazole) (PBD), especially when the quantum dot (QD) concentration is reduced. Sample e, which contains no QDs, acts as a control to underscore the absence of QD-associated emission.

Figure 4.20 illustrates the energy-level alignment for a CdSe/ZnS quantum dot light-emitting diode, with work functions of ITO and PEDOT:PSS, along with electron affinities (EA) of TPD, PBD, and the CdSe QDs, sourced from established literature. Within these devices, hole transport commences at the ITO anode, facilitated by the PEDOT:PSS layer, subsequently migrating through the hole-transporting TPD matrix before being delivered to the QDs. Conversely, electrons are conveyed from the aluminum cathode to the electron-transporting layer PBD, establishing a complementary charge flow architecture.

QDs in PL solutions have demonstrated quantum efficiencies surpassing 50% (Hines and Guyot-Sionnest 1996), with their emission wavelengths tunable between 470 and 630 nm, a characteristic closely linked to their particle size synthesis. Despite this, the performance efficiency of QD LEDs has historically lagged behind that of organic LEDs. Prior research employed QD multilayer films, comprising 10–20

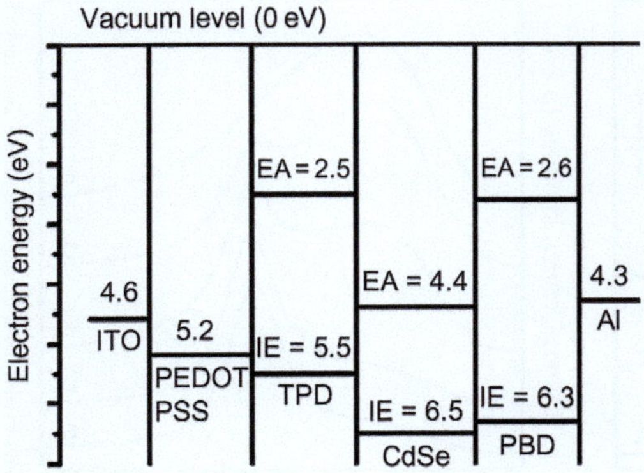

Figure 4.20 Schematic representation of the energy level alignment in multilayered CdSe/ZnS quantum dot light-emitting diodes (QDLEDs), detailing the electronic structure integral to their operation. *Source:* Zhao et al. (2004) / AIP Publishing LLC.

layers, serving dual purposes as electron transporters and light emitters. However, these structures suffered from inadequate electrical conductivity, causing an imbalance in charge injection and capping the luminous efficiency at below 0.10 cd/A.

Additionally, the presence of high pinhole defect densities within the multilayer QD architecture contributed to low manufacturing yields and inconsistent device performance. Recent advancements in QD LED research point toward the potential for planar light emission systems. A pioneering approach entails the fabrication of organic-inorganic hybrid QD light-emitting diodes (QD LEDs). This strategy capitalizes on the varied properties of organic materials in conjunction with the outstanding optical and electrical attributes of inorganic materials. By utilizing a single-layer QD as the sole emitter, this method addresses the challenges linked to multilayer structures. In the structure of a hybrid QD LED featuring a singular QD layer, the organic layer takes on the responsibility of charge-carrier transport. It channels the charge carriers to the QD monolayer, where light emission takes place. Consequently, the function of the QD is optimized exclusively for emission.

Figure 4.21 includes an inset detailing the schematic of a single-layer QD-based sandwiched hybrid QLED, as reported by Coe et al. (Coe et al. 2002). The devices under examination are designated as Device I (Figure 4.21a) and Device II (Figure 4.21b). The quantum dots were encapsulated with TOPO and trioctylphosphine (TOP), yielding a photoluminescence efficiency of 22% and 62% for the solution-contained QDs. The absorption peak and PL emission manifest at distinct wavelengths, specifically 545 and 561 nm, respectively. These spectral features are characteristic of CdSe quantum dots (QDs) that possess a 3.8-nm core diameter and are encapsulated by 1.5 monolayers of ZnS. These QDs were integrated into a chloroform-TPD blend. The resulting mixture was uniformly deposited onto an ITO coated glass substrate via spin coating, yielding a 35-nm thick TPD layer accompanied by a precisely regulated QD monolayer. In the case of Device I, a 40-nm thick Alq_3 layer was deposited through thermal evaporation. Subsequently, a 75-nm cathode composed of Mg:Ag with a 10:1 weight ratio, along with a final 50-nm Ag capping layer, was fabricated. In contrast, Device II (Figure 4.21b) incorporates an additional 10-nm layer of TAZ prior to the deposition of the 40-nm Alq_3 layer, introducing a modification to the structure for enhanced performance.

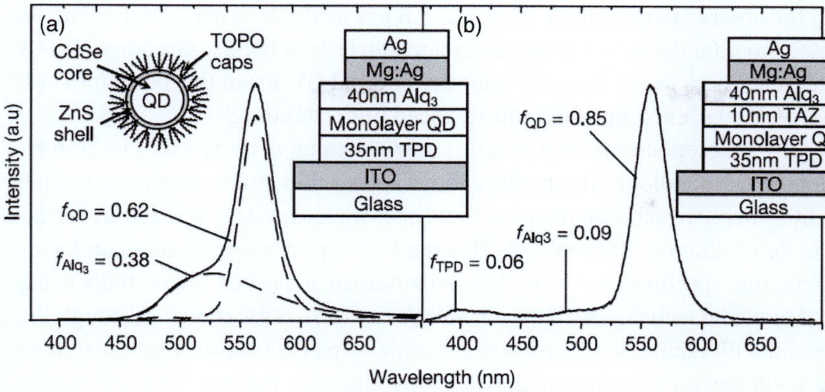

Figure 4.21 Spectra of electroluminescence and architectural configurations for the quantum dot light-emitting diodes, designated as Device I (a) and Device II (b), are presented. The electroluminescent profiles are dissected by dashed lines, representing the spectral contributions from Alq_3 and the quantum dot entities. An accompanying inset depicts an illustrative representation of a core-shell structured QD, highlighting the specialized architecture intrinsic to these devices. In the context of these LEDs, QDs denote quantum dots, while LED refers to light-emitting diodes, emphasizing their optoelectronic nature. *Source:* Coe et al. (2002) / Springer Nature.

In Figure 4.21, the electroluminescence results of the QD LED are presented. The spectral analysis exhibits distinct peaks at 562 and 400 nm, along with a broad secondary peak that is centered around the 530 nm wavelength. These features correspondingly originate from the emissions of the QDs, TPD, and Alq_3 components. The dashed lines in Figure 4.21a illustrate the spectral decomposition into contributions from Alq_3 and the QDs, with a quantification method that yields emission fractions, denoted as fx. Specifically, it is determined that fQD is approximately 0.62 and fTPD is around 0.38. Conversely, Device II exhibits a more confined EL spectrum, with adjusted fractions of fQD at 0.85, fTPD at 0.06, and $fAlq_3$ at 0.09. This narrowing is attributed to the inclusion of a 10-nm TAZ layer, which enhances QD emission due to its structural effect, albeit with a 72% coverage leading to reduced Alq_3 emission, suggested by the low $fAlq_3$ value. The diminished contribution from TPD (fTPD = 0.06) might be linked to inefficient TPD molecular transport for QD excitation within Device II, potentially influenced by the partial TAZ coverage hindering hole transport to Alq_3, thereby suppressing its emission.

4.4.3 Quantum Efficiency

QY serves as a quantitative metric that expresses the ratio of the number of photons emitted to the number of photons absorbed. It can be categorized into two types: internal QY and external QY. The internal QY reflects the intrinsic efficiency of a material. It is determined by calculating the total number of photons emitted following the absorption of a single photon. However, due to the high refractive indices of semiconductor materials, which lead to significant internal reflection at critical angles, a mere fraction of internally emitted photons escape for detection, thereby defining the observed external QY. This necessitates meticulous measurement of the external QY to infer the intrinsic efficiency accurately. In scholarly discourse, unless specified, references to QY typically infer the internal QY. From this point forward, we denote QY by η, encompassing this nuanced understanding.

Due to the inherent complexity in accurately assessing η, there exists no universally adopted methodology. Among the various approaches, one exemplar technique is presented here, which circumvents the direct measurement of external intrinsic QY to determine the internal QY. This method capitalizes on a semiconductor heterostructure, specifically a GaAs/AlAs superlattice, deposited sequentially using advanced epitaxial techniques. The internal QY of the SL is ascertained through the analysis of the PL excitation spectrum of the underlying GaAs layer. The researchers propose a simple yet effective model. They hypothesize that the SL's QY can be deduced from the intensity ratio (R) of two specific PL signals emanating from the GaAs layer. These signals are labeled as I_a and I_b, and they are associated with excitations that exceed and are below the excitonic transition energy of the SL, respectively. The mathematical formulation for GaAs PL intensity is then given as follows:

$$I_{\mathrm{GaAs}} \propto I_{\mathrm{exc}}[1 - \exp(-\alpha_{\mathrm{GaAs}} l)]\eta_{GaAs} \tag{4.43}$$

where the excitation energy density, I_{exc}, is considered, with α_{GaAs} representing the absorption coefficient specific to GaAs. The sample's thickness is denoted by l, while η_{GaAs} signifies the internal QY of GaAs, and p quantifies the proportion of photons that successfully escape the sample. Below the superlattice's excitonic transition energy, I_{exc} aligns with the incident light intensity, I_{inc}. Conversely, above this transition, I_{exc} accounts for the direct incident light not absorbed by the SL. The fraction that constitutes the indirect contribution to the SL's PL intensity, emerging from processes beyond direct excitation, is then

$$I_{\mathrm{exc}} = I_{\mathrm{exc}}^{\mathrm{dir}} + I_{\mathrm{exc}}^{\mathrm{ind}} \propto I_{\mathrm{inc}} \exp(-a_0 d) + I_{\mathrm{inc}}[1 - \exp(-\alpha_0 d)\eta_{\mathrm{SL}}(1 - p_0)] \tag{4.44}$$

where α_0 denotes the absorption coefficient at the SL's excitonic transition, η_{SL} signifies the internal QL of the superlattice, and p_0 represents the fraction of PL photons that escape the crystal lattice, a quantity governed by the critical angle for total internal reflection. Consequently, the determination of the internal QY for the SL becomes

$$\eta_{\mathrm{SL}} = \frac{R - \exp(-\alpha_0 d)}{(1 - p_0)[1 - \exp(-\alpha_0 d)]} \tag{4.45}$$

Under the condition that α_0 multiplied by the sample depth d far exceeds unity ($\alpha_0 d \gg 1$), the superlattice QY (SLQY) closely approximates the ratio R, with R defined as less than I_a divided by I_b. However, in scenarios where the PL intensity of GaAs exhibits a nonlinear dependence on the excitation power, and when reabsorption and potential photon reexcitation phenomena come into play, deducing the QY solely from the ratio R becomes intricate. It is imperative to maintain equivalent excitation power levels during the measurement of PL intensities from GaAs to ensure accuracy. Adopting this protocol, the SLQY can be reliably inferred from the ratio of PL intensities, R, adhering to the outlined methodology. Notably, at elevated

Table 4.2 Quantum yield of various nanomaterial.

Materials	Measurement method	Temperature (°C)	Quantum yield
GaAs/AlAs	Ratio of emission intensities	10–40 K	0.88–0.91
Alq_3 crystal	Photocalorimetric	2 K	0.76 ± 0.03
Film	Photocalorimetric	2 K	0.79 ± 0.03
Alq_3 film	—	RT	0.32
CdSe/ZnSe film	—	RT	0.85
CdSe film	—	RT	0.05–0.10
PPy film	Integrating sphere to collect all the emitted light	—	0.31 ± 0.05
PPV film	—	RT	0.27 ± 0.03
BEHP PPV film	—	0	0.06
PHTP film	Integrating sphere to collect all the emitted light	—	100

QY values, the impact of photon reexcitation must be meticulously accounted for to avoid inaccuracies in the determination.

In the investigation of the superlattice, the authors ascertained an SLQY, denoted η_{SL}, ranging between 0.88 and 0.91. Utilizing a substantial R value of 0.86 from the GaAs PLE spectrum for η_{SL} computations yields results in close agreement with empirical findings. However, beyond a temperature of 40K, this methodology encounters limitations due to a significant decline in the GaAs PL intensity, rendering it impractical for determining η under such thermal conditions.

Table 4.2 compiles the measured quantum yields (QYs) for various materials, highlighting the substantial influence of temperature on these values. Notably, for poly(2-(2′,5′-bis(2-ethylhexyloxy)phenyl)-1,4-phenylenevinylene (BEHP-PPV) films, an examination reveals that η experiences a significant reduction to half its initial value as the temperature escalates from 0 to 144 °C. Similarly, in the case of aluminum tris(8-hydroxyquinoline), or Alq_3, film, the QY diminishes from 0.79 at 2 K to 0.32 at ambient conditions, illustrating the critical role thermal agitation plays in disrupting the radiative decay process of its S1 state. Despite CdSe films with grain sizes between 25 and 30 nm exhibiting QYs within a modest range of 0.05–0.10, the introduction of a ZnSe/CdSe superlattice structure sees a remarkable increase in QY to 0.85, demonstrating the potential for architectural enhancements to mitigate the temperature-dependent efficiency loss (Figure 4.22).

In their pursuit of further understanding the QY in nanomaterials, Dabbousi and colleagues conducted measurements on (CdSe)ZnS nanocomposites encapsulated with TOPO. As previously noted, these nanoparticles feature a ZnS shell of 0.31 nanometers in thickness enveloping a CdSe core with a diameter ranging between 4.2 and 10 nm. Figure 8.11 presents findings that illustrate an initial rise in the PL QY as the ZnS coating layer thickness increases, culminating in a peak of 50% at

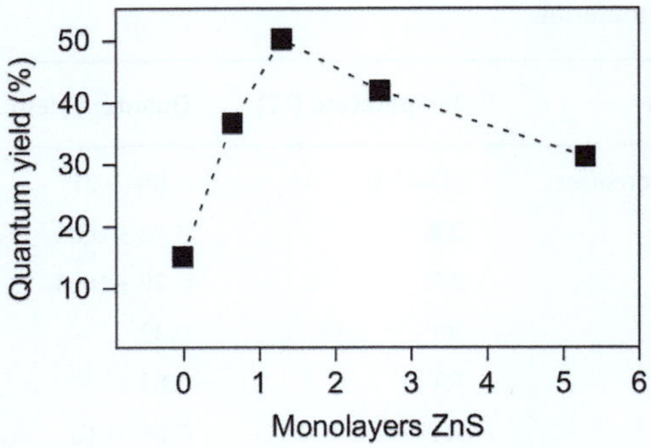

Figure 4.22 The photoluminescence quantum yield is depicted as a function of ZnS layer thickness. An increment in PL intensity is observed with the progressive addition of ZnS, achieving a notable 50% enhancement at a coverage of B1.3 monolayers, followed by a gradual decrease beyond this point. *Source:* Royal Society of Chemistry.

a coverage of 1.3 monolayers, before experiencing a systematic decline with further increases in thickness. The rationale behind this trend constitutes an intriguing aspect of the study.

The authors conducted an in-depth analysis of the XPS outcomes for the composite nanomaterials. Their findings revealed that, after 80 hours of air exposure, the absence of SeO_2 lines in the X-ray diffraction (XRD) pattern attests to the effective protection of the selenium (Se) core by the ZnS layer against oxidation. Supplementary findings derived from TEM and SAXS studies substantiated the notion that CdSe core was efficiently passivated once the thickness of the ZnS layer attained 1.3 monolayers. As the ZnS coating thickens, TEM structural analyses illustrate a loss of uniformity in the ZnS layer, with a preferential [002] growth orientation aligning with the wurtzite structure of the CdSe, leading to the formation of irregular QDs. Contrasting with thinner coatings, thicker ZnS layers promote the development of a fluffy, flake-like morphology, potentially due to crystal defects. The 12% lattice mismatch at the CdSe/ZnS heterointerface induces interfacial strain, generating crystalline defects such as threading dislocations and low-angle grain boundaries in the ZnS shell. These defects act as non-radiative recombination centers, leading to a progressive decline in QY (η) with increasing ZnS shell thickness, underscoring the critical role of structural imperfections in optoelectronic performance.

4.5 Light Scattering Spectroscopies

Light scattering constitutes a fundamental phenomenon occurring when electromagnetic radiation encounters a medium populated with particles or traverses interfaces bearing structural irregularities or defects. This noninvasive methodology is a comprehensive suite of techniques highly adaptable for characterizing a broad spectrum of macromolecules and particles suspended in solution. By leveraging the interaction of light in the visible spectrum with solvated or suspended entities, these techniques afford an incisive, label-free means to assess nanomaterials and biomaterials, elucidating their dimensions, morphology, internal architecture, and interplay without causing any damage. They uniquely harness the optical response of particles in suspension to unravel their physicochemical attributes. Unlike conventional analytical techniques that necessitate calibration standards, light scattering operates as an absolute measurement technique, imparting it with a distinct advantage. While alternatives like mass spectrometry, membrane osmometry, and sedimentation equilibrium can determine molecular weights, only light scattering comprehensively embraces the analysis of diverse macromolecular assemblies and their oligomeric forms. Central to its significance is its ability to probe the solution behavior of macromolecules, a feature pivotal in numerous technological advancements where the manipulation and exploitation of light are paramount.

4.5.1 Light Scattering Principles and Techniques

Dynamic light scattering (DLS) constitutes a spectroscopic methodology of gentle interrogation, enabling the assessment of (i) the scattered light intensity emanating from nanoparticles and concurrently (ii) the dimensions of particles dispersed within a liquid medium, spanning a nanoscale range from 1 to 1000 nm.

The analysis of light scattered through its interaction with matter yields insights into the sample's physical characteristics. In the late 19th century, John Tyndall (Tyndall 1869) conducted pioneering studies on light scattering experiments. It is noteworthy that for particles smaller than one-tenth of the incident wavelength (λ), the scattering process is isotropic, a phenomenon referred to as Rayleigh scattering. Conversely, Gustave Mie's theory (Mie 1908) explicates that when particles exceed this size threshold, the scattering becomes angle-dependent, characterizing inelastic scattering events.

Particles dispersed within a liquid undergo stochastic displacements known as Brownian motion. The fundamental premise of DLS measurements relies on the Doppler effect principle, wherein the frequency of scattered light is modulated according to the velocity of the particle. Consequently, particles of smaller size exhibit heightened mobility within the medium, leading to a more rapid fluctuation in the intensity signal, as depicted in Figure 4.23a, b. This technique quantifies the temporal signature of these intensity fluctuations, which is intricately linked to the particles' diffusion coefficient.

The spectral characteristics of light scattering are not only shaped by the random thermal movement of the scattering sites but are also profoundly impacted by the count of these sites present within the observation volume. In the case of opaque colloidal systems, where multiple scattering events abound, this complicates obtaining trustworthy measurements, thereby limiting the accuracy of such analyses.

Hence, leveraging the temporal autocorrelation function (ACF, denoted as $G(t)$ in Figure 4.23c and d) enables the extraction of particle radius information from the light diffusion intensity spectrum. Through experimental investigation of colloidal ACFs, a conventional approach employs the correlation between the diffusion coefficient and the spectral profile, yielding insights into the particles' characteristics:

$$G = A + Be^{-2k\tau} \tag{4.46}$$

The relationship k, equivalent to Dq^2, is established, where q, the light scattering vector, is determined by experimental configuration as $4\pi n\sin(\theta/2)/\lambda$, with n representing the refractive index and θ the scattering angle.

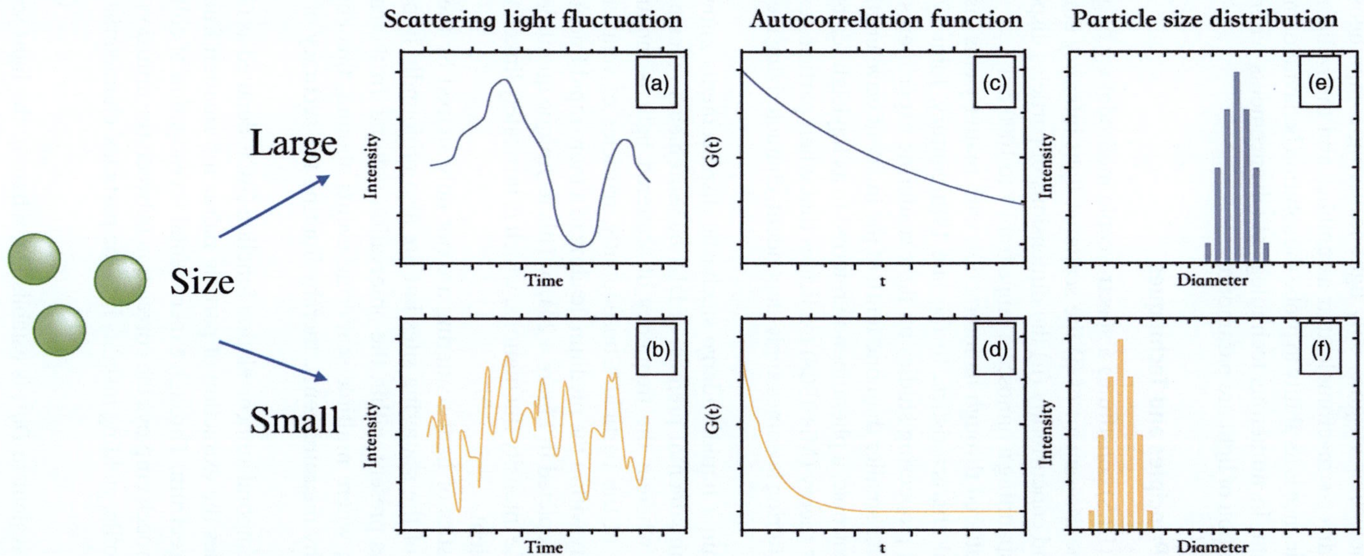

Figure 4.23 The impact of particle size on the variability of scattered light intensity is illustrated in figures (a) and (b), demonstrating how smaller particles induce more rapid intensity fluctuations. These fluctuations are systematically analyzed through the autocorrelation function (ACF), as shown in figures (c) and (d), where the ACF captures the temporal correlation of these intensity changes, yielding insights into particle dynamics. Subsequently, the derived PSD, presented in figures (e) and (f), encapsulates the collective outcome of these analyses, revealing the spectrum of particle sizes present within the system.

For the determination of hydrodynamic radii, employment of the Stokes–Einstein relation is imperative; this equation elucidates the correlation between the diffusion coefficient, D, and the radius of spherical particles, denoted as R_H, within a fluid medium:

$$D = \frac{k_B T}{6\pi \mu R_H} \tag{4.47}$$

The dynamic viscosity of the medium, denoted as μ, and the medium's temperature, T, are crucial parameters, with the Boltzmann constant, $k_B = 1.3806 \times 10^{-23}$ J* molecule^{-1} K^{-1}, collectively informing the calculations.

It is imperative to acknowledge that the hydrodynamic radius, R_H, represents a theoretical construct. Within the computations of DLS, particles are idealized as rigid spheres, a simplification that deviates from the reality of suspensions. This radius effectively characterizes the particle's dynamic interaction within the fluid environment rather than reflecting its actual geometry.

Nanoparticle samples in their authentic form exhibit a spectrum of particle sizes, necessitating the utilization of a PSD that can illustrate size metrics through intensity, volume, or count-based representations. This permits the computation of mean sizes within each peak. However, it is crucial to recognize that transitioning from an intensity-based distribution, which serves as the fundamental metric, to volume- or number-based distributions may introduce substantial inaccuracies post-conversion, as shown in Figure 4.23(e) and (f).

To characterize the size distribution of samples, the Z-average, in conjunction with the polydispersity index (PI), serves as a pivotal DLS method, which not only quantifies the distribution's breadth but also represents the intensity-weighted mean of the hydrodynamic radii across the entire sample ensemble. In the case of a monodisperse sample, the congruence between the Z-average and the single peak's mean underscores its homogeneity. Conversely, polydisperse samples yield a Z-average that encapsulates complexity, while the intensity distribution reveals multiple peaks, each with distinct mean and dispersion. Subsequently, the reported sizes from DLS analyses herein are derived as the mean from the intensity-weighted distribution.

In the realm of DLS, a sophisticated ensemble of components operates in unison to ascertain the intensity ACF of the scattered light, subsequently yielding insights into the particulate size distribution within a solution. Referencing Figure 4.24, the intricate instrumentation employed in this technique is delineated, highlighting its architectural and functional components.

a) **Coherent Illumination Source:** The employment of a laser in DLS serves as the genesis of monochromatic radiation, with wavelengths spanning from 532 to 633 nm, ensuring spectral purity. This light, in a state of phase congruence, is directed upon the sample matrix. Upon interaction with particulate matter, it undergoes scattering in a diffuse pattern, the resultant photons being meticulously captured by the detection system. The paramount importance of the laser's coherence and unwavering stability cannot be overstated, as they underpin the precision of measurements.

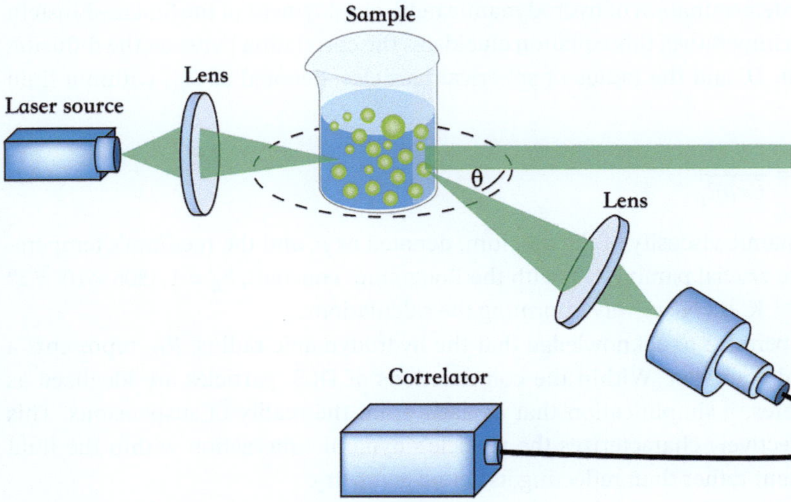

Figure 4.24 Dynamic light scattering.

b) **Optical Configuration:** The system incorporates a monochromatic laser beam, which is meticulously aligned to impinge upon the sample held within a cuvette. Upon encountering the suspended particles, the laser radiation undergoes scattering in a comprehensive angular span. Over time, the detection of this scattered light at a predefined angle becomes pivotal, as it subsequently feeds into the calculation of diffusion coefficients and particle dimensions via the esteemed Stokes–Einstein relation. The photons, having been scattered, are meticulously channeled onto a light-sensitive receptor, such as a photomultiplier tube (PMT) or an avalanche photodiode (APD), to quantify the intensity of the scattered light, thereby ensuring the acquisition of critical data.
c) **Detection and Signal Processing:** The photodetector, in a precise transduction role, transforms incident photons into electrical signals, the magnitude of which is congruent with the scattered light's intensity. These signals are subsequently amplified and subjected to intricate electronic processing, ultimately leading to the extraction of an autocorrelation-derived metric that delineates the temporal variations in the intensity of the scattered light.
d) **Correlation Analyzer:** Employing a digital correlator, the computation of the ACF for the scattered light's intensity was undertaken. This involved a meticulous analysis of the fluctuations over time in the scattered intensity, wherein the instrument ascertained the correlation coefficients between intensity measurements separated by variable temporal offsets.
e) **Advanced Data Processing Suite:** The instrument is governed by dedicated software, which orchestrates the system's operations, facilitates the acquisition of data from the photodetector, and conducts instantaneous analyses to yield the intensity ACF, thereby integrating data capture with computational interpretation.

f) **Sample Enclosure:** The specimen is contained within a specialized receptacle, referred to as a cuvette, fabricated from either quartz or glass. These materials are selected for their propensity to mitigate spurious light scattering and to reduce background noise, thereby ensuring data integrity.
g) **Temperature Stabilization:** To ensure the integrity of DLS experiments, a regimen of temperature control is incorporated into numerous DLS setups, guaranteeing a uniform sample temperature throughout the measurement protocol. The accuracy of these measurements hinges critically on precise thermal management due to the temperature-sensitive nature of particle diffusion rates. Over a predefined temporal interval, the detection of scattered light serves as a proxy for tracking particle dynamics, revealing intensity variations that are temporally dependent. Specifically, particles of lesser size exhibit more rapid intensity fluctuation signatures compared to their larger counterparts, which yield greater amplitude oscillations between intensity peaks and troughs. Utilizing the intensity profile, a correlation function is derived, which constitutes a mathematical encapsulation of these fluctuation patterns. This function, in turn, facilitates the computation of the translational diffusion coefficient by comparing the light intensity at any given time, denoted as t, with its phase-shifted equivalent after a delay period.
h) **DLS Application Protocol:** The deployment of DLS involves a systematic procedure, entailing the following steps. First, meticulous sample preparation, which often necessitates filtration and suitable dilution to minimize interference; subsequently, the transfer of these samples into a cuvette or arranging them within a microtiter plate for multisample analysis. The subsequent step requires positioning the cuvette or plate within the DLS apparatus. Data acquisition can then be executed through either manual intervention or automated protocols, ensuring efficient and controlled experimentation.

The integration of DLS equipment encompasses high-precision optical elements, keenly sensitive detectors, and intricate data analysis algorithms. This cohesive assembly enables the precise determination of PSDs in liquid suspensions by analyzing intensity fluctuations in scattered light, a phenomenon rooted in Brownian motion. Collectively, these elements facilitate profound understanding of the characteristics and dynamics within nanoscale entities, proteins, liposomes, and broader colloidal architectures, yielding insights of considerable scientific value.

4.5.2 Application of Light Scattering in Chalcogenide Nanomaterials

DLS holds a pivotal role in numerous industrial and scientific domains, where the precise understanding of PSDs and solution dynamics is imperative. The advent of lipid nanoparticle (LNP) technologies, exemplified by the prominent SARS-CoV-2 vaccines—BNT162b2 from Pfizer-BioNTech and mRNA-1273 by Moderna—has spotlighted DLS's indispensable contribution to nanomedicine, accentuating its role in ensuring the quality and characterization of these novel therapeutic modalities (Jin et al. 2022). This manuscript delves into the breadth of DLS applications, highlighting its versatility as an analytical tool. Beyond merely determining particle size and distribution, its capability to assess zeta potential and additional colloidal attributes, including molecular weight, positions DLS as an invaluable instrument across diverse fields.

Within the field of materials science, the widespread application of DLS is evident in the comprehensive analysis of colloidal systems, nanoscale particles, and polymeric materials. This technique offers distinct advantages in industries ranging from paint and pigment manufacturing to food, ceramics, cosmetics, beverages, and personal care, where it surpasses destructive analytical methods, including microscopic examination, electrical sensing-based Coulter counters, hydrodynamic and field-flow fractionation techniques, particle sizing methods utilizing disc centrifuges, size-exclusion chromatography, and alternative scattering-based analytical procedures. Notably, Lim et al. (Lim et al. 2013) employed DLS to investigate the size distribution and colloidal stability of magnetic nanoparticles, comparing its outcomes with those of transmission electron microscopy (TEM) and dark-field microscopy (DFM) (Hassan et al. 2015). Their comparative study highlighted the nuanced strengths and limitations of DLS in accurately assessing magnetic nanoparticle (MNP) sizes. The measurement of zeta potential, furthermore, plays a pivotal role in advanced material synthesis through electrophoretic deposition (Besra and Liu 2007), as well as in mineral processing techniques like flotation, where the surface chemistry, influenced by water layer orientation and the electrochemical double layer, with the point of zero charge being a decisive factor, is crucial for controlling the separation process (Kamble et al. 2022).

5

Chalcogenide Compounds for Optical Communications

5.1 Introduction

Chalcogenide glasses constitute a category of amorphous solid materials. Their primary constituents are chalcogens, including sulfur (S), selenium (Se), and tellurium (Te). These chalcogens form covalent bonds with other elements like germanium (Ge), arsenic (As), and antimony (Sb). From the standpoint of chemical composition, chalcogens occupy Group 16 (VIA) in the periodic table. They exhibit a wide range of valence states and distinctive electronic structures, which serve as the foundation for the unique properties of chalcogenide glasses. For instance, sulfur atoms have six electrons in their outermost shell and can form complex network structures by forming covalent bonds with other elements when forming glass.

From a structural perspective, chalcogenide glasses lack the highly ordered arrangement of atoms extending over long distances that is characteristic of crystalline materials. Instead, they exhibit regions where atoms are arranged in a relatively ordered fashion over short distances. Taking the common Ge–S system as an example, germanium atoms typically form tetrahedral coordination structures with four sulfur atoms, and these tetrahedra are interconnected through shared sulfur atoms, thereby forming a three-dimensional glass network. Unlike traditional oxide glasses, where silicon-oxygen tetrahedra are the basic structural units, the network structure of chalcogenide glasses is relatively loose, which endows them with some unique physical and chemical characters.

Chalcogenide glasses exhibit a series of special characters in optics, electricity, and thermodynamics. In terms of optics, their prominent feature is good transmissibility in the infrared band, with the transmission band of some chalcogenide glasses extending to 12 μm or even longer (C. G. Lin et al. 2018). This characteristic stems from the vibration modes of chemical bonds between atoms in chalcogenide glasses, which result in low absorption and scattering losses when infrared light propagates through them. Additionally, chalcogenide glasses exhibit a substantial nonlinear optical coefficient. Their refractive index and other optical characteristics undergo marked alterations when exposed to intense light. This property holds immense significance in domains like the manipulation of optical signals and the development of optical switches within the realm of optical communication

Chalcogenide Nanophotonics, First Edition. Tun Cao, Ying Su, Yinghan Li, and Hongze Gao.

systems (Zhang et al. 2019). Electrically, chalcogenide glasses have certain semiconductor characters, and their electrical conductivity can be regulated by altering the composition and external conditions (such as light and temperature). Thermally, chalcogenide glasses have a relatively low coefficient of thermal expansion and good thermal stability, which helps maintain the stability of material characters under different environmental temperatures.

The exceptional chemical constituents, atomic arrangement patterns, and performance attributes of chalcogenide glasses are precisely what render them indispensable core materials in the realms of optical communications and sensing. These glasses are well-suited to fulfill the requirements of contemporary technology for high-performance optical and sensing functional materials.

The history of chalcogenide glass is also very much worth telling. In the early 20th century, the field of materials science was in a stage of vigorous development. Scientists were not satisfied with the characters of traditional oxide glasses and actively explored new types of glass systems. Chalcogen elements (such as sulfur, selenium, and tellurium) gradually came into the sight of researchers due to their special electronic structures. At that time, people initially realized that the glasses formed mainly with these chalcogen elements might have unique characters. Early experimental attempts mainly focused on synthesizing simple chalcogenide glass samples and exploring their formation conditions. For example, by melting and mixing elements such as sulfur and selenium with some metal or semimetal elements at high temperatures and then cooling them rapidly, attempts were made to obtain an amorphous glass structure.

However, due to the limitations of experimental equipment and technology at that time, the quality of the prepared chalcogenide glass samples was low, with a large number of defects, making it difficult to conduct in-depth research on their characters. As time passed into the middle of the 20th century, the research methods in materials science continued to be enriched and improved. Technologies such as X-ray diffraction and spectral analysis began to be applied to materials research. Scientists used these technologies to conduct systematic research on the structure and basic physical and chemical characters of chalcogenide glasses. It was found that chalcogenide glasses have structural characteristics that are completely different from those of traditional oxide glasses. In oxide glasses, structural units such as silicon oxygen tetrahedrons are the basis for constructing the glass network, while in chalcogenide glasses, chalcogen elements form complex network structures through covalent bonds. This unique structure endows chalcogenide glasses with special optical, electrical, and thermal characters. Particularly noteworthy is that chalcogenide glasses exhibit good transmittance in the infrared band. This characteristic makes them show great potential in the field of infrared optics and attracts more researchers to engage in research in this field. After that, chalcogenide glasses ushered in great development.

During the period spanning from the mid-20th century to the close of the 20th century, as semiconductor technology advanced at a brisk pace, materials science encountered fresh opportunities and formidable challenges (Tanaka K. et al., 2018). The prospective application worth of chalcogenide glasses within the optoelectronic domain started to be thoroughly investigated. During this period, scientists discovered that chalcogenide glasses have a special response to optical signals. For example, after being irradiated by light, the optical characters of some chalcogenide glasses will change reversibly. This characteristic makes them suitable as optical storage media. In early optical storage technologies, chalcogenide glasses were used to make rewritable optical discs. By irradiating with a laser, the crystalline and amorphous structures of the chalcogenide glass thin film are changed to achieve data writing, reading, and erasing, bringing new changes to the field of information storage.

At the same time, chalcogenide glasses also began to show their potential in simple optoelectronic devices. For example, photodetectors made of chalcogenide glasses can convert optical signals into electrical signals and have a relatively fast response speed. During this period, the application research of chalcogenide glasses in optoelectronic devices laid a solid foundation for their large-scale application in the fields of optical communication and sensing in the future. In addition, with the deepening of theoretical research, people have a clearer understanding of the energy band structure and electron transition mechanism of chalcogenide glasses, providing theoretical guidance for the optimization of material characters. Regarding the preparation procedures, techniques like the chemical vapor phase deposition (CVD) process and thermal evaporation have progressively been adopted and refined. These methods can more precisely control the glass composition and structure, and prepare chalcogenide glass materials with higher quality and more stable performance, further promoting the process of chalcogenide glasses from laboratory research to practical application.

In the 21st century, as information technology experiences an exponential growth, the necessity for advanced optical communication and sensing technologies has escalated dramatically, and chalcogenide glass has consequently opened up a more expansive development horizon. Within the realm of photonic communication, the quest for communication systems capable of achieving extended-range, high-velocity, and substantial-volume data transmission has driven researchers to persistently enhance the performance characteristics of optical fibers made from chalcogenide glass. Through enhancing the production procedure, for instance, employing the modified chemical vapor deposition (MCVD) approach integrated with distinctive doping methodologies, the attenuation of the fabricated chalcogenide-based optical fibers operating within the mid-wavelength infrared spectrum has been significantly diminished, thereby facilitating the transmission of optical signals over ultra-long distances.

Concurrently, chalcogenide glass holds a pivotal position in the amalgamation of optical communication devices. Optical devices such as optical modulators and optical switches made of chalcogenide glass are integrated with other optoelectronic devices on the same chip, greatly improving the integration degree and working efficiency of the optical communication system. In the field of sensing, the application of chalcogenide glass has also been greatly expanded. With the combination of nanotechnology, micro-electromechanical system (MEMS) technology, and chalcogenide glass, a series of high-performance sensors has been developed. To illustrate, gas detecting devices constructed using chalcogenide glass nanowires exhibit a heightened adsorption capability for particular gas molecules, owing to the exceedingly vast specific surface area of these nanowires. They can achieve ultra-high sensitivity detection of harmful gases by detecting the changes in optical characters caused by the adsorption of gas molecules, and the detection limit can reach the ppb (parts per billion) level. In the aspect of biosensing, chalcogenide glass is combined with biometric technology to construct sensor devices for detecting specific biomarkers in living organisms. By modifying biometric molecules such as antibodies and aptamers atop the exterior face of chalcogenide glass, and using the optical response of chalcogenide glass to the interaction of biomolecules, rapid and accurate detection of biomarkers can be achieved, providing a powerful tool for early disease diagnosis, personalized medicine, etc. In addition, with the continuous deepening of basic research on chalcogenide glass, new chalcogenide glass systems are constantly being developed, and their performance is also constantly improving, creating conditions for its application in more fields.

Chalcogenide glass is typically applied within the realms of optical communication and sensing. In the field of optical communication, the main applications are chalcogenide glass optical fibers and optical modulators. The most prominent advantage of chalcogenide glass optical fibers in optical communication is their extremely low loss in the mid-infrared band (2–12 μm). Traditional silica optical fibers have high loss in this band on account of the absorption and scattering features of the material itself, which limits the long-distance transmission of optical signals. In contrast, chalcogenide glass optical fibers can effectively reduce signal loss, allowing optical signals to be transmitted over longer distances without the need for relay amplification. For example, in some special long-distance communication scenarios, such as the expansion of specific bands in submarine optical cable communication and the construction of long-distance communication links in remote areas, chalcogenide glass optical fibers can significantly improve communication efficiency and reliability and reduce the cost of building relay stations.

At the same time, chalcogenide glass optical fibers also have a relatively large nonlinear coefficient, and this characteristic gives them unique advantages in realizing wavelength conversion of optical signals, generation of supercontinuum spectra, etc., and can be used to expand the bandwidth and functions of optical communication systems. Optical modulators are key devices for realizing optical signal modulation in optical communication systems. Chalcogenide glass performs excellently in the manufacture of optical modulators. Physical characters such as the electro-optic effect and acousto-optic effect of chalcogenide glass can be utilized to modulate optical signals. By applying an external electric field or acoustic wave field, the refractive index of chalcogenide glass is changed, and thus, the rapid modulation of parameters such as the intensity, phase, and frequency of the optical signal is achieved.

In sophisticated optical communication frameworks, exemplified by ultra-rapid communication networks operating at 100 Gbps and beyond, optical modulators are mandated to exhibit exceptionally swift response velocities and minimal driving power. Chalcogenide glass optical modulators can meet these requirements. Their response speed can reach the picosecond level, and the driving voltage is relatively low, which helps to improve the performance of the entire optical communication system and reduce energy consumption (Gupta N, et al., 2022).

In the field of sensing, it is mainly applied to gas sensors and biosensors. Chalcogenide glass gas sensors are based on their sensitive response characteristics to specific gas molecules. When specific gas molecules come into contact with the surface of chalcogenide glass, physical adsorption or chemical reactions will occur, resulting in changes in the optical characters of chalcogenide glass, such as changes in refractive index and absorption spectrum. By detecting these changes in optical characters, the detection of gas types and concentrations can be achieved. For instance, in the realm of environmental monitoring, there is a need to detect hazardous airborne substances, like sulfur-containing dioxide and nitrogen-based oxides. Gas sensors constructed with chalcogenide glass exhibit notable merits, including a heightened sensitivity, a rapid response tempo, and excellent selectivity. Their sensitivity can reach extremely low concentration levels, and they can promptly detect trace changes in harmful gases in the environment. Moreover, through the optimized design of the composition and surface modification of chalcogenide glass, specific detection of different gases can be achieved, effectively avoiding interference from other gases. In the field of biosensing, chalcogenide glass can be used to construct various types of biosensors. By utilizing the optical characters of chalcogenide glass and combining biometric technology, specific biomarkers in living organisms can be detected. For example, in disease diagnosis, the detection of tumor markers, pathogen nucleic acids, etc. By immobilizing biometric molecules on the surface of chalcogenide glass, when the target biomarkers in the biological sample specifically bind to them, it will cause changes in the optical signals of chalcogenide glass, such as changes in fluorescence intensity and surface plasmon resonance signals. This detection method has the characteristics of high sensitivity, high specificity, and rapid detection. In the field of point-of-care testing (POCT), biosensors based on chalcogenide glass are expected to achieve the development of miniaturized and portable detection devices, providing strong support for on-site rapid diagnosis, which is of great significance for the prompt identification and management of diseases.

In summary, chalcogenide glass has traversed a protracted developmental trajectory, evolving from its initial exploratory phase to its present widespread utilization across the domains of optical communication and sensing technologies. Figure 5.1 shows several typical applications of chalcogenide glass.

With the continuous progress of technology, its performance has been continuously optimized, and the application fields have also been continuously expanded. In the foreseeable future, it is anticipated to assume a pivotal role across a broader spectrum of domains, thereby significantly propelling the advancement of science and technology and fostering social progress. The optical characters of chalcogenide glass can be divided into infrared transmittance and nonlinear optical characters. One of the most remarkable optical characters of chalcogenide glass is its excellent transmittance in the infrared band. Its transmittance range can cover the mid-infrared and even the far-infrared region, and the transmission band of some

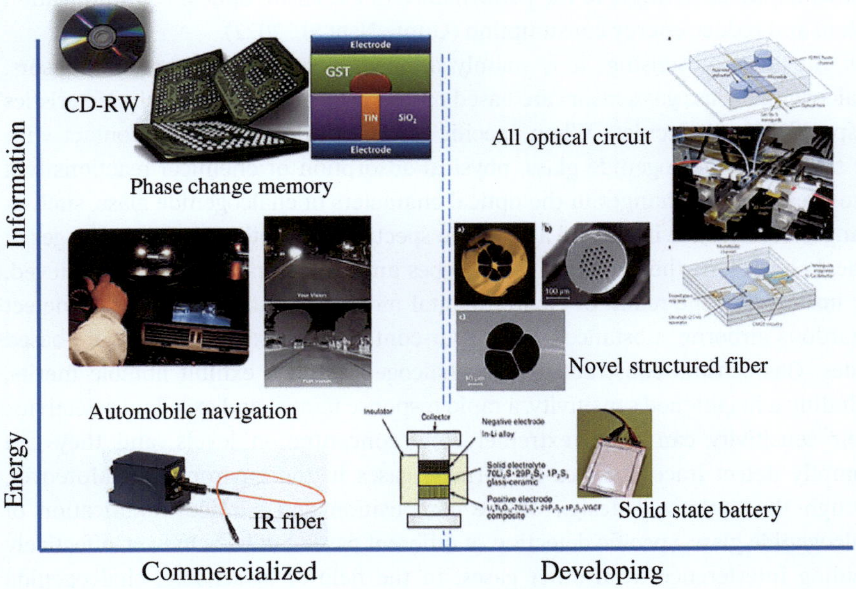

Figure 5.1 Several typical applications of chalcogenide glass. *Source:* Lin et al. (2018) / with permission of Elsevier.

chalcogenide glasses can extend to 12 μm and longer. This characteristic originates from the vibrational pattern of the chemical linkages within the atomic structure.

In chalcogenide glass, there exists a poor degree of frequency correspondence between the vibrational frequencies of these chemical linkages and those of infrared light. Consequently, during the transmission of infrared light, it experiences negligible absorption and scattering losses. Compared with traditional oxide glasses, chalcogenide glass has obvious transmittance advantages in the infrared band, providing a broad space for applications in the field of infrared optics. For example, in infrared imaging systems, chalcogenide glass can be used to manufacture optical components such as lenses and windows, which can effectively improve the system's ability to collect and transmit infrared signals and enhance the imaging quality. Chalcogenide glass has a relatively large nonlinear optical coefficient. Under the action of strong light, its optical parameters, such as refractive index, will change significantly. This nonlinear optical effect encompasses various types, such as second harmonic generation and the optical Kerr effect. For instance, in the optical Kerr effect, when a high-intensity laser passes through chalcogenide glass, the refractive index of the glass changes in relation to the light intensity. This change allows for the modulation of the phase and polarization state of the optical signal. This property is highly valuable in the field of optical communication and has applications in optical signal processing, optical switching, and optical limiting.

In optical signal processing, the nonlinear optical characteristics of chalcogenide glass can be utilized to perform functions such as wavelength conversion and supercontinuum generation. This capability expands the bandwidth and enhances the signal processing abilities of optical communication systems. As for the preparation process of chalcogenide glass optical fibers, it can be divided into the CVD method and the melt drawing method. The CVD method is one of the commonly used methods for preparing high-quality chalcogenide glass optical fibers. First, gaseous precursors containing chalcogen elements (such as sulfur, selenium, tellurium) and other glass-forming elements (such as germanium and arsenic) are transported to the reaction area through a chemical vapor reaction. Under high-temperature conditions, these gaseous precursors undergo chemical reactions to form solid glass deposits. By precisely controlling parameters such as the flow rate, temperature, and pressure of the reaction gases, the composition and structure of the glass can be precisely controlled. Then, the deposited glass preform is drawn to make chalcogenide glass optical fibers. The CVD method can prepare chalcogenide glass optical fibers with uniform composition and high purity, and is suitable for large-scale production of high-performance optical fiber products. Regarding the melt drawing method, first, the chalcogenide glass raw materials mixed in a certain proportion need to be put into a high-temperature furnace for melting. At high temperature, the raw materials are fully mixed and form a uniform glass melt. Then, through a specific drawing device, the glass melt is drawn into filaments at a certain speed, that is, chalcogenide glass optical fibers. During the drawing process, it is essential to strictly control process parameters such as temperature and drawing speed to ensure the uniformity of the diameter and the stability of the optical fiber's performance. The melt drawing method has a relatively simple process and low cost, but there are certain challenges in controlling the uniformity of the glass composition and the quality of the optical fibers compared with the CVD method. In order to enable readers to have a better understanding of chalcogenide compound optical communication technology, this chapter will provide a detailed introduction from three aspects: the optical characters of chalcogenide compound optical fibers, the preparation process of chalcogenide glass optical fibers, and their applications.

5.2 Optical Characters of Chalcogenide Glasses

5.2.1 Linear Refractive Index

1) **Basic Concept of Linear Refractive Index:**

The linear refractive index is a crucial physical quantity that describes how light propagates in a medium. When light transitions from one medium to another, refraction occurs, and the refractive index dictates the extent of this refraction. For chalcogenide glasses, the linear refractive index is defined as the ratio of the speed of light in vacuum to the speed of light in the chalcogenide glass. Snell's law, a well-known principle, describes how light propagates at the interface between materials with different refractive indices. This highlights the critical role of the linear refractive index in the field of optics. In chalcogenide glasses, the linear refractive index influences the path of light propagation, the phase changes that occur, and the interactions with other optical elements. This property is vital for the use of chalcogenide glasses in various applications, including optical communication, optical sensing, and infrared optics.

2) **Influencing Factors of the Linear Refractive Index of Chalcogenide Glasses:**
   a) **Influence of Chemical Composition:**

The Role of Glass Formers: Chalcogenide glasses are primarily composed of chalcogen elements, such as sulfur (S), selenium (Se), and tellurium (Te), along with other glass-forming elements like germanium (Ge), arsenic (As), and antimony (Sb). The choice of glass formers significantly affects the linear refractive index of the glass. For instance, in the commonly studied Ge-S system in chalcogenide glasses, germanium atoms form covalent bonds with sulfur atoms to create the glass network (Gong et al. 2008). As the germanium content increases, the number of (Ge-S) bonds within the glass network also rises. Because germanium atoms have a relatively large atomic mass and a more complex electron cloud distribution, the electronic polarizability of the glass increases. This leads to an increase in the linear refractive index. Research has shown that within a certain range, for every 10% increase in the germanium content of germanium disulfide glass, the linear refractive index in the visible light spectrum can increase by approximately 0.05.

The Influence of Chalcogen Elements: The characters of chalcogen elements themselves also have an extremely important influence on the linear refractive index. With an increment in the atomic number, from sulfur to tellurium, the atomic (Adam J. I. et al., 2014) radius gradually increases, and the polarizability of the electron cloud is enhanced. In the same glass system, replacing sulfur with tellurium will significantly increase the linear refractive index of the glass. In the (As–S–Te) glass system, when the amount of tellurium is gradually increased while keeping the ratios of the other elements relatively stable, the linear refractive index of the glass in the infrared range can rise from approximately 2.5 at low tellurium content to more than 3.0 at high tellurium content. This increase occurs because the larger electron cloud of tellurium atoms is more susceptible to distortion under an external electric field (with light acting as an alternating electric field). As a result, the medium's response to light is enhanced, leading to a higher refractive index.

The Influence of Doped Elements: Introducing appropriate doped elements into chalcogenide glasses can finely tune their linear refractive index. When a small amount of indium (In) is introduced into Ge–Se glass as a dopant, the indium atoms integrate into the glass's network structure. This incorporation alters the distribution of the electron cloud and modifies the bond strength within the original network. Due to the unique valence state and electronic structure of indium atoms, an optimal doping range of 1–3 atomic percent can decrease the linear refractive index of the glass in the near-infrared region by approximately 0.02–0.05. This is because the structure of the glass network after doping is optimized, reducing the scattering centers of electrons, decreasing the energy loss during the propagation of light, and thus changing the refractive index.

b) **Influence of Temperature:**

The impact of temperature on the linear refractive index of chalcogenide glasses is quite complex and primarily occurs due to changes in the glass's microscopic structure and the vibrational state of its atoms. As the temperature increases, the thermal vibrations of the atoms intensify, slightly increasing the interatomic distance and leading to a decrease in the glass's density. According to the Lorentz-Lorenz formula, a reduction in density decreases the number of atoms per unit volume, which, in turn, results in a lower linear refractive index (Danto S. et al., 2013). Experimental measurements have found that for common glasses, within the temperature range from room temperature to 200 C, for every 10 C increase in temperature, the linear refractive index in the visible light band decreases by approximately 0.0005. In addition, temperature changes may also lead to the relaxation process of the internal structure of the glass, further affecting the arrangement of atoms and the distribution of the electron cloud, thus having an impact on the linear refractive index. At high temperatures, weak bonds in the glass network structure may break and recombine. This structural change will, to a certain extent, change the electronic polarizability of the glass, and thus change the linear refractive index.

c) **Influence of Pressure:**

When external pressure is applied to glass, the distances between atoms decrease, causing the interatomic interactions to become stronger. This change also affects the distribution of the electron cloud, leading to a modification in the electronic polarizability of the glass. As a result, the linear refractive index of the glass is impacted as well. Studies have shown that when pressure is applied to chalcogenide glasses at room temperature, for every increase of 1 atmosphere in pressure, there will be a small but measurable change in the linear refractive index. For example, for Ge–As–Se glass, when the pressure increases from 1 atmosphere to 100 atmospheres, its linear refractive index in the near-infrared band increases by approximately 0.0001. The effect of pressure on the linear refractive index is particularly important in certain optical applications. For example, in high-pressure optical sensors, the way in which the linear refractive index of chalcogenide glasses varies with pressure can be used to accurately measure pressure levels.

3) **Measurement Methods of the Linear Refractive Index of Chalcogenide Glasses:**

a) **Minimum Deviation Angle Method:**

The minimum deviation angle method is a well-established technique for measuring the refractive index of glass. This method relies on the principle of light refraction in a triangular prism (Cao, Y. et al., 2010). When a beam of parallel light passes through a triangular prism, refraction occurs, resulting in a deviation angle. By rotating the triangular prism, the position where the deviation angle is minimized is found, and the deviation angle at this time is called the minimum deviation angle. For chalcogenide glasses, it is first necessary to process them into the shape of a triangular prism and then measure them on a high-precision spectrometer. This method has a relatively high measurement accuracy and can reach a relatively low magnitude. However, it has high requirements for sample processing, and the measurement process is rather cumbersome, requiring precise adjustment of the optical path and measurement angle.

b) **Ellipsometry Method:**

The ellipsometry method is a technique for measuring the refractive index of materials by utilizing the polarization characteristics of light. When a beam of polarized light strikes the surface of chalcogenide glass, the optical anisotropy of the glass—meaning even macroscopically isotropic glasses have some level of optical anisotropy at the microscopic level—causes a change in the polarization state of the reflected light. By analyzing the changes in the polarization state, specifically the ellipsometric angle (φ) and the phase difference (Δ), and combining these measurements with an appropriate optical model, such as the Fresnel formula, we can deduce the linear refractive index of the glass. The advantage of the ellipsometry method is that it has a high measurement accuracy, which can reach the magnitude of ($10-^4$), and it causes little damage to the sample surface and does not require complex processing of the sample. However, this method has high requirements for the measuring instrument, and it is necessary to be equipped with a high-precision polarized light generator and detector. Moreover, the processing of the measurement data requires certain mathematical models and calculation methods.

c) **Interference Method:**

The interference method uses the interference principle of light to measure the linear refractive index of chalcogenide glasses. Common interference methods include the Michelson interferometer method and the Mach–Zehnder interferometer method. Taking the Michelson interferometer as an example, the chalcogenide glass sample is placed in one of the optical paths of the interferometer. Since the refractive index of the glass is different from that of air, the optical path of the light in this optical path will change, resulting in the shift of the interference fringes. The interference method has a relatively high measurement accuracy and has relatively low requirements for the shape of the sample. It can measure thin-film or bulk samples. However, this method has high requirements for the experimental environment, and it is necessary to strictly control factors such as the temperature, humidity, and vibration of the environment to avoid interference with the interference fringes.

4) **The Importance of Linear Refractive Index in Chalcogenide Glass Applications:**

a) **Optical Communication Field:**

Fiber Optic Communication: In optical communication, chalcogenide glass optical fibers have garnered significant attention due to their low-loss characteristics in the mid-infrared range. The linear refractive index is vital for optical fiber communication. First, the waveguide properties of optical fibers depend on the refractive index difference between the core and the cladding materials. By carefully controlling the composition of these materials in chalcogenide glass optical fibers, it's possible to achieve an appropriate linear refractive index difference. This difference enables efficient transmission of optical signals within the core while minimizing signal loss into the cladding. For instance, when designing chalcogenide glass optical fibers for mid-infrared communication, the refractive index of the core is typically designed to be slightly higher than that of the cladding. The refractive index difference generally falls between 0.01 and 0.05 to ensure that optical signals are effectively confined and can propagate within the core. Second, the dispersion characteristics of the linear refractive index—specifically, how the refractive index changes with optical wavelength—significantly impact the bandwidth and quality of signal transmission in optical communication. Chalcogenide glass optical fibers have unique dispersion characteristics in specific bands. By reasonably designing the linear refractive index distribution of the optical fibers, dispersion compensation can be achieved. Thus, in long-distance and high-speed optical communication, the broadening and distortion of signals can be avoided, and the communication capacity and reliability can be improved.

Optical Integrated Devices: In the field of optical integration, chalcogenide glasses can be used to fabricate various optical integrated devices, such as optical modulators, optical switches, and wavelength division multiplexers. The linear refractive index in these devices determines the propagation path and the mode of interaction of optical signals. Take an optical modulator as an example. External electric fields, magnetic fields, and thermal fields can alter the linear refractive index of chalcogenide glasses. This change allows for the modulation of optical signals' intensity, phase, or polarization state. In an electro-optic modulator, the linear electro-optic effect (known as the Pockels effect) of chalcogenide glasses is employed. When an electric field is applied, the linear refractive index of the glass changes, resulting in a shift in the phase of the light traveling through it. This mechanism enables the modulation of optical signals. For high-quality optical modulators, it is required that chalcogenide glasses have a large linear electro-optic coefficient and stable linear refractive index characteristics to ensure efficient and low-loss modulation of optical signals (Anonymous 2024a).

b) **The Field of Optical Sensing:**

Gas Sensing: Gas sensors that use chalcogenide glasses operate by detecting changes in the linear refractive index when these glasses interact with specific gas molecules. When gas molecules are adsorbed onto the surface of the chalcogenide glass, they cause physical or chemical changes to the glass. This interaction results in a shift in the linear refractive index, allowing for the measurement of gas concentrations. By employing optical methods (such as surface plasmon resonance technology and fiber optic interferometry), this change in refractive index can be precisely measured, allowing for high-sensitivity detection of gas concentrations. For instance, when detecting hydrogen sulfide gas, active sites on the surface of chalcogenide glass will chemically react with hydrogen sulfide molecules, resulting in changes to the electron cloud distribution and structure of the glass surface layer, thus altering the linear refractive index. Experiments have shown that for certain specific compositions of chalcogenide glass, the linear refractive index change near the surface can reach a magnitude of 10^{-4}–10^{-3} when the concentration of hydrogen sulfide gas varies between 1–100 ppm (Zhao, X. J. et al., 2017). By using an appropriate optical sensing system, this minute change in the refractive index can be accurately detected, thereby achieving high-sensitivity detection of hydrogen sulfide gas.

Biosensing: In the field of biosensing, chalcogenide glasses can be used to construct biosensors for detecting biomolecules, cells, and other biological substances. The linear refractive index also plays a crucial role in biosensing. For instance, in surface plasmon resonance (SPR)-based biosensors, chalcogenide glass serves as the sensing substrate. When biomolecules undergo specific binding on their surface, the refractive index distribution near the glass surface changes, thereby affecting the resonance condition of the surface plasmon wave. By monitoring changes in the resonance wavelength or resonance angle, the presence and concentration of biomolecules can be indirectly detected. Because the interaction between biomolecules and the glass surface is very weak, it is essential to have a high detection accuracy for changes in the linear refractive index. Typically, this requires the ability to detect refractive index changes as small as 10^{-6} or even lower. Therefore, the stable and precisely measurable linear refractive index characters of chalcogenide glasses are crucial for constructing high-sensitivity biosensors.

In summary, the linear refractive index is a key optical parameter of chalcogenide glasses, influenced by various factors, and is of utmost importance in the preparation, performance research, and broad application fields of chalcogenide glasses. A thorough understanding and precise control of the linear refractive index of chalcogenide glasses are essential for further expanding their applications in cutting-edge technological areas such as optical communication, optical sensing, and infrared optics (Gong and Zhao 2009). As materials science and optical technology continue to evolve, research into the linear refractive index of chalcogenide glasses will deepen. This progression holds promise for the development of higher-performance chalcogenide glass materials and related optical devices.

5.2.2 Infrared Transmission Characteristics

Within the vast family of optical materials, chalcogenide glasses stand out with their exceptional infrared transmission characteristics, becoming a key material in numerous infrared optical application fields. From the early exploration of chalcogenide glasses to their widespread application in cutting-edge technological fields such as optical communication, infrared imaging, and sensing technology, research into their infrared transmission characteristics has continuously deepened. This not only drives the development of materials science but also provides robust support for the innovation of related industries. Understanding the infrared transmission characteristics of chalcogenide glasses is crucial for optimizing material performance, expanding the scope of applications, and developing new types of infrared optical devices.

1) **The Principle of Infrared Transmission Characteristics:**
   a) **Basic Concepts:**

The infrared transmission characteristic describes the transmission ability of a material for electromagnetic waves in the infrared band, which is usually quantified by the transmittance. Transmittance (T) refers to the ratio of the intensity of infrared light passing through the material (I) to the intensity of the incident infrared light (I_0), that is, $T = I/I_0 \times 100\%$. For chalcogenide glasses, their infrared transmission characteristics stem from the interaction between the internal atoms and molecules of the glass and the infrared light. When infrared light strikes chalcogenide glass, some of the light is absorbed, some is scattered, and the rest passes through the glass. Chalcogenide glasses demonstrate effective transmission in the infrared range, which is closely linked to microscopic processes such as the vibration of chemical bonds and electron transitions within the material.

   b) **Chemical Bond Vibration and Infrared Absorption:**

Sulfur glass is mainly composed of a combination of sulfur elements (sulfur S, selenium Se, tellurium Te) and other glass-forming elements (germanium Ge, arsenic As, antimony Sb, etc.) through covalent bonding. In the infrared band, these bonds undergo vibrational absorption. Different chemical bonds have specific vibration frequencies. When there is a match between the frequency of infrared light and the vibration frequency of the chemical bond, resonance absorption takes place. This phenomenon leads to the dissipation of light energy and a decrease in transmittance. For example, in the common Ge–S system, sulfur glass, the vibrational absorption of the Ge–S bond mainly occurs in a specific frequency range of the mid-infrared band. Through the design and adjustment of the glass composition, the type and number of chemical bonds can be changed, so as to regulate the position and intensity of the infrared absorption peak, and optimize the infrared penetration performance.

   c) **The Influence of Electronic Transition on Infrared Absorption:**

In addition to chemical bond vibrations, electron transitions in sulfur glasses can also have an effect on infrared absorption. In the glass, the electrons are in different energy level states. When infrared light illuminates a material, photons with appropriate energy levels can excite electrons, enabling them to transition from a lower energy state to a higher energy state. During this process, the photons' energy is absorbed, which subsequently causes a diminishment in the intensity of the light. However, compared with some semiconductor materials, the electron transition absorption of sulfur glass in the infrared band is relatively weak, which is one of the reasons for its better infrared transmittance characteristics. It is the band structure of sulfur glass that governs the high-energy requirement for electron transition. In the infrared band (generally 0.76–1000 μm), most of the photons are not enough to cause the electron transition, thus reducing the absorption loss of light.

2) **Factors affecting the infrared penetration characteristics of sulfur system glass:**
   a) **The Influence of the Chemical Composition:**

The role of the glass-forming bodies: Different glass formers significantly influence the infrared transmittance characteristics of sulfur glass. For instance, in the Ge–S–Se system, germanium (Ge) atoms play a crucial structural role in the glass network. As the Ge content increases, the glass network structure becomes more compact, leading to stronger chemical bonds. This change affects the absorption and scattering properties of the glass. Research has shown that higher Ge content can reduce the number of defects and non-bridging oxygen atoms within the glass, which in turn decreases light scattering loss and improves infrared transmittance. However, when the Ge content is too high, it may cause the displacement and broadening of the chemical bond vibration absorption peak of the glass, which will adversely affect the infrared penetration. For example, in the Ge_xS_{1-x} glass, when x increases from 0.2 to 0.3, the infrared transmittance in the 3–5 μm band can increase from 70% to 75%, but continues to increase x to 0.4, the transmittance decreases at some wavelengths.

The influence of sulfur group elements: Sulfur element is the core component of sulfur glass, and its type and content have a great impact on the infrared transmittance characteristics. From sulfur (S) to selenium (Se) to tellurium (Te), the atomic radius gradually increases, and the electronegativity gradually decreases. This leads to changes in the chemical bonds they form with other elements. In the infrared band, the positions of different sulfur elements are different. For example, the vibrational absorption peaks of S—S bonds are typically found in the shorter wavelength region of the infrared spectrum, while the peaks of Te—Te bonds shift toward longer wavelengths. In the As—S—Te glass system, as the amount of Te is gradually increased, the transmittance of the glass in the long-wave infrared range (8–12 μm) also rises. This occurs because the vibrational absorption peak of the Te—Te bond moves toward longer wavelengths, avoiding this range and thereby reducing absorption loss. However, it is important to note that increasing the Te content may decrease the thermal stability of the glass, which imposes stricter requirements on the preparation process.

The influence of the doping elements: The introduction of appropriate doped elements into the sulfur glass allows for the fine regulation of its infrared penetration characters. For example, by doping a small amount of indium (In) into the Ge–Se glass, the indium atoms will enter the glass network structure, changing the electron cloud distribution and chemical bond state of the network. The right amount of doping can decrease the defect density in glass and reduce the scattering of light, thereby improving infrared transmittance to some extent. Research shows that the infrared transmittance of Ge–Se glass in the 2–4 μm range is optimal when the doping level of In is around 3% to 5%.

However, the species and content of doped elements need to be precisely controlled, as excessive doping may introduce new absorption centers or change the structural stability of the glass, with negative effects on infrared penetration.

b) **The Influence of the Microstructure:**

Glass network structure: The microscopic network structure of sulfur glass has a profound influence on its infrared penetration characteristics. The glass network is connected by various chemical bonds, and the order and uniformity of its structure can affect the light propagation therein. In the uniform and orderly glass network, the scattering loss of light is small, which is conducive to the transmittance of infrared light. Conversely, if there are many defects, holes, or uneven regions in the glass network, light scattering occurs during propagation, leading to reduced transmittance. For example, by optimizing the preparation process, the network structure of sulfur glass is more uniform, and by reducing the internal stress concentration area, the infrared transmittance can effectively improve the transmittance. The sulfur glass film prepared by CVD has a more uniform network structure and a higher transmittance in the infrared band than the bulk glass prepared by the traditional melting method.

Non-bridge oxygen atoms and defects: The non-bridge oxygen atoms are a defect structure in the continuity and symmetry of the glass network. The electron cloud distribution around the non-bridge oxygen atoms is different from the bridge oxygen atoms, which makes it easy to form a local absorption center and produce additional absorption loss to the infrared light. In addition, other defects in the glass, such as vacancies and impurity atoms, can also cause light scattering and absorption. Reducing the number of non-bridge oxygen atoms and defects is an important way to improve the infrared penetration characteristics of sulfur-bound glass. By conducting high-purity treatment of raw materials and optimizing the temperature and pressure control in the preparation process, the concentration of defects in glass can be effectively reduced, and the infrared penetration performance can be improved. For example, when preparing As_2S_3 glass, using high-purity raw materials and melting in an inert atmosphere can significantly reduce the number of non-bridge oxygen atoms in the glass, increasing its infrared transmittance in the 3–6 μm band from 60% to more than 75%.

c) **The Influence of the Preparation Process:**

Melting method: The melting method is one of the most common techniques for preparing sulfur-series glass. In this process, various parameters like temperature, insulation time, and cooling rate significantly affect the infrared penetration characteristics of the glass. During the high-temperature melting phase, if the temperature is excessively high or the insulation time is too long, it can lead to the volatilization and oxidation of glass components. This may introduce impurities and defects, ultimately reducing the infrared transmittance of the glass. For example, when preparing Ge–As–Se glass, too high a melting temperature will volatilize the Se element, causing the glass composition to deviate from the expectation, resulting in a new absorption peak in the infrared band, and the transmittance decreases. The slow cooling rate may lead to the phenomenon of crystal evolution inside the glass, and the existence of crystals will cause the scattering of light, which is also not conducive to the penetration of infrared light. Therefore, optimizing the process parameters of the melting method and selecting the appropriate melting temperature, insulation time, and cooling rate are crucial to obtaining the sulfur glass with high infrared transmittance performance.

CVD: CVD has unique advantages in preparing high-quality sulfur glass films. By accurately controlling the flow rate, reaction temperature, and pressure of the gaseous precursor, the glass film with uniform composition and dense structure can be prepared, thus obtaining good infrared transmittance performance. However, small fluctuations in reaction temperature and pressure during the CVD process can impact the growth rate and quality of the glass film, subsequently affecting its infrared penetration characteristics. For example, when using CVD to prepare GeS_2 glass film, if the reaction temperature is unstable, the distribution of Ge—S bonds in the film will be uneven, the scattering loss will increase in the infrared band, and the transmittance will be reduced. Therefore, in the process of CVD preparation, high-precision control equipment is needed to ensure the stability of process parameters to ensure the infrared penetration performance of sulfur glass film.

The sol–gel method: The sol–gel method is a preparation technique based on solution chemistry, conducted at low temperatures to prevent the volatilization of components and the introduction of impurities during high temperatures. However, the sulfur glass prepared by the sol–gel method may have some characteristics in the microstructure, such as high porosity and a relatively loose network structure, and these factors have a certain influence on the infrared transmittance characteristics. The presence of pores can cause increased scattering of light and decrease the transmittance. However, by optimizing the sol–gel process, such as selecting the appropriate precursor, controlling the hydrolysis and condensation reaction conditions, and post-processing, the porosity, the microstructure of glass, and the infrared penetration performance can be improved. For example, high-temperature annealing of Ge–S glass prepared by the sol–gel method can make the glass network structure more compact and reduce the pore size, thus improving its transmittance in the infrared band.

3) **Measurement Method of Infrared Penetration Characteristics of Sulfur-series Glass:**
   a) **Fourier Transform Infrared Spectroscopy (FTIR):**

Fourier transform infrared spectroscopy is a widely utilized technique to measure the infrared transmittance characteristics of sulfur glass. This method uses the interferometer to modulate the infrared light, then detects the interference map and Fourier transforms the infrared absorption spectrum of the glass and then calculates the transmittance. During the measurement, the sulfur glass sample is thin and placed in the sample pool of the FTIR spectrometer to scan a certain infrared range (e.g., 400–4000 cm^{-1}, corresponding to the wavelength range of 2.5–25 μm). The FTIR spectrometer can accurately measure the transmittance value at different wavelengths and draw the curve of the transmittance change with wavelength. This method has high measurement accuracy and can obtain the penetration information of glass at multiple wavelengths at the same time, which is very effective for studying the infrared absorption characteristics and penetration performance of sulfur glass.

For example, the peak position and intensity of the various chemical bonds in the sulfur glass can be clearly observed by the FTIR spectrometer, thus analyzing the influence of the glass composition and structure on the infrared penetration (Singh 2015).

b) **Spectrophotometer Method:**

The spectrophotometer method is commonly used to measure the infrared transmittance of sulfur series glass. In this process, infrared light emitted by a light source passes through the sample. The light that successfully transmits through the sample is detected and converted into an electrical signal. This signal is then amplified and processed to obtain the transmittance value. Unlike FTIR spectroscopy, spectrophotometers generally can only measure the transmittance at a specific wavelength. When measuring the infrared transmittance of sulfur glass, choose an appropriate wavelength range and detector. For example, for measurements in the mid-infrared band (2–5 μm), a lead sulfide (PbS) detector can be used to draw a transmittance curve by measuring the transmittance at different wavelengths. The spectrophotometer method is relatively simple to operate and has a lower equipment cost, but measures relatively few wavelength points, which is less comprehensive than FTIR spectroscopy.

c) **Infrared Imaging Method:**

The infrared imaging method can intuitively observe the penetration of sulfur glass in the infrared band. This method uses the infrared camera to image the infrared light through the sulfur glass, and reflects the transmittance difference of different areas of the glass through the image gray value. During the measurement, a sulfur-tied glass sample was positioned between an infrared light source and an infrared camera to capture an infrared image through the glass. By analyzing and processing this image, we can determine both the average transmittance of the glass and the distribution of transmittance across its surface. The infrared imaging method effectively and intuitively assesses the uniformity of infrared penetration in sulfur glass, which is crucial for evaluating large-area sulfur glass samples or optical components. For example, when detecting the infrared penetration performance of the sulfur glass window plate, infrared imaging can clearly find out whether there are defects or uneven areas inside the glass, which show abnormal gray values in the image.

4) **The Importance of Infrared Penetration Characteristics in the Application of Sulfur System Glass**

a) **Optical Communication Field:**

Medium-infrared optical fiber optic communication: In the field of optical communication, sulfur glass fiber exhibits low loss and good infrared transmittance in the mid-infrared band (2–12 μm), which makes it an ideal material for mid-infrared fiber communication. In this band, many gas molecules have characteristic absorption peaks, and remote sensing and communication can be realized using the infrared penetration characteristics of sulfur glass fiber. For example, in the fields of the petrochemical industry and environmental monitoring, the concentration and leakage of harmful gases in the pipeline can be monitored in real time. The high infrared transmittance of sulfur glass fiber ensures the low loss of optical signal in long-distance transmission, and improves the reliability and sensitivity of communication. At the same time, its unique dispersion characteristics in the mid-infrared band can realize the dispersion compensation through the reasonable design of optical fiber structure and refractive index distribution, and meet the requirements of high-rate and large-capacity optical communication.

Optical signal processing: Sulfur-based glass also plays an important role in light signal processing. Optical modulator, optical switch, and other optical devices based on sulfur glass use their infrared transmittance characteristics and optical nonlinear effect to realize the modulation and control of optical signals. In the light modulator, the optical characters of the sulfur glass, including infrared transmittance and refractive index, are changed by external electric field, magnetic field, or thermal field, so as to realize the modulation of the light signal intensity, phase, or polarization state. In the infrared band, it can be modulated efficiently without significant loss of optical signal energy, which provides strong support for high-speed optical communication and processing of optical information. For example, in a high-speed optical communication network, a sulfur glass-based optical modulator can realize optical signal modulation at a rate of 100 Gbps and above to meet the increasing demand for communication bandwidth.

b) **Infrared Imaging Field:**

Infrared thermal imaging: In the field of infrared thermal imaging, sulfur glass is widely used in the production of infrared lenses, windows, and other optical components because of its excellent infrared penetration characteristics. Infrared thermal imaging technology, by detecting the infrared radiation emitted by the object and sulfur glass, can effectively pass through the infrared light radiated by the object and focus it on the detector to form a clear image. In military, security, industrial testing, and other fields, infrared thermal imaging equipment needs to work in harsh environments. Sulfur glass has good chemical stability and mechanical characters, which can meet the requirements of optical components in these application scenarios. For example, in military reconnaissance, a thermal imager equipped with a sulfur-based glass infrared lens can clearly detect the thermal signal of a target object at night or in harsh environments such as smoke, providing an important basis for combat decisions. The high infrared transmittance and low scattering characteristics of sulfur glass ensure the high resolution and clarity of infrared thermal imaging.

Infrared spectroscopy imaging: Infrared spectroscopy imaging technology combines infrared spectroscopy analysis with imaging technology to simultaneously capture both spatial and spectral information about objects. Sulfur glass serves as an optical window and a transmission medium in infrared spectral imaging systems, and its infrared penetration characteristics are crucial to the system's performance. Different substances exhibit unique absorption spectra in the infrared region, allowing for the identification and analysis of an object's chemical composition and structure through spectral analysis of the infrared light that passes through the sulfur glass. This technology is widely utilized in fields such as materials science and biomedicine. For example, in biomedical research, it can be detected nondestructively by analyzing the absorption and spectrum changes of tissues in the infrared band, diagnosing diseases, and studying the structure and function of biomolecules. The high precision infrared transmittance performance of sulfur glass ensures the accuracy and reliability of infrared spectral imaging.

c) **Optical Sensing Field**

Gas sensing: The gas sensor made from sulfur glass functions by utilizing changes in the infrared transmittance characteristics of the glass when specific gas molecules are present. When gas molecules adsorb onto the surface of the sulfur glass, they cause physical or chemical changes to the glass surface, which leads to variations in infrared transmittance. By accurately measuring these changes in transmittance using optical methods—such as infrared absorption spectroscopy or surface plasmon resonance technology—it is possible to achieve highly sensitive detection of gas concentrations. For instance, when detecting hydrogen sulfide (H_2S) gas, the active sites on the surface of the sulfur glass react with H_2S molecules. This interaction alters the vibrational state of the chemical bonds in the surface layer of the glass, thus affecting its absorption and penetration characteristics in the infrared spectrum. Experiments show that for some specific components of sulfur glass, when the concentration of H_2S gas changes in the range of 1–100 ppm, the infrared transmittance at a specific wavelength can reach 1–10%. This small transmittance change can be accurately detected by the appropriate optical sensing system, so as to achieve a highly sensitive detection of H_2S gas.

Biosensing: In the field of biosensing, sulfur-based glass can also be used to build biosensors for the detection of biomolecules, cells, and other biological substances. The use of the infrared penetration characters of sulfur glass is combined with the specific interaction of biomolecules with the glass surface. For example, in biosensors based on surface plasmon resonance (SPR), sulfur glass serves as the sensing substrate. When the biomolecule is specifically bound to its surface, it will change the refractive index and infrared transmittance characteristics near the glass surface, thus affecting the resonance conditions of the surface plasmon wave. The presence and concentration of biomolecules can be detected indirectly by monitoring the change in the resonance wavelength or resonance angle. Because the interaction between biomolecules and the glass surface is very weak, the detection accuracy of the change in infrared transmittance characteristics is very high. The stable and accurately measurable infrared transmittance characteristics of sulfur glass are crucial for the construction of highly sensitive biosensors. Biosensors based on sulfur glass have great application prospects in disease diagnosis and food safety detection.

In conclusion, the infrared transmittance of sulfur glass is one of the key characters in many fields. In principle, its infrared penetration characteristics are close to the microscopic processes such as chemical bond vibration and electronic transition inside the glass.

5.2.3 Photosensitivity Characteristics

As a kind of important optical material, sulfur glass's unique optical characters show great application potential in optical storage, optical communication, optical sensing, and many other fields. Photodarkening, photobleaching, and photoinduced anisotropy are important manifestations of the photosensitivity characters of sulfur glass. A deep understanding of these characters will not only help to reveal the relationship between the microstructural changes inside the sulfur glass and the optical characters, but also provide a solid theoretical basis for the development of new photofunctional devices (Hubert M. et al., 2011). With the continuous development of materials science and optical technology, the study of sulfur glass has increasingly attracted attention and become one of the hot research fields.

1) **Photoinduced Darkening and Photoinduced Bleaching:**
   a) **Phenomenon Description:**

Light Darkening: Light-induced darkening is the phenomenon in which the transmittance of sulfur glass gradually decreases when exposed to specific wavelengths of light, causing the material to become darker in color. This change is observed macroscopically as an increased absorption of light by the glass, characterized by a shift of the absorption edge toward longer wavelengths, a process commonly referred to as a "redshift." For example, the common AS_2S_3 sulfur glass can be observed in a short time (a few minutes to a few minutes) under blue or ultraviolet light. In the visible light band, the transmittance may decrease from about 70% to 50% or even lower (Masaki A. et al., 1993), while the color of the glass gradually changes from the original light yellow to dark yellow or even tan.

Photo-bleaching: Photoinduced bleaching is the opposite of photoinduced darkening, which refers to the optical performance of sulfur glass under another wavelength of light, or after some thermal annealing treatment, gradually recovers, the transmittance increases, and the absorption edge moves to the short wavelength direction, that is "blue shift." For example, for As_2S_3 glass that is darkened by blue light, illuminated with near-infrared light, or heated to a certain temperature (such as 100–150 °C) and maintained for a period of time, the transmittance of the glass will gradually rise, the color becomes lighter, and it gradually returns to the initial state.

   b) **Generation Mechanism:**

Theory of Structural Change: At present, the mechanism of photodimination and photobleaching of sulfur glass is not fully defined, but it is generally believed that it is closely related to the structural changes inside the glass. A widely accepted theory is the model of photoinduced structural changes. In sulfur glasses, various chemicals bond such as Ge—S and As—S. When light is exposed, the photon energy is absorbed by the glass, leading to the breaking and reorganization of some chemical bonds. Take As_2S_3 glass as an example, in the process of photoshading, the light makes the As—S bond break, producing sulfur radical (S) and As atomic clusters. These radicals and atomic clusters would recombine to form new structures, such as S—S bonds or As—As bonds. The formation of these new structures changes the electron cloud distribution and band structure of the glass, enhancing the absorption of the glass to light, which leads to light-induced ening. In the process of photobleaching, the energy provided by another wavelength of light irradiation or thermal annealing makes these newly formed structures change again, the chemical bonds return to a relatively stable initial state, and the optical characters of the glass are restored.

Defect Formation and Elimination: Another mechanism is related to the formation and elimination of defects in the glass. During photodarkening, electron–hole pairs generated by photoexcitation migrate inside the glass. Caves interact with anions in the glass network (such as S^2) to form defect centers that can absorb light at specific wavelengths, leading to photoshading. In the photoinduced bleaching process, through the irradiation of the other light, the electronic state of these defects is changed, the defect is eliminated, the absorption of the glass decreases, and the transmittance increases. For example, in the Ge–S series glass, it is found that the Ge—Ge weak bond defects are generated during the photoinduced darkening process, and these defects are the important causes of the enhanced light absorption (Daniel V. Hahn et al. 2005). However, during photoinduced bleaching, these Ge—Ge weak bond defects gradually disappear, and the optical characters of the glass are restored.

c) **Influencing Factors:**

The Wavelength and The Intensity of the Light: The wavelength and intensity of light significantly influence the processes of photodarkening and photobleaching in sulfur glass. Light at various wavelengths possesses different photon energies, and photodarkening or photobleaching occurs only when the photon energy aligns with the bond energy of the internal chemical bonds within the glass. In general, short-wavelength light (such as ultraviolet and blue light) is more likely to cause light darkening because of its higher photon energy. For example, in As–Se glass, irradiation with 365 nm UV light can trigger light dimming more quickly than blue light with 450 nm. The intensity of light also plays an important role, with higher light intensity providing more photons and accelerating the photochemical reaction process, thus accelerating the rate of photodarkening. However, very high light intensity may lead to adverse phenomena such as overheating inside the glass, affecting the performance of the glass. In the process of photoinduced bleaching, it is also necessary to choose the appropriate wavelength of light, as its photon energy can promote the transformation of the new structure formed in the process of light-induced shading. For example, the Ge–As–S glass with 800–1000 nm.

Glass composition: The chemical composition of glass is one of the key factors affecting the characteristics of photodimming and photobleaching. Different glass formers and sulfur elements can lead to different chemical bond characters and network structures, which can affect their response to light (Shen et al. 2015). In the Ge–S–Se system, the Se content increases. This is because the Se—Se bond has relatively low bond energy and is more likely to break and recombine under the action of light. In addition, doping elements can also have an impact on photoinduced darkening and photoinduced bleaching. For example, the doping of a small amount of thallium (Tl) into the As_2S_3 glass can significantly increase the photoinduced dimming rate of the glass, while also changing the behavior of the photoinduced bleaching. It is shown that the doping of Tl may introduce new defect levels and promote electronic transitions during photoexcitation, thereby affecting the photosensitive characters of the glass.

Temperature: Temperature also has an important influence on the photodarkening and photobleaching process of sulfur glass. In the process of photoshading, appropriately increasing the temperature can accelerate the rate of photochemical reaction, because the temperature will increase the thermal motion of atoms, which is conducive to the breaking and reorganization of chemical bonds. However, too high a temperature may lead to irreversible changes such as thermal decomposition of glass. The role of temperature is even more pronounced during photoinduced bleaching (Nie and Wang 2018). Thermal annealing is one of the common photoinduced bleaching methods, and by heating the photodarkened glass to a certain temperature and maintaining it for some time can provide sufficient energy to restore the structure inside the glass to its initial state. For example, for AS–S glasses, with thermal annealing around 120 °C, significant photobleaching can be observed after several hours, and the transmittance of the glass returns to near the initial value.

d) **Study of Cases and Applications:**

Optical Storage Applications: The photodarkening and photobleaching characteristics of sulfur glass have important applications in the field of light storage. Data can be written, and by controlling the irradiation area and intensity of light, different degrees of dark areas can be formed in the glass, and these areas can represent different information coding. Data can be erased and modified by photobleaching (Li and Liu 2016). For example, in the study of erasable optical discs, sulfur glass is used as the storage medium, including blue light writing data (photodimming), and near-infrared light wiping data (photobleaching). This optical storage technology has the advantages of high storage density and fast data transmission rate, and is expected to become an important development direction of the next generation of optical storage technology.

Optical Sensing Applications: In the field of optical sensing, the photodarkening and photobleaching characters of sulfur glass can be used to detect changes in specific environmental factors. For example, the sensitivity of phototodarkening to certain gas molecules allows one to develop gas sensors. When specific gas molecules are adsorbed on the surface of sulfur glass, they will change the optical characters of the glass surface and affect the rate and degree of photoshading (Dai and Xu 2020). Gas concentration can be detected by monitoring the change in glass transmittance in the photoshading process. In biosensing, it is also possible to detect photoinduced dimming interactions with photoinduced bleaching characters. For example, when the biometric molecule is fixed on the surface of sulfur glass, when the target biomolecule is combined with the recognition molecule, it will cause the microenvironmental changes on the glass surface, and then affect the process of photodarkening and photobleaching. High-sensitivity detection of biomolecules can be achieved.

2) **Photoanisotropy:**
   a) **Phenomenon Description:**

Photoinduced anisotropy refers to the phenomenon that the optical characters (such as refractive index and birefringence) of sulfur glass differ in different directions under the irradiation of light. When linearly polarized light passes through sulfur glass, the glass exhibits optical anisotropy related to the polarization direction of the incoming light. Specifically, the refractive index of the glass differs depending on whether it is aligned parallel or perpendicular to the polarization of the incident light (Anonymous 2024b). This difference in refractive index results in birefringence. This birefringence can be observed and measured by some optical experimental methods, such as polarized optical interference experiments. For example, by placing a sulfur glass sample treated with phototropic anisotropy between the orthogonal polaroid sheets, the appearance of interference stripes can be observed when light is transmitted, and the spacing and intensity of the stripes are related to the degree of birefringence of the glass.

   b) **Generation Mechanism:**

Molecular Orientation Theory: The generation mechanism of photoinduced anisotropy is mainly related to the orientation change of molecules or atomic groups in the glass. In the sulfur series glass, various chemical bonds and molecular structure units exist. When the linearly polarized light is illuminated, the electric field vector of the light will act on the molecules or atomic groups in the glass to change their orientation. In the case of a sulfur-bound glass containing a long-chain molecular structure, these long-chain molecules are gradually oriented along the polarization direction of the light. This change of molecular orientation causes differences in the electron cloud distribution and chemical bond characters in different directions, thus causing the anisotropy of the refractive index. For example, in Ge–As–S glass, when exposed to linearly polarized light, the As–S chain structure is gradually arranged parallel to the polarization direction of the light, making the refractive index differ between parallel and perpendicular to the polarization direction, producing birefringence.

Photoinduced Structural Changes with the Anisotropy: Similar to the photodimination process, photoinduced anisotropy may also be accompanied by structural changes inside the glass (Nie and Wang 2018). Photoexcitation leads to the breakage and reorganization of chemical bonds, and such structural changes may differ in different directions, leading to optical anisotropy. For example, under the irradiation of light, some regions of the glass may form atomic clusters or defect structures with specific orientations, making the optical characters of the glass different in different directions. It is found that in some sulfur glasses, directional Ge—Ge bond or As—As bond aggregation regions are generated, in which the response to light varies in different directions, which leads to the anisotropy of the refractive index.

c) **Influencing Factors:**

The Polarization Characters of Light: The polarization direction and degree of polarization are the key factors affecting the photoanisotropy of sulfur glass. Only linearly polarized light can effectively induce anisotropy in glass, and the higher the polarization, the more obvious the induced anisotropy effect. When circularly polarized light or unpolarized light is irradiated, because the electric field vector of light is evenly distributed in all directions to make the directional orientation of molecules or atomic groups in the glass, there is no obvious anisotropic light. Moreover, the irradiation time of light also affects the degree of photoinduced anisotropy. With increasing irradiation time, molecules or groups inside the glass have more time to adjust their orientation, and birefringence gradually increases. However, when the irradiation time reaches a certain degree, the photoinduced anisotropy may tend to saturate, and the birefringence will no longer increase significantly (Treacy et al. 1979).

Glass Composition and Microstructure: The chemical composition and microstructure of glass have an important influence on the photoinduced anisotropy. Different glass systems have different responses to light and anisotropic ability to produce anisotropy due to their different chemical bond characters and molecular structure units. In some sulfur glasses with complex molecular structures, obvious photonic anisotropy is more likely. For example, in sulfur-bound glasses containing a large number of long-chain or layered structures, these structural units are more prone to orientation change under light, resulting in large birefringence. In addition, the defects and impurities in the glass can also affect the photoinduced anisotropy. The presence of defects may change the propagation path and electric field distribution of light in the glass, which in turn affects the orientation process of molecules or atomic groups. Studies have shown that the sensitivity of photoinduced anisotropy is improved in sulfur glass containing appropriate defects, but too many defects may lead to increased scattering of light and mask the effect of photoinduced anisotropy.

Temperature and Stress: Temperature and stress also have some effects on the photoinduced anisotropy of sulfur glass. In the process of photoinduced anisotropy, appropriately increasing the temperature can accelerate the orientation adjustment speed of molecules or atomic groups, thus enhancing the photoinduced anisotropy effect (Yang et al. 1987). However, the high temperature may cause the hot movement of the internal structure of the glass to be too intense, which is not conducive to the directional arrangement of molecules, but weakens the photoinduced anisotropy. Moreover, the externally applied stress also affects the photoinduced anisotropy of the glass. Stress can change the microstructure and chemical bond state inside the glass, couple to the action of light, and affect the orientation process of molecules or atomic groups. For example, under tensile stress, molecules or atomic groups in sulfur glass are more likely to orient along the stress direction and superposition with the photoinduced orientation, further enhancing the photoinduced anisotropy.

d) **Study of Cases and Applications**

Light Polarization Device: Photoinduced anisotropy of sulfur glasses has important applications in the preparation of photopolarized devices. Light polarization components, such as a wave sheet and polarization beam splitter, can be made by photoinduced anisotropy. For example, by controlling the irradiation conditions of the light, a specific birefringence can be induced in the sulfur-bound glass film. This results in the creation of a wave sheet that can precisely control the polarization state of the light. Wave sheets based on sulfur glass systems offer advantages such as fast response speeds and dynamic adjustments. They are widely used in fields like optical communication and laser technology. In optical communication systems, these wave sheets can adjust the polarization state of optical signals, thereby enhancing signal transmission quality and improving resistance to interference.

Optical Information Handling: In the field of optical information processing, the photoinduced anisotropy can be used to realize the modulation and storage of optical signals (Hari et al. 2000). Optical signals can be diffracted and modulated by the photoanisotropic birefringent grating formed in a sulfur-bound glass. For example, using linearly polarized light to write a birefringent grating in a sulfur glass, diffraction occurs when another light is exposed to the grating, and the intensity and polarization state of the diffracted light are modulated by the grating's birefringence. This optical modulation mode can be used for applications such as optical signal encoding, decoding, and image processing. In addition, the phototropic anisotropy can also be used for optical data storage. By recording the birefringence information in different directions in the glass, high-density optical data storage can be realized, and the data storage capacity and reading speed can be improved.

The important photoshading, photobleaching, and photoinduced anisotropy of sulfur glass have shown great application potential in many fields. From the phenomenon description to the generation mechanism, to the influencing factors and practical applications, the study of these characteristics continues to deepen. Photoshading and photobleaching enable the reversible regulation of optical performance by altering the structure of glass and eliminating defects. These processes hold significant application potential in the fields of optical storage and sensing. Additionally, photoinduced anisotropy creates variations in the optical properties of glass by changing molecular orientation and inducing structural modifications. This offers new technological opportunities in the development of optical polarization devices and optical information processing (Zakery and Elliott 2003). However, there are still some problems in the study of these photosensitive characters, such as the mechanism not being fully defined, and the characteristic regulation of the complex glass system not being accurate enough. In the future, as materials science, optical technology, and computational methods continue to advance, we can expect a deeper understanding of the photosensitive properties of sulfur glass. This will enable more precise regulation and broader applications of the material, ultimately driving technological innovation and development in related fields.

5.2.4 Nonlinear Characteristic

Due to its unique chemical composition and microstructure, sulfur glass shows rich and excellent nonlinear optical characteristics, which occupy an important position in the field of modern optical technology. These nonlinear characters make the sulfur glass have great application potential in many frontier technology fields such as optical communication, optical signal processing, optical imaging, and optical sensing. A thorough examination of the nonlinear properties of sulfur glass is crucial for advancing optical technology. This research is key to developing new optical devices and addressing the increasing demand for high-speed, efficient, and high-capacity optical information processing (Bruno et al. 2004).

1) **Glass Series With High Nonlinear Optical Characters (example):**
   a) **Ge–As–S Series of Glass:**

Ge–As–S glasses are a typical class of sulfur glasses with high nonlinear optical characters. Among them, Ge, as a glass network formation body, builds the glass network structure by forming covalent bonds with sulfur elements (such as S). The introduction of As further modulates the electronic cloud distribution and chemical bonding characters of the network. We have shown that glasses exhibit significant nonlinear optical responses in $Ge_xAs_yS_{1-x-y}$ systems when x and y are in the appropriate$_x$ proportional range. For instance, $Ge_{0.25}As_{0.35}S_{0.4}$ The glass has a high nonlinear refractive index (n_2) in the near-infrared band, and its n_2 value can reach the magnitude of 10^{18}–10^{17} m^2/W. This high nonlinear refractive index gives the glass great advantages in optical signal modulation and wavelength conversion applications in optical communication. In terms of optical signal modulation, fast modulation of optical signal intensity, phase, and polarization state can be realized by using its nonlinear optical characters through the interaction between light and glass, satisfying the requirements of high-speed optical communication for signal processing speed.

   b) **As–Se Series of Glass**

The As–Se glass is also a sulfur glass system with prominent nonlinear optical characters (Keiji Tanaka 2002). Due to the large atomic radii of the Se atoms and the abundant electron clouds, the As—Se bonds have a unique electronic structure and polarization characters. As_2Se_3 Glass shows good nonlinear optical performance in the mid-infrared band. In some experimental studies, measurements of the nonlinear optical parameters of As_2Se_3 glass in the mid-infrared band reveal a relatively low two-photon absorption coefficient, and the third-order nonlinear polarizability is high. This characteristic makes As_2Se_3 glass valuable in the applications of light confinement and supercontinuum generation in the mid-infrared band. For example, in light-limiting applications, when a high-intensity mid-infrared laser irradiates the As_2Se_3 glass, the glass transmittance rapidly decreases with increasing light intensity due to its nonlinear optical characters, thus protecting the rear-end optical elements from strong light damage.

c) **Ge–S–Te Series of Glass**

The Ge–S–Te series of glasses combines the benefits of both the Ge–S and Ge–Te systems, showcasing distinctive characteristics in the realm of nonlinear optics. The addition of Te alters the vibrational modes within the glass network structure and its chemical bonds, consequently influencing the nonlinear optical properties of the material. In Ge–S–Te glasses, the nonlinear refractive index steadily increases as the content of tellurium rises. For example, in the specific composition of the $Ge_{0.3}S_{0.5}Te_{0.2}$. In the glass, the nonlinear refractive index can reach 10^{16} m^2/W in the long-wave infrared band of 10–12 μm. This high nonlinear refractive index makes the glass have potential applications in the frequency conversion of long-wave infrared light and optical parametric oscillation. In the process of optical parametric oscillation, the nonlinear optical characters of Ge–S–Te glass can convert the frequency of long-wave infrared light into other required wavelengths, which expands the application range of long-wave infrared light (Hirohisa Kanbara et al. 1997).

2) **Two-photon Absorption Spectra:**
   a) **Principle:**

Two-photon absorption is a nonlinear optical process of third order. In this process, when a material is illuminated by high-intensity laser light, molecules or atoms can simultaneously absorb two photons, causing them to leap from the ground state to an excited state. The occurrence of this process requires the satisfaction of both the energy conservation and the momentum conservation conditions. For sulfur glasses, the two-photon absorption process involves electronic transitions of atoms or molecules inside the glass. Due to the band structure characteristics of sulfur glasses, electrons can transition from the valence band to the conduction band by two-photon absorption at a suitable laser wavelength and intensity. The two-photon absorption process can be described by the two-photon absorption coefficient (β), and β represents the light intensity decay coefficient due to the two-photon absorption per unit length.

b) **Measurement Method:**

The common method for measuring two-photon absorption spectra of sulfur glass is the Z-scan technique (which will be detailed in the subsequent Z-scan section). In the Z-scan technique used for two-photon absorption measurements, a sulfur-bound glass sample is positioned along the optical axis of a focused laser beam. The change in light intensity as it passes through the sample is measured by moving the sample along the optical axis (Petkov and Ewen 1999). When the sample is located at the focal point, a noticeable trough in the light intensity is observed, which is indicative of enhanced two-photon absorption. By analyzing the change in the penetrating light intensity as the sample is repositioned, we can calculate the two-photon absorption coefficient. Additionally, pump-detection techniques can also be employed to measure the two-photon absorption spectra. In the pump-detection experiment, the two-photon absorption is used to excite the sample with high intensity (David Furniss, et al., 1999), and then the weak detection light is used to detect the optical characters changes after the two-photon absorption, and the two-photon absorption spectral information is obtained by analyzing the signal changes of the detection light.

c) **Study of Cases and Applications:**

In the two-photon absorption study of Ge–As–S glasses, it is found that the two-photon absorption coefficient shows a specific pattern with the wavelength change. Studies show that near 1.55 μm wavelength, $Ge_{0.2}As_{0.3}S_{0.3}$, the two-photon absorption coefficient of the $_5$ glass is about on the order of 10^8 cm/W. This two-photon absorption characteristic has potential applications in the optical storage field. Three-dimensional light storage can be realized in the two-photon absorption process. Using a high-resolution laser beam, information is recorded by two-photon absorption, generating photoinduced structural changes at specific locations inside the glass. Compared with the traditional 2D optical storage, 3D optical storage has higher storage density and better information security. In the field of biological imaging, novel two-photon fluorescence imaging techniques based on two-photon absorption characters can be developed. Two-photon absorption of a specific wavelength laser by sulfur glass is used to emit fluorescence from fluorescent material modified inside or on the glass to achieve high-resolution imaging of biological samples.

3) **Degeneration and Four-wave Mixing (DFWM):**
   a) **Principle:**

Degenerate four-wave mixing is an important third-order nonlinear optical process. In this process, three laser beams of the same frequency (degenerate) (called pump light) simultaneously illuminate the sulfur glass sample, producing a fourth light (called signal light) through nonlinear interaction. This process satisfies both energy conservation and momentum conservation (Z. H. Zhou, et al. 1998). In the sulfur glass, the pump light interacts with the electron cloud inside the glass, causing the nonlinear polarization of the electron cloud, and then produces a new polarization field, which radiates the signal light. The frequency of the signal light matches that of the pump light; however, its direction of propagation and phase are influenced by the nonlinear optical characteristics of both the pump light and the glass. The degenerate four-wave mixing process can be described using the third-order nonlinear polarizability (χ^3), which indicates the strength of the glass's nonlinear response to the optical field. (Hirohisa Kanbara, et al. 1997).

b) **Experimental Device and Measurement:**

In degenerate four-wave mixing experiments, the primary components used include a laser source, a beam splitter, mirrors, and a focusing lens. The laser beam generated by the laser source is divided into three pump beams with the help of the beam splitter. These beams are then focused onto a sulfur glass sample at specific angles and with precise phase relationships. By adjusting factors such as the intensity, angle, and phase of the pump light, the degenerate four-wave mixing process can be optimized to enhance the intensity of the resulting signal light. Measuring the intensity and phase of the mixed signal light provides valuable information, such as the third-order nonlinear polarizability of the sulfur glass. For example, the nonlinear refractive index of the glass can be calculated indirectly, by measuring the phase difference between the signal light and the pump light.

c) **Application:**

Simplified four-wave mixing plays a crucial role in optical communication. In light signal processing, phase conjugation of the light signal can be accomplished through degenerate four-wave mixing. The phase-conjugated light produced by this process compensates for phase distortions caused by optical fiber dispersion and nonlinear effects during transmission. As a result, it enhances both the transmission quality and distance of the optical signal. In the field of optical imaging, degenerate four-wave mixing can be used for the construction of optical phase-conjugate microscopy. The illumination of the sample by the phase-conjugate light generated by degenerate four-wave mixing can eliminate interference factors such as scattering and aberrations in the sample and improve the imaging resolution and contrast. In quantum optics, the degenerate four-wave mixing process can be used to generate entangled photon pairs, providing an important light source for studies such as quantum communication and quantum computing (K. Petkov, et al., 1999).

4) **Z scan:**
   a) **Principle:**

The Z-scanning technique is a widely used method for measuring the nonlinear optical properties of materials. In a Z-scan experiment, a sulfur glass sample is positioned along the optical axis of a focused laser beam, and the sample is moved along this axis (the z direction). When the sample is located far from the focus, the intensity distribution of the laser beam is relatively uniform, causing the light absorption and refraction of the sample to primarily exhibit a linear optical response. As the sample gradually approaches the focus, the laser beam intensity gradually increases, and the nonlinear optical effects begin to appear. When the sample is in the focal position, the nonlinear optical effect is most significant because the laser intensity reaches the maximum. By measuring the change curve of the light intensity through the sample, the characteristics of nonlinear absorption and nonlinear refraction can be obtained. If the material has a nonlinear absorption process such as two-photon absorption, the light intensity curve will appear in the focus position; if the material has a nonlinear refraction, the light intensity curve will have an asymmetric distribution of peak and valley values before and after the focus.

b) **Experimental Steps and Data Analysis:**

When conducting Z-scan experiments, the first step is to prepare a high-stability laser light source with a wavelength that aligns with the nonlinear optical response of the sulfur glass. The laser beam is then focused on the sample using a lens, and the sample's movement along the optical axis is precisely controlled using a high-precision displacement device (Xu Yan-Tao et al. 2015). During this movement, the light intensity passing through the sample is measured in real time with a detector. To enhance measurement accuracy, the experimental data are typically processed multiple times and averaged. In the data analysis stage, the measured optical intensity data are fitted to calculate key nonlinear optical parameters, such as the two-photon absorption coefficient and nonlinear refractive index of the material. Commonly used fitting models include numerical models based on nonlinear absorption and nonlinear refraction theories, where model parameters are adjusted to achieve the best fit to the experimental data (Chen Jiang et al. 2016).

c) **Advantages and Applications:**

The Z-scanning technique has the advantages of high measurement accuracy, a relatively simple experimental device, and the ability to measure the nonlinear absorption and nonlinear refraction of materials simultaneously. In sulfur glass research, Z-scanning technology is widely used to screen glass materials with high nonlinear optical characters. The Z-scan measurements of sulfur-bound glass with different compositions and structures can rapidly assess its nonlinear optical characters for the optimized design of materials. In light-limited materials research, Z-scanning technology can be used to measure the light-limiting characteristics of sulfur glass at different light intensities and evaluate its potential in the field of strong light protection. In optical device development, Z-scanning technology can be used to measure the nonlinear optical performance of sulfur-bound glass under practical application conditions, providing data support for device performance optimization and reliability evaluation.

5) **Tertiary Harmonic Generation (THG), Optical Kerr Shutter (OKS):**
   a) **Third Harmonic Generation (THG):**

Principle: Teraharmonic generation is a third-order nonlinear optical process. When high-intensity laser light (at the base frequency) illuminates sulfur glass, it generates light at three times the base frequency, known as the third harmonic. This phenomenon occurs due to the third-order nonlinear polarizability of the glass. In this process, the fundamental frequency light interacts with the electron cloud within the glass, causing it to oscillate nonlinearly. This oscillation results in a polarized field that emits third-harmonic light. The generation of the third harmonic satisfies the principles of energy conservation and momentum conservation. Additionally, its efficiency depends on several factors: the third-order nonlinear polarizability of the glass, the cube of the intensity of the base frequency light, and the length of the glass (Ya'nan Sun et al. 2015).

Apply: Tertiary harmonic generation has important applications in the field of optical frequency conversion. The laser in the infrared band can be converted to light in the visible or ultraviolet band by selecting the appropriate sulfur-bound glass material and controlling the parameters of the base frequency light. For example, in some laser processing applications, an ultraviolet laser is used for high-precision material processing, but the existing laser sources may be mainly infrared band. By using the third-harmonic generation characteristics of sulfur glass, the infrared laser can be converted into an ultraviolet laser to meet the processing requirements. In the field of optical imaging, third-harmonic imaging images the sample, which can provide different contrast information than conventional imaging techniques, and helps study the microstructure and optical characters of the sample.

b) **Optical Kerr Shutter (OKS)**

Principle: The optical Kerr shutter is based on the optical Kerr effect, which is a manifestation of the third-order nonlinear optical characters of the sulfur-bound glass. When the high-intensity pump light on the sulfur glass changes proportionally to the pump intensity, namely $n = n_0 + n_2 I$, where n_0 is the linear refractive index, n_2 is the nonlinear refractive index, and I is the pump intensity. In an optical Kerr shutter, a beam of detection light hits the sulfur glass at a specific angle. When the pump light is not illuminated, the transmittance of the glass to the detected light is low; when the pump light is illuminated, the refractive index of the glass changes due to the optical Kerr effect, and the polarization state and propagation characteristics of the detected light also change, thus significantly increasing the transmittance of the detected light, and realizing the rapid switching control of the detected light (Jasbinder S. S., et al., 1995).

Apply: Optical Kerr shutter is widely used in the field of ultrafast optics. In the ultrafast laser pulse measurement, the optical Kerr shutter can be used to select the laser pulse at a specific time point to achieve an accurate measurement of the temporal characteristics of the laser pulse. In the field of optical communication, an optical Kerr shutter can be used to construct high-speed optical switches to realize fast routing and exchange of optical signals. The optical Kerr shutter made from sulfur glass can achieve high-speed optical switching at picosecond or even femtosecond levels, thanks to its high nonlinear refractive index and rapid response speed. This capability meets the demands of high-speed optical communication for efficient signal processing.

6) **Prediction of the Nonlinear Refractive Index:**
   a) **Theoretical Model**

Classical Theoretical Model: Some classical theoretical models are widely used in predicting the nonlinear refractive index of sulfur-bound glasses. Among them, the modified models based on the Lorenz–Lorenz formula are more commonly used. The model treats the glass as a medium composed of atoms or molecules with a certain polarizability, and calculates the nonlinear refractive index by considering the electronic structure and chemical bond characters of the glass. In sulfur-bound glasses, the polarizability of different atoms or molecules is determined according to the chemical composition of the glass, and then the modified Lorentz–Lorentz formula is used to calculate the linear and nonlinear refractive index of the glass. However, classical theoretical models often ignore the complexity of the internal glass structure and the quantum mechanical effects, and there are some limitations to the nonlinear refractive index prediction of some complex sulfur-glass systems.

Quantum Mechanics Model: To more accurately predict the nonlinear refractive index of a sulfur-bound glass, a quantum-mechanical model is introduced

(Chen et al. 2011). The quantum mechanical model calculates the polarizability and nonlinear response of the electron cloud by solving the Schrodinger equation and considering the quantum states and interactions of the electrons inside the glass. For example, density functional theory (DFT) can be used to calculate the electronic structure and optical characters of sulfur glass. In DFT calculation, the atomic structure and electron interaction of the glass are described by a specific functional form, and the electron density distribution and energy level structure of the glass are obtained through iterative calculation, and then the nonlinear refractive index is calculated. The quantum mechanical model can consider the microstructure and electronic interactions inside the glass more precisely, but with a large computational amount and high requirements on computational resources.

b) **Analysis of Influencing Factors:**

Chemical Composition: The chemical composition of sulfur glass is a key factor affecting its nonlinear refractive index. Different glass formers and sulfur elements can lead to different electronic structures and chemical bonding characters, which can affect the nonlinear refractive index. As mentioned above, in the Ge–As–S series glasses, with the increase of Ge content, the network structure of the glass is more compact, the polarizability of the electron cloud changes, and the nonlinear refractive index changes accordingly. Moreover, the type and content of doped elements have a significant effect on the nonlinear refractive index (Chen Yu, et al. 2011). Doping of some transition metal elements can introduce new electronic energy levels, change the electronic transition characters of the glass, and thus affect the nonlinear refractive index.

Micromechanism: The microstructure of the glass, such as the uniformity of the network structure, the orientation of chemical bonds, and the distribution of defects, can also have an influence on the nonlinear refractive index. In the nonlinear optical glass, the nonlinear optical response is relatively stable, and the prediction of the nonlinear refractive index is relatively accurate. However, when numerous defects or heterogeneous regions exist in the glass, light scattering and additional nonlinear interactions occur during the propagation process, further complicating the prediction of the nonlinear refractive index. For example, defects such as non-bridge oxygen atoms and vacancies in glass can change the electron cloud distribution, affect the nonlinear polarizability, and then affect the nonlinear refractive index.

c) **Research the Current Situation and Challenges:**

At present, some progress has been made in the prediction of the nonlinear refractive index of sulfur glass, but it still faces many challenges. On the one hand, although the theoretical model is constantly developing and improving, there is still some deviation between the model and the actual situation for some complex sulfur glass systems. On the other hand, the comparison and validation of the experimental measurement data and the theoretical prediction results are not sufficient, and further high-precision experimental measurement work should be carried out to provide more accurate data support for the optimization of the theoretical model. In addition, as new sulfur glass materials continue to be developed, predicting their nonlinear refractive index quickly and accurately has become a significant focus of current research. In the future, it will be essential to integrate theoretical calculations, experimental measurements, and machine learning techniques to enhance the prediction of the nonlinear refractive index for sulfur glass. This approach aims to improve the accuracy and reliability of these predictions, guiding the design and application of these materials.

The nonlinear characters of sulfur glass cover many important aspects, from glass series with high nonlinear optical characters, to nonlinear optical processes such as two-photon absorption spectra, degenerate four-wave mixing, Z scanning, third-harmonic generation and optical Kerr shutter, and predictions of nonlinear refractive index. These characteristics show great application potential in many fields, such as optical communication, optical storage, optical imaging, and optical sensing. By conducting a thorough study of the principles, measurement methods, applications, and predictive technologies related to the characteristics of sulfur glass materials, we can not only advance their development and use but also inspire new ideas and methods for innovation in optical technology. However, there are still challenges in understanding the nonlinear properties of sulfur glass. For instance, our comprehension of the nonlinear optical mechanisms in complex glass systems is not sufficiently deep, and there is a need to enhance the consistency between theoretical predictions and experimental results. In the future, it is necessary to further strengthen basic research, combine with interdisciplinary integration, and continuously explore and discover (Li et al. 2015).

5.3 Preparation Process of Sulfur System Glass Fiber

5.3.1 Purification of the Sulfur-series Glass Material

Sulfur glass materials are widely used in optical communication, infrared optics, optical storage, and many other fields because of their unique optical and electrical characters. However, impurities in the material can seriously affect their performance, such as impurities may increase optical absorption and scattering, reducing optical transmittance, and interfering with electrical conduction characteristics. Therefore, purification has become the key link to obtain high-quality sulfur glass materials, which plays a decisive role in improving the comprehensive performance of sulfur glass and expanding its application scope (Lingqi Li et al. 2017).

5.3.2 Several Methods for the Preparation of Sulfur Glass Fiber

Because of its excellent optical performance in the infrared band, such as low loss and high nonlinearity, it has shown great application potential in optical communication, optical sensing, infrared imaging, and other fields. The key to preparing high-quality sulfur glass fiber is to choose a suitable and efficient preparation method. Tube and rod drawing method, extrusion method, and double crucible drawing technology are commonly used preparation methods, and they have unique process characteristics and scope of application, respectively (J. Keirssea, et al., 2017).

1) **Prefabricated Rod Drawing Technique:**
   a) **Pipe-rod Method:**
      i) Principle

The core principle of the tube rod method is to use the close coordination and melting fusion between the glass pipe and the glass rod. First, a high-quality sulfur glass tube was prepared as the cladding material, while a sulfur glass rod matching the inner diameter of the glass tube was prepared as the fiber core material. During the drawing process, the glass rod is inserted into the glass tube, making the two melt and fuse at high temperature through heating, and then pulled into an optical fiber with a specific structure and size through the drawing device. In this process, the composition difference between the glass tube and the glass rod determines the different optical characteristics of the cladding and fiber core of the final fiber.

ii) Flow path

Preformed rod preparation: Sulfur glass tubes and glass rods were prepared by CVD and the melting method. For glass pipes, ensure that the inner diameter and the pipe wall thickness are uniform; for glass rods, the diameter matches the inner diameter of glass tubes accurately, and the chemical composition of both meets the design requirements of the target optical fiber. For example, in the preparation of sulfur glass fiber for infrared optical communication, the glass tube can choose a sulfide-based glass system, while the glass bar uses sulfur glass containing specific doped elements to achieve the required fiber core optical performance.

Assembly and Heating: Carefully insert the glass rod into the glass tube to ensure that both are coaxial. The assembled prefabricated rods are then placed in the wire drawing furnace and heated at a certain temperature rate to the sulfur glass softening temperature range (generally in a few hundred degrees Celsius to more than 1000 C, depending on the composition of the glass). In the heating process, the temperature distribution should be accurately controlled to ensure that the glass pipe and the glass rod are evenly heated.

Wire Drawing: When the prefabricated rods reach a suitable softening state, they are pulled down at a steady speed through a drawing device. The drawing speed needs to be accurately adjusted according to the viscosity of the glass and the diameter of the target fiber, generally between a few meters to dozens of meters per minute. During the drawing process, the precise diameter monitoring equipment is used to monitor the fiber diameter in real time, and the drawing speed is automatically adjusted through the feedback control system to ensure the uniformity of the fiber diameter.

iii) Advantages and disadvantages and applications

Merit: The pipe-rod process is relatively simple, easy to understand, and operate. By choosing glass tubes and glass rods with different compositions, it is possible to design the fiber cladding and core structure flexibly to meet the specific optical

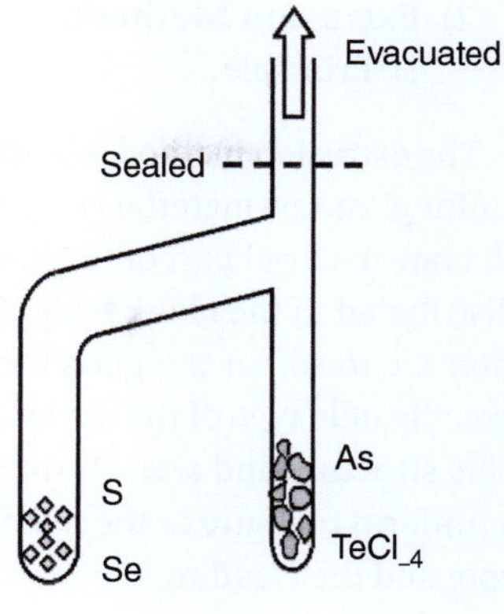

Figure 5.2 Quartz pipe structure. *Source:* Xu Yan-Tao et al. (2015) / Editorial Department of Journal of Inorganic Materials.

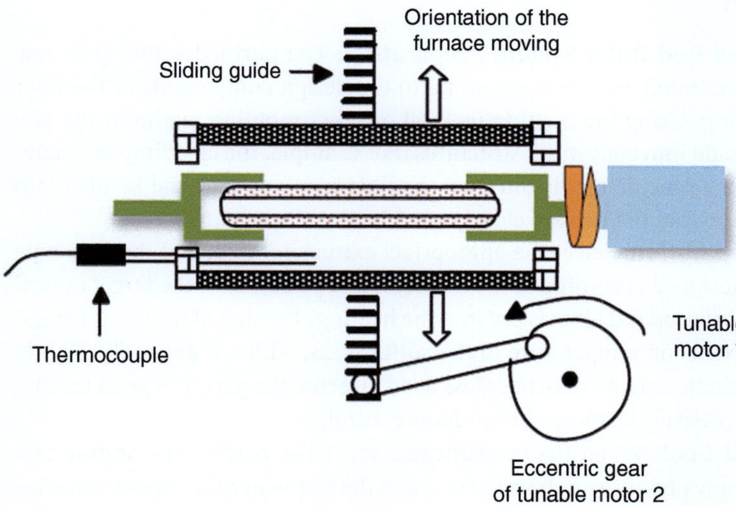

Figure 5.3 Schematic diagram of the rotary furnace structure. *Source:* Xu Yan-Tao et al. (2015) / Editorial Department of Journal of Inorganic Materials.

performance requirements of various applications. At the same time, the method has relatively low requirements for the equipment, and the cost is relatively controllable.

Shortcoming: Due to the small eccentricity of the glass tube and glass rod in the assembly process, the cladding and core of the final fiber are not concentric, which affects the uniformity of the optical performance of the fiber. In addition, during the melting and fusion process, defects such as bubbles and impurities may be generated at the interface, reducing the quality of the fiber.

Apply: It is suitable for some application fields with relatively low optical fiber performance requirements and cost sensitivity, such as short-distance infrared optical transmission and simple optical sensing. In some infrared monitoring systems in the industrial field, the sulfur glass fiber prepared by the tube rod method can meet the basic requirements of signal transmission and sensing, and the cost is low, and it has good economy (Figures 5.2 and 5.3).

2) **Extrusion Method:**
   i) Principle

The extrusion method is based on the principle of hot extrusion molding. First, the sulfur glass raw material is made into a prefabricated rod blank body, which contains the raw material part of the fiber core and the cladding, and both are concentrically distributed in the blank body. The prefabricated rod body is then put into the extrusion die to soften the glass body at high temperature and squeeze it out from the specific hole type of the die by applying pressure to form an optical fiber with a specific structure and size. During the extrusion process, the glass blank body receives a uniform pressure in the mold, thus ensuring the structural uniformity of the fiber core and the cladding.

ii) Flow path

Prefabricated Rod Billet System Preparation: The purified sulfur glass raw materials are accurately matched according to the design components of the fiber core and cladding. Using the melting method or other molding methods, the raw materials are made into concentric structures. For example, the cladding raw material is formed in a specific mold, and then the fiber core raw material is filled into the tube to ensure that the two are closely combined and concentric.

Squeeze Preparation: Select the appropriate extrusion mold, and the hole type of the mold is designed according to the diameter and structure of the target optical fiber. The prefabricated rod billet is put into the heating chamber of the extruder and heated to the extrusion temperature of the sulfur glass, which is generally slightly higher than the softening point of the glass to ensure that the glass has good fluidity but does not excessively soften and lose shape control.

Squeeze and Cool: When the prefabricated rod billet reaches the appropriate temperature, apply pressure on the billet through the piston or other pressure device to squeeze it from the mold hole type to form an optical fiber. During extrusion, the pressure is stable to ensure the diameter of the fiber. The extruded optical fiber is quickly cooled through the cooling device to prevent its deformation. The cooling mode can be air-cooled or water-cooled, specifically according to the characteristics of the optical fiber and production speed.

iii) Advantages, disadvantages, and applications

Advantages: The extrusion method can effectively ensure the concentricity of the fiber core and cladding of the optical fiber, because the whole glass body is under uniform pressure during the extrusion process, and the structural stability is good. The fibers prepared by this method have high surface quality and controllable size accuracy, which is beneficial to improve the optical performance uniformity and reliability of the fibers. At the same time, the extrusion method is suitable for mass production, and the production efficiency is relatively high.

Disadvantages: Extrusion equipment cost is high, needs to be equipped with a special extruder and high-precision mold. The process control requirements are very strict, and the small fluctuations of extrusion temperature, pressure, and speed may

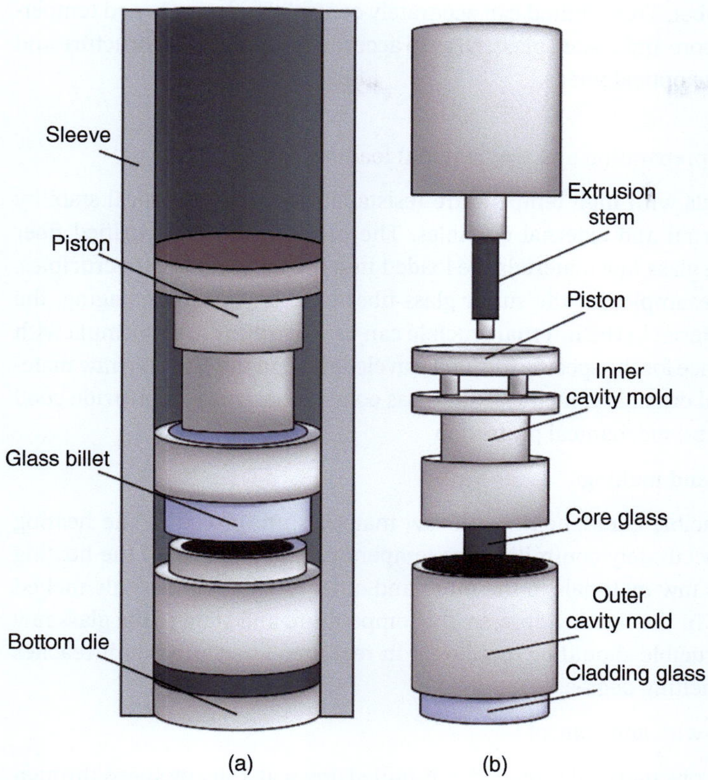

Figure 5.4 Installation of the two extrusion modes (a) DSE, (b) ISE.

affect the quality of optical fiber. In addition, due to the limitations of the mold, it is difficult to prepare optical fibers with special structures or complex shapes.

Applications: It is widely used in fields with high optical fiber performance requirements, such as long-distance optical communication, high-end optical sensing, etc. In the case of the communication of submarine optical cable, it is necessary to use high-quality and low-loss sulfur glass optical fiber. The optical fiber prepared by the extrusion method can meet its strict requirements on optical performance and structural stability, ensuring the reliability of the signal in long-distance transmission (Bruno et al. 2004, 2014) (Figure 5.4).

2) **Double Crucible Pulling Technology:**
   a) **Principle:**

The double crucible pulling technology adopts two concentric crucibles; the internal crucible holds the fiber core glass raw materials, and the external crucible holds the cladding glass raw materials. During the pulling process, the glass raw materials in the two crucibles are melted simultaneously by heating. Then, using the specially designed wire drawing device, make the fiber core glass in the inner crucible and the cladding glass in the outer crucible flow out simultaneously, and fuse at the exit to form the optical fiber. This method can accurately control the flow rate and temperature of the fiber core and coated glass, so as to accurately regulate the structure and performance of the optical fiber.

b) **Flow Path:**

i) Crucible preparation and raw material loading

Choose materials with high temperature resistance and good chemical stability to make the internal and external crucibles. The precise ratio and purified fiber core and cladding glass raw materials are loaded into the inner and outer crucibles, respectively. For example, for the sulfur glass fiber used for infrared imaging, the fiber core raw material in the internal crucible can use the sulfur glass formula with a high transmittance for the specific infrared wavelength, and the cladding raw material in the external crucible can choose the glass composition that can provide good optical isolation and mechanical protection.

ii) Heating and melting

The double crucible components with raw materials are placed in the heating furnace, and by accurately controlling the temperature distribution of the heating furnace, the glass raw materials in the inner and outer crucibles are evenly melted at the same time. In the heating process, the temperature and state of the glass raw material in the crucible should be monitored in real time to ensure that it reaches the appropriate melting degree.

iii) Pull the wire and control

When the glass raw material is melted, it is pulled down at a steady speed through the drawing device. The drawing device is equipped with a high-precision flow control and temperature monitoring system. This system can adjust the outflow speed and temperature of both the fiber core and the cladding glass in real time, ensuring uniformity during the fusion process. Additionally, diameter monitoring equipment is utilized to track the fiber diameter in real time. Through a feedback control system, adjustments are made to the drawing speed and glass flow, maintaining consistent fiber diameter throughout the process.

c) **Advantages, Disadvantages, and Applications**

Advantages: The double crucible pulling technology can realize the independent control of the fiber core and the cladding glass, and can accurately adjust the composition and proportion of both, so as to prepare the optical fiber with a complex optical structure and high performance. The fibers prepared by this method have high concentricity, clear interfaces, and stable optical characters. At the same time, due to the real-time adjustment of glass flow and temperature, the process fluctuations in the production process have a good adaptability, and product quality consistency is good.

Disadvantages: The equipment is complex and expensive, so it needs to be equipped with high-precision double crucible components, a heating system, flow control, and a temperature monitoring device. The technical level of the operator is very high, and the parameter adjustment and monitoring in the production process

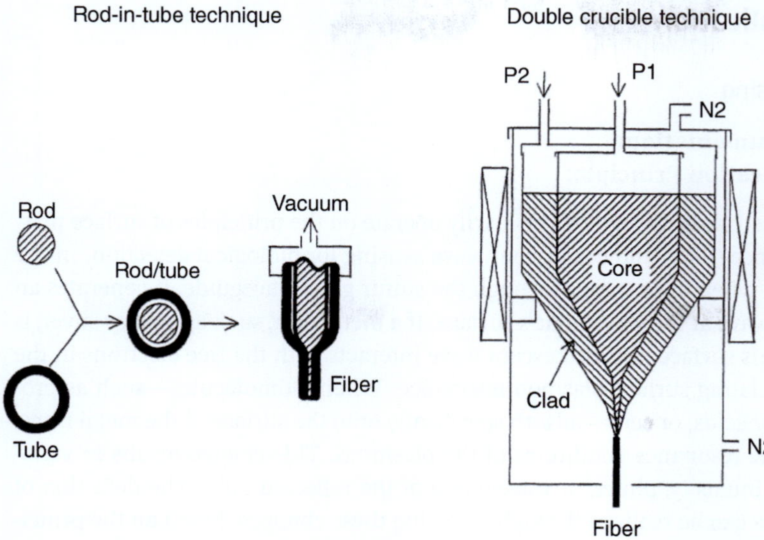

Figure 5.5 Bar-tube and double crucible processes used for fiber stretching. *Source:* Jasbinder Singh Sanghera et al. (1995) / SPIE.

require rich experience and professional knowledge. Moreover, the production efficiency of double crucible pulling technology is relatively low, limiting its large-scale application.

Applications: It is mainly used in the high-end fields with extremely high requirements for optical fiber performance, such as ultra-high-speed optical communication, high-resolution optical sensing, and quantum optics. In the ultra-high-speed optical communication system, the optical fiber needs to have extremely low loss and accurate dispersion characteristics. The sulfur glass fiber prepared by double crucible pulling technology can meet these strict requirements and guarantee the realization of high-speed and stable optical signal transmission. In quantum optical experiments, this fiber demonstrates excellent optical uniformity and stability. Additionally, the fiber prepared with this technology effectively reduces the loss of quantum signals (Figure 5.5).

Pipe and rod method, extrusion method, and double crucible pulling technology have their advantages and disadvantages in the preparation of sulfur glass fiber. The pipe-rod method has a simple process and low cost, but fiber concentricity and quality limitations; the extrusion method can ensure the fiber structure stability and surface quality, suitable for mass production, but the equipment and process requirements are high. The double crucible pulling technology can accurately control the fiber core and cladding, and the optical fiber prepared has excellent performance, but the equipment is complex, high-cost, and has low production efficiency. In the future, with the continuous progress of technology, it is expected to optimize and improve the existing preparation method, and more efficient and advanced preparation technology will be developed to further improve the performance and application value of sulfur glass fiber.

5.4 Application

5.4.1 Sensing

1) **Bioinstrumentation:**
   a) **Application Principle:**

Sulfur-based glass waveguides primarily operate on the principles of surface plasmon resonance (SPR) and evanescent wave sensing for biological detection. In the case of SPR, when light travels through the sulfur glass waveguide, it generates an evanescent wave at the waveguide's surface. If a metal film, such as gold or silver, is applied to this surface, the evanescent wave interacts with the free electrons in the metal, stimulating surface plasmon resonance. When biomolecules—such as proteins, nucleic acids, or cells—adsorb specifically onto the surface of the metal films, they alter the resonance conditions of the plasmons. This change results in variations in the intensity, phase, or wavelength of the reflected light. The detection of biomolecules can be realized through detecting these changes. Based on the principle of evanescent wave sensing, the biomolecules on the surface of the waveguide interact with the evanescent wave to change the propagation characteristics of light in the waveguide, such as the absorption, scattering, or phase change of light, so as to realize biological detection.

   b) **Advantages:**
      i) High sensitivity

Very low concentrations of biomolecules can reach picomolar (pM) or even lower concentration levels, which is critical for early disease diagnosis and trace biomarker detection. For example, in the early diagnosis of cancer, very trace tumor markers in the blood can be detected, providing a basis for early intervention.

      ii) Strong specificity

By modifying specific biorecognition molecules (such as antibodies and nucleic acid aptamers) on the surface of the waveguide, the target biomolecules can be specifically identified, reduce the interference of other biomolecules, and improve the accuracy of detection. For example, in pathogen detection, specific viruses or bacteria can be accurately identified to avoid misdiagnosis.

      iii) Real-time monitoring

It can monitor the binding process of biomolecules in real time and obtain the kinetic information of biomolecular interaction, such as binding rate and dissociation rate, which is helpful to deeply study the function and action mechanism of biomolecules. In drug development, the binding of drug molecules and target biomolecules can be monitored in real time, and drug efficacy can be evaluated.

   c) **Specific Examples:**
      i) Protein detection

In immunodetection, antibodies against specific proteins are fixed on the surface of the sulfur-bound glass waveguide. When the biological sample containing the target protein flows through the waveguide surface, the protein specifically binds to the antibody and causes SPR signal changes, thus achieving quantitative protein detection. For example, to detect the C-reactive protein in human serum for inflammation monitoring.

ii) Nucleic acid testing

Using the principle of nucleic acid hybridization, the probe complementary to the target nucleic acid sequence is fixed on the waveguide surface, and the target nucleic acid in the sample is hybridized with the probe to change the optical characters of the waveguide surface and realize nucleic acid detection. In the novel coronavirus nucleic acid test, the virus nucleic acid can be detected quickly and accurately to help the epidemic prevention and control.

1) **Fluid Detection:**
   a) **Application Principle:**

Sulfur-based glass waveguides use their sensitivity to liquid refractive index changes in liquid detection. When a liquid comes into contact with the surface of a waveguide, its refractive index can influence how light propagates within the waveguide. According to the theory of light propagation in waveguides, particularly mode coupling theory, changes in the refractive index of the liquid will affect the mode distribution and the propagation constant of the light traveling through the waveguide. By detecting these changes in optical parameters, it is possible to deduce the refractive index of the liquid, which can then be used to analyze the composition and concentration of the liquid. In addition, if the waveguide surface modification has a functional material with selective adsorption or chemical reaction of a specific liquid component, the qualitative and quantitative detection of a specific component can also be realized by detecting the change of the light signal.

   b) **Advantages:**
      i) Noninvasive detection

Without the complex pretreatment of the liquid samples, the waveguide can be tested in contact with the liquid, which reduces the sample contamination and loss, and is suitable for the detection of precious or perishable liquid samples. In the detection of biological fluids, blood and urine can be directly tested to avoid the cumbersome sample preparation process.

ii) High resolution

It can accurately detect the small change of liquid refractive index, corresponding to the detection of liquid concentration and other parameters, and high-resolution detection can be realized. In chemical analysis, the concentration changes of trace components in mixed solutions can be accurately detected for quality control and compositional analysis.

iii) Multiparametric detection

By combining different optical detection methods (such as interferometry and spectral analysis), various parameters of liquid, such as refractive index, absorption spectrum, etc., which is helpful to fully understand the characters of the liquid. In food detection, food water content, sugar concentration, and additive composition can be detected at the same time.

c) **Specific Examples:**
   i) Solution concentration detection

In the chemical laboratory, the concentration of solute in solution is detected using a sulfur glass waveguide. If the concentration of the salt solution is detected when the concentration of the salt solution changes, the refractive index change causes a change in light propagation characteristics in the waveguide. The method can accurately determine the concentration of light intensity or phase of salt solution by measuring the light intensity or phase change.

   ii) Water quality monitoring

In environmental monitoring, it is used to detect the pollutant concentration in water bodies. For example, to detect the concentration of heavy metal ions in water, by modifying the materials with selective adsorption of heavy metal ions on the surface of the waveguide, when the water sample containing heavy metal ions flows through the waveguide surface, the adsorption of heavy metal ions leads to the change of the optical characters of the waveguide, so as to realize the detection of the concentration of heavy metal ions and ensure the safety of water quality.

b) **Gas Detection:**
   i) **Application Principle:**

Sulfur glass waveguide in gas detection is mainly based on the change of the waveguide's optical characters. When gas molecules are adsorbed on the waveguide surface, the optical parameters will change the refractive index and absorption characteristics of the waveguide surface. The interaction between gas molecules and the surface of the waveguide can alter the distribution of the electron cloud on that surface, which may affect the refractive index. Additionally, certain gas molecules absorb light at specific wavelengths, increasing light absorption within the waveguide. By detecting changes in the optical parameters of the waveguide, we can improve both the specificity and sensitivity of detection on its surface.

b) **Advantages:**
   i) Quick response

The adsorption and desorption process of gas molecules on the waveguide surface is relatively fast, which can achieve a rapid response to the change of gas concentration, and can be used to monitor the dynamic change of gas concentration in the environment in real time. In the monitoring of industrial waste gas emission, gas leakage or excessive concentration can be found in time.

   ii) High sensitivity

Very low concentrations of gas can be detected, and the detection limit of some harmful gases, such as formaldehyde and hydrogen sulfide, can reach pb (one in one billion) level, which is conducive to the early detection of harmful gas pollution in the environment, and ensure human health and environmental safety.

iii) Portability

Sulfur glass waveguide devices are small and lightweight, and easy to integrate into portable devices for rapid field detection. In air quality testing, portable gas detection equipment can be developed to facilitate air quality monitoring in different places.

c) **Specific Examples:**

i) Hazardous gas detection

In the indoor air quality test, the concentration of formaldehyde gas is detected. In the sulfur system, glass waveguide surface modification has specific adsorption and reaction of formaldehyde materials. When the indoor air contains formaldehyde, formaldehyde molecules adsorb on the surface of the waveguide, causing the optical characters of the waveguide to change. Through the detection of light signal changes, the concentration of formaldehyde can be quantitatively detected to ensure the safety of the indoor living environment.

ii) Industrial gas detection

In chemical production, the concentration of chlorine gas and ammonia gas is detected. Use the sensitivity of the waveguide to these gases to monitor the gas concentration in the production process in real time, prevent gas leakage from causing safety accidents, and optimize the production process to improve production efficiency.

5.4.2 Nonlinear Effects and All-light Treatment

1) **Optical Switch Angle**

a) **The Principle of the Sulfur System Glass Photoswitch:**

The sulfur glass photoswitch is based on the optical Kerr effect in its nonlinear optical effects. The optical Kerr effect means that the refractive index of a material changes with the light intensity, i.e., $n = n_0 + n_2 I$, where n is the refractive index of the material, n_0 is the linear refractive index, n_2 is the nonlinear refractive index coefficient, and I is the light intensity. In the sulfur-system glass light switch, there is usually a control light and a signal light. When the intensity of the control light is sufficiently high, it causes a significant change in the refractive index of the sulfur-bound glass. By designing suitable waveguide structures, such as the Mach–Zehnder interferometer (MZI) structure, the phase of the signal light will change due to the glass refractive index when passing through the sulfur glass waveguide. In the MZI structure, the signal light is divided into two beams, each propagating in different arms, one of which contains a sulfur-bound glass waveguide. When the control light is applied to the sulfur glass waveguide, the phase of the signal light in the arm is changed so that the interference state of the two signal lights in the interference region is changed, thus realizing the switching control of the signal light. For example, when the control light is not on, the two-signal light interference is long, the signal light can pass smoothly; when the control light is on and reaches a certain intensity, the sulfur glass refractive index changes due to the change in the light phase of the signal, the signal light is blocked, complete the operation of the optical switch.

b) **Advantages:**

i) High-speed response

The optical Kerr effect of sulfur glass is very fast, and can achieve refractive index change in the order of 10^{-12} seconds or even femtosecond 10^{-15} seconds, which enables the optical switch based on sulfur glass to achieve ultra-high-speed switching operation. In the high-speed optical communication network, it can be quickly switched to meet the needs of high-speed data transmission and exchange, and greatly improve the communication efficiency.

ii) Low power consumption

Compared with traditional electro-optic or mechanical optical switches, sulfur glass optical switches use the nonlinear interaction of light and material, without a complex electrical or mechanical drive device, thus reducing power consumption. In large-scale optical networks, the low-power consumption characteristic of many optical switches helps to reduce the energy consumption of the entire system and reduce the operating costs.

iii) Good integration

Sulfur glass can be prepared into a waveguide structure integrated with other optical devices through a variety of micro-nano processing technologies, such as lithography and electron beam etching. This enables the optical switch to be easily integrated with other optical components, such as optical waveguides, optical detectors, and optical amplifiers, on the same chip, realizing the high integration of the optical signal processing system, reducing the system volume, and improving reliability.

c) **Application Examples:**

The sulfur glass optical switch plays a key role. The OXC system needs to flexibly switch the optical signals between different optical fiber lines to realize the routing and distribution of optical signals. The optical switch based on sulfur glass can switch the optical signal at high speed and low power consumption to ensure the efficient operation of the optical communication network. For example, in an OXC node containing multiple input and output ports, the sulfur-bound glass optical switch can quickly switch the optical signal of a specific input port to the specified output port according to the real-time requirements of the network, ensuring the accurate transmission of data and the stability of the network (Figure 5.6).

2) **Optical Nonlinear Angles Used for all-optical Signal Processing:**

a) **The Application of the Optical Nonlinear Principle in all-optical Signal Processing**

i) Four-wave mixing (FWM)

In sulfur glass, four-wave mixing occurs when multiple beams of light with different frequencies are simultaneously propagated in the waveguide. Based on the third-order nonlinear polarizability, the input three beams (frequencies omega_1, omega_2, and omega_3, respectively) generate a fourth light through

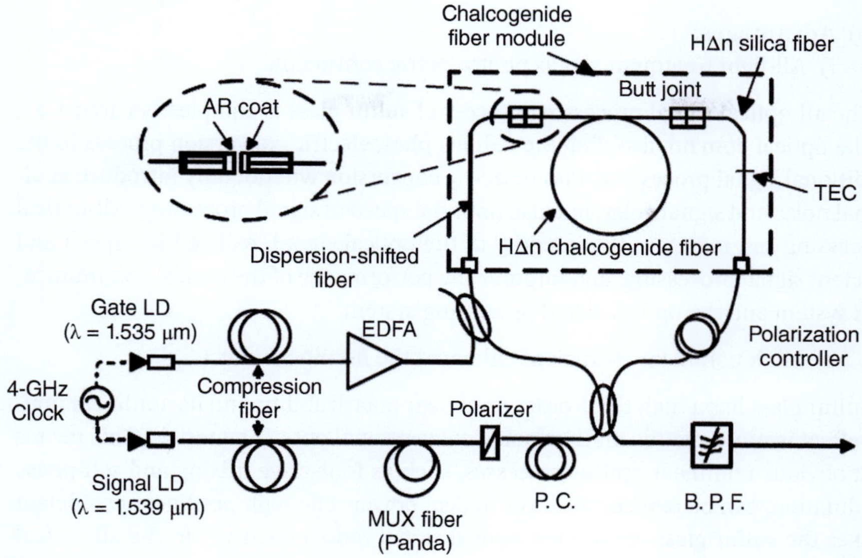

Figure 5.6 Experimental setup of the all-optical switch NOLM using As_2S_3-based optical fibers. *Source:* Adapted from Masaki Asobe et al. (1997).

nonlinear interaction, whose frequency omega_4 satisfies the energy conservation relationship omega_4=omega_1 + omega_2-omega_3. In all-optical signal processing, four-wave mixing can be used for wavelength conversion. For example, in an optical communication system, when it is necessary to convert light of a particular wavelength to meet different transmission or processing requirements, by controlling the frequency and power of the input light, the signal light can be converted to the desired wavelength, so as to realize the flexible conversion of the optical signal between different wavelength channels and improve the utilization of wavelength resources in the optical communication system.

ii) Self-phase modulation (SPM) and cross-phase modulation (XPM)

Self-phase modulation refers to the alteration of the refractive index of sulfur glass, which leads to the modulation of the light signal itself. Cross-phase modulation occurs when the intensity change of one light signal affects the phase of another light signal due to the nonlinear properties of sulfur glass. In all-optical signal processing, these phenomena can be utilized for the modulation and demodulation of optical signals. For instance, in optical soliton communication, optical solitons can be formed by balancing self-phase modulation with fiber dispersion. Additionally, the self-phase modulation characteristics of sulfur glass can be leveraged to generate and control optical solitons. In terms of optical signal demodulation, cross-phase modulation can be used to extract phase information from optical signals carrying information and realize the demodulation of optical signals, which has important applications in coherent detection of high-speed optical communication.

b) **Advantages:**
   i) All-light treatment avoids photoelectric conversion

The all-optical signal processing process of sulfur glass is completely carried out in the optical domain, avoiding the tedious photoelectric conversion process in the traditional signal processing. Photoelectric conversion will not only introduce additional noise and signal delay, but also limit the speed of signal processing. All-optical processing can maintain the integrity of the optical signal, realize high-speed and efficient signal processing, and improve the performance of the optical communication system and the optical signal processing system.

   ii) High nonlinear coefficient enhances the nonlinear effect

Sulfur glass has a high third-order nonlinear polarizability, and its nonlinear optical effect is more significant than some conventional optical materials. This means that obvious nonlinear optical processes, such as four-wave mixing and self-phase modulation, can be realized at lower optical power. The high nonlinear coefficient makes the sulfur glass control the optical signal more effectively in the all-optical signal processing, reduces the requirement for optical power, and is conducive to realizing the all-optical signal processing system with low power consumption and high performance.

   iii) Wide band applicability

Sulfur glass has good optical nonlinear performance in a wide wavelength range (from visible to infrared band). This enables the all-optical signal processing technology based on sulfur glass to adapt to the optical signals of different wavelengths and meet the needs of various application scenarios. In the field of optical communication, optical signals of different wavelengths can be processed simultaneously; in optical sensing, the nonlinear processing of different wavelengths can expand its application scope.

c) **Application Examples:**
   i) All-optical wavelength conversion and multiplexing

In the dense wave division multiplexing (DWDM) optical communication system, it is necessary to multiplex multiple optical signals of different wavelengths in an optical fiber, and de-multiplex wavelength conversion at the receiving end. The four-wave mixing effect of sulfur glass can realize efficient all-light wavelength conversion, convert the signal light of different wavelengths into the required wavelength, and facilitate the transmission and processing in a DWDM system. By accurately controlling the optical power and wavelength in the sulfur system glass waveguide, the simultaneous conversion and multiplexing of multiple wavelength channels can be realized, which greatly improves the capacity and flexibility of the optical communication system.

   ii) All-optical logic operation

A full optical logic gate can be constructed to realize full optical logic operation. For example, all-optical door circuits or gates based on four-wave mixing and cross-phase modulation. In the all-optical computing system, these all-optical logic gates can perform logical operations on optical signals to realize data processing and calculation. Due to the fast-computing speed and low energy consumption of all-optical logic, it is expected to promote the development of high-speed and low-power all-optical computing technology in the future.

With its unique optical and physical characters, sulfur glass has shown excellent application value in many cutting-edge science and technology fields. In the field of optical switching, the sulfur glass optical switch based on the optical Kerr effect changes the glass refractive index by controlling light, and controls the signal light with the help of a Mach-Zehnder interferometer. Its high-speed response characteristics can complete the refractive index change in a picosecond or even a femtosecond, is suitable for high-speed optical communication network, can quickly switch optical signal path, can greatly improve communication efficiency, has low power consumption and no complex drive device, can reduce energy consumption and operation cost, has good integration with other optical devices and high integration of optical signal processing system, and can reduce the system volume and enhance reliability.

The optical nonlinear effect of sulfur glass plays a key role in all-optical signal processing. Four-wave mixing can be used for wavelength conversion, realizing the light conversion of different wavelengths and improving the cross-phase modulation can be used for modulation and demodulation of optical signals, such as an important application in optical soliton communication and coherence detection. All-optical processing avoids the noise and delay of photoelectric conversion. Its high nonlinear coefficient makes the nonlinear effect significant even under the low light signal. With an applicable band-wide range, it can meet a variety of application scenarios for different wavelength light signal processing requirements, such as in dense wave points multiplexing optical communication systems of all optical wavelength conversion and the use, and all optical computing systems of all optical logic operation.

In addition, sulfur glass also performs well in the field of biological, liquid, and gas detection. In biological testing, surface plasmon resonance and evanescent wave sensing principles enable high-sensitivity, specific, and real-time monitoring of bio molecules. They can be applied to protein and nucleic acid detection, supporting early disease diagnosis and drug development.

In fluid testing, these principles (or specific detection technologies based on them) achieve non-invasive, high-resolution multi parameter detection by leveraging their sensitivity to changes in liquid refractive index and interactions with modified functional materials. This allows for solution concentration detection and water quality monitoring.

In gas detection, they (or gas sensors based on these principles) enable the detection of harmful and industrial gases by exploiting the optical property changes caused by gas adsorption. Their fast response, high sensitivity, and portability help ensure environmental safety and stable industrial production.

6

Integrated Optics and On-Chip Photonic Devices of Sulfide Compounds

6.1 Introduction

Chalcogenide materials, renowned for their exceptional optical properties and reversible phase transition capabilities, have emerged as a transformative force in integrated photonics. Phase-change materials (PCMs) such as Sb_2Se_3, Ge–Sb–Te, and $Ge_2Sb_2Te_5$ (GST) exhibit non-volatile behavior, ultrahigh refractive index contrast ($\Delta n > 1.0$), and sub-nanosecond switching speeds, positioning them as cornerstones for next-generation photonic devices. These materials enable dynamic reconfiguration of optical functionalities through external stimuli—electrical, thermal, or optical pulses—transitioning between amorphous and crystalline states. This unique attribute unlocks unprecedented opportunities for low-power, high-density photonic memory and adaptive optical systems, addressing critical limitations of conventional static silicon photonic platforms. In the realm of optical data storage, chalcogenides have already revolutionized commercial technologies, such as Blu-ray discs, where the contrast in reflectivity between amorphous and crystalline states encodes binary data. Recent advancements extend this paradigm to non-volatile photonic memory, where PCMs integrated with silicon waveguides enable localized modulation and persistent storage of optical signals. Such capabilities are pivotal for emerging applications like optical computing and photonic neural networks, where ultrafast, energy-efficient data processing is paramount. Beyond storage, chalcogenides are reshaping dynamic displays through structural color engineering. For instance, Ge–Sb–Se–Te-based nanostructures exploit optical interference effects in phase-tunable metasurfaces, enabling pixel-level color switching with high resolution and minimal power consumption—a breakthrough for on-chip displays in augmented reality (AR) and ultracompact photonic systems.

The convergence of chalcogenide PCMs with integrated photonics has ushered in a transformative era for reconfigurable optical systems and next-generation photonic devices. Chalcogenide alloys, notably GST and Sb–Se systems, exhibit extraordinary optoelectronic properties—including non-volatile, reversible phase transitions between amorphous and crystalline states—enabling unprecedented control over light-matter interactions at subwavelength scales. This unique combination of ultrafast switching kinetics (sub-ns), high refractive index contrast

Chalcogenide Nanophotonics, First Edition. Tun Cao, Ying Su, Yinghan Li, and Hongze Gao.

($\Delta n > 1.0$), and complementary metal oxide semiconductor (CMOS)-compatible processing positions chalcogenides as a cornerstone for developing adaptive photonic integrated circuits (PICs).

Central to this technological revolution is the paradigm-shifting application of PCMs in optical data storage and color pixel displays. Phase-change optical storage leverages the non-volatile refractive index modulation of GST films to achieve multilevel, diffraction-limited data encoding, surpassing conventional charge-based memory in energy efficiency (pJ/bit) and storage density (>1 Tb/in^2). Concurrently, chalcogenide-based structural color displays exploit tunable plasmonic resonances and Mie scattering effects to realize full-color gamut reproduction without organic dyes, offering enhanced stability (>10^6 switching cycles) and sub-micron pixel resolution—a critical advancement for AR visors and ultra-high-definition microdisplays.

This chapter systematically explores the fundamental physics, fabrication methodologies, and device architectures underpinning chalcogenide-enabled photonic systems. Through heterogeneous integration strategies that bridge silicon photonics with chalcogenide functional layers, we demonstrate practical implementations spanning non-volatile optical memories, wavelength-selective switches, and bio-inspired neuromorphic photonic networks. The chapters dissect these breakthroughs while addressing persistent challenges in cyclability, thermal cross-talk mitigation, and nanoscale phase uniformity—charting a roadmap toward energy-efficient, intelligent photonic systems that redefine the boundaries of optical computing, sensing, and visualization technologies.

6.2 Chalcogenide Photonic Memory

Storing information has been pivotal in humanity's advancement towards sophisticated knowledge and technology. From ancient petroglyphs to modern semiconductor transistors, the underlying objective has consistently been to preserve information indefinitely. The rapid evolution of semiconductors and magnetic storage media over the past century, coupled with the transition from analog to digital formats, has enabled us to achieve storage densities exceeding 10 Gb per square millimeter. This transition not only improved storage capabilities but also facilitated data processing through arithmetic and logic operations at speeds surpassing 1 GHz. Ongoing advancements in semiconductor technology and nanofabrication have fueled the development of systems based on processing transistors, leading to highly efficient, portable, and successful digital computing devices.

However, current technologies predominantly adhere to the digital computing framework proposed by Von Neumann, which distinctly separates data storage and processing units. This architecture, while efficient, creates a significant challenge known as the "Von Neumann bottleneck," where data must be constantly transferred between these two units, thus slowing down the process and creating a bottleneck in the data communication channel. To mitigate this, various solutions have been proposed, with fiber optic interconnects being the most prevalent in today's data centers. These interconnects utilize light to transfer data at the highest possible speeds, reducing the time data spends in transit.

Despite these advancements, the core operations within data centers, between processors and memory banks, still rely on electrical signals. This necessitates electro-optical conversions at both ends, which incurs an energy penalty. Innovations such as on-chip optical links and scalable 3D silicon photonic-electronic integrated circuits are emerging, yet the energy demands of electro-optical conversions remain a significant challenge, especially with the growing computational requirements of contemporary data centers. These centers also face issues with heat dissipation, both internally and externally, due to the energy-intensive nature of their operations, necessitating advanced cooling systems.

A highly viable solution to tackle these obstacles lies in designing architectures where memory components can be directly read using optical methods. This approach bypasses the need for an intermediate electro-optical conversion step, thereby mitigating excessive heat generation. Another effective tactic involves embedding computational processes directly within the memory structure, which curtails unnecessary data transfers and mirrors the brain's remarkable computational efficiency. These advancements pave the way for non-Von Neumann architectures, which could potentially eradicate the memory access bottleneck and usher in a new era of more streamlined data processing systems. The advancement of photonic memories, particularly those capable of multilevel storage, assumes paramount importance in this landscape, especially for analog, brain-inspired computing frameworks.

Electronic and photonic devices are essential to the operation of modern electronic systems, serving to manipulate electrons and photons for information transmission. Electronic devices, which work with electrons and their associated fields, are integral to the creation of memories, oscillators, displays, amplifiers, and detectors within electronic circuits. Similarly, photonic devices utilize photons to transmit information through optical fibers and waveguides, performing comparable functions to their electronic counterparts. The rivalry between electronic and photonic technologies has intensified recently, as both fields have seen significant advancements. Electronics and photonics have both experienced rapid growth, notably in fiber transmission rates (as per Keck's law) and semiconductor storage densities (according to Moore's law). Currently, photon-based information transmission is often favored because it offers a higher frequency bandwidth, providing advantages over electronic systems. While electronic systems excel in miniaturization and integration, reaching scales smaller than the diffraction limit of light, they are also highly effective in data storage and processing. This capability is supported by advanced transistor-transistor logic and the use of field-programmable gate arrays, which allow for versatile and reconfigurable functionalities suitable for a variety of applications. In contrast, to remain competitive, photonic devices must evolve to become more reconfigurable and reprogrammable, achieving higher levels of integration and miniaturization. This necessitates the development of materials with tunable optical properties. This review delves into state-of-the-art photonic memories, which incorporate rewritable mechanisms into traditional computing frameworks.

We explore devices and architectures that support data storage with optical readout, whether in free-space, on chips, or through optical fibers. Our examination covers a variety of architectures that enable access to optical memory cells, including random memory addressing and content-based addressing. We utilize wavelength-division multiplexing (WDM) to expand into multibit two-dimensional memory arrays. Additionally, we discuss optical architectures for digital, in-memory, and neuromorphic computing, emphasizing the photonic memories that support these technologies. Our goal is to provide insights into emerging computing scenarios that could benefit from photonic memory arrays and outline a comprehensive strategy for integrating tunable nanophotonics into data storage and computing environments.

Finally, we consider the future potential of photonic memories, addressing the challenges they face and the research opportunities ahead. Unlike electronic memories, photonic memories rely on light-matter interactions due to the non-interacting nature of photons under normal conditions. This review also categorizes photonic memories based on their material mechanisms, the nature of optical measurement, and performance metrics, distinguishing between volatile and non-volatile types based on their speed and power needs. Phase Change Materials (PCMs) are highlighted as a promising solution for reprogramming photonic circuits. When PCMs like GST are exposed to optical or electrical pulses, they undergo a reversible change between crystalline (SET) and amorphous (RESET) states, significantly altering their optical properties and electrical resistivity. This variability is pivotal to the adaptability of PCM-based photonic systems, which can endure over 10 billion switching cycles with switch times under 500 ps and can be finely tuned through interface engineering.

6.2.1 Design of Chalcogenide-Based Photonic Memory

PCMs like $Ge_2Sb_2Te_5$ (GST) revolutionizes through their unique combination of non-volatile phase transitions and substantial refractive index modulation ($\Delta n > 1$), enabling reconfigurable optical switches, memory arrays, and neuromorphic computing systems. Their operational advantages stem from abrupt optical property changes during phase transitions, multistate stability, reversible optical/electrical switching capability, and theoretical endurance reaching 10^{15} cycles, complemented by sub-nanosecond response times and minimal energy requirements (~1.2 aJ/nm^3). Standard deposition and patterning techniques ensure seamless integration with existing CMOS processes, while materials like GST and Ag-In-Sb-Te (AIST) benefit from established electronic memory manufacturing infrastructure. However, practical implementation challenges persist, including sensitive phase transition control where improper power or duration can cause material damage, and actual cycling performance (10^9–10^{12} cycles) falling short of theoretical limits-making them unsuitable for high-frequency applications but ideal for reconfigurable systems like photonic field programmable gate array (FPGAs) where frequent switching isn't required. The inherent stochasticity in multilevel operation due to unpredictable atomic rearrangement after each phase change can be addressed through discrete PCM segment architectures, where each transmission level corresponds to specific patch states. While GST exhibits problematic optical losses in visible/near-IR ranges, requiring careful device engineering at telecom wavelengths (1550 nm), emerging wide-bandgap alternatives promise low-loss operation across broader spectral ranges. This comprehensive analysis explores PCM material properties, device optimization strategies, and application-specific performance tradeoffs, while identifying key research challenges in switching reliability, loss reduction, and scalable integration-providing a roadmap for future developments in hybrid material systems and novel PCM discovery for next-generation silicon photonics.

6.2.2 Design of Photon Memory Based on Sulfide Compounds

PCMs are chalcogenides that can exist in two distinct micro-structural forms: crystalline and amorphous, both stable at room temperature. These forms can be rapidly interchanged within nanoseconds using thermal pulses generated by either electrical or optical methods. Specifically, converting a crystalline PCM to an amorphous state involves heating it above its liquidus temperature (T_l) using Joule heating via a brief voltage or laser pulse lasting tens of nanoseconds (Lencer et al. 2011). The rapid cooling from the melted region to its surroundings causes the material to transition into the amorphous state, a process known as RESET. Although crystallization is thermodynamically favored, it is slowed by low atomic mobility. To induce the transition of the PCM from its amorphous state back to the crystalline form, the PCM is subjected to a temperature range situated between the glass transition temperature (T_g) and T_l. This is achieved by applying a relatively long-duration pulse, which can extend for several hundred nanoseconds. Over time, this process facilitates atomic diffusion. This particular transformation is commonly known as the SET process.

Phase changes significantly alter optical constants and electrical resistivity due to variations in the electronic bandgap, creating distinct differences in properties between the two states. These characteristics make PCMs ideal for memory applications; initially, they were used in rewritable compact disks and later in random access memory. Selection criteria for effective PCMs have historically emphasized low electrical resistivity for efficient Joule heating in memory cells (Lankhorst et al. 2005) and high optical absorption for laser writing in disks. Researchers identified most functional PCMs within the Ge–Te–Sb ternary phase diagram (Figure 6.1b). The first group of materials is situated along the pseudobinary GeTe-Sb_2Te_3 tie line and includes compounds such as $GeSb_2Te_4$ and $Ge_2Sb_2Te_5$. Later developments included other groups, such as Ge_xSb_1-x and doped Sb_2Te, which were designed to enhance switching speed (Chen et al. 2006).

In recent years, integrating PCMs with nanophotonic structures has revolutionized reconfigurable integrated photonics. For these applications, traditional criteria for selecting PCMs, such as those used in random access memory and rewritable disks like GST, are no longer adequate (Chen et al. 2006). Low optical loss has become critical due to the proximity of optical modes to PCMs in nanophotonic waveguides. For instance, GST experiences significant band-to-band absorption in the visible to near-infrared spectrum, as well as free carrier absorption in the mid

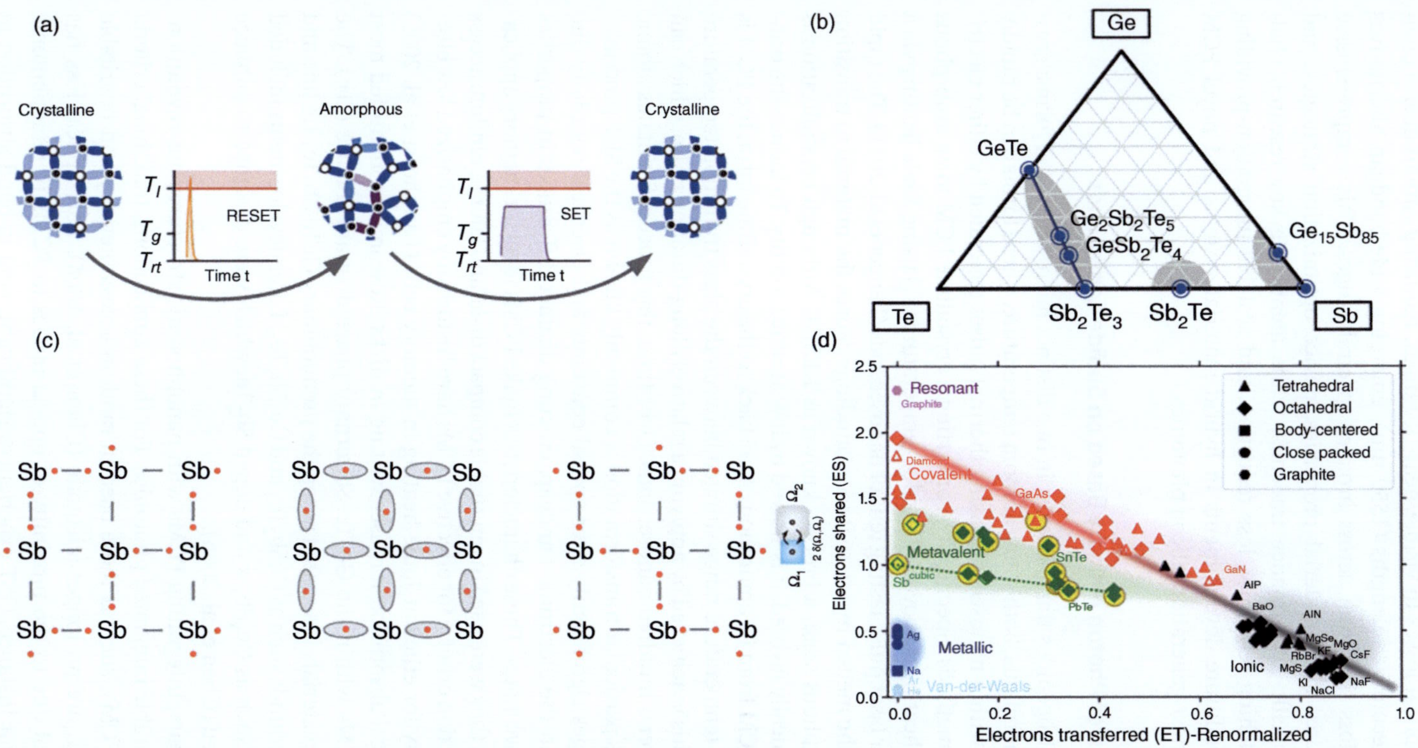

Figure 6.1 Phase change materials for integrated photonic applications. (a) The conditions for reversible phase transition of PCMs and the bonding characteristics in amorphous and crystalline states. (b) Ge–Te–Sb Ternary phase diagram, where most of the functional PCMs for random access memory applications were found. (c) Resonant bonding in crystalline PCMs. (d) Classifying the bonding types based on the number of shared electrons and transferred electrons in crystalline materials. Most PCMs are found to be metavalent. *Source:* Fang et al. (2021) / John Wiley & Sons.

Table 6.1 Comparison of SB_2S_3, SB_2Se_3. GST, GeTe, and GSST.

	633 nm				1310 nm				1550 nm			
	Δn	Δk	$\frac{\Delta n}{k_a}$	$\frac{\Delta n}{k_c}$	Δn	Δk	$\frac{\Delta n}{k_a}$	$\frac{\Delta n}{k_c}$	Δn	Δk	$\frac{\Delta n}{k_a}$	$\frac{\Delta n}{k_c}$
GeTe	1.15	1.50	1.28	0.48	1.86	0.42	30.16	3.90	1.83	0.24	53.94	6.79
GST	0.16	2.54	0.08	0.04	2.82	1.20	5.93	1.69	2.74	1.07	137.00	2.51
GSST	0.62	1.75	0.77	0.24	1.81	0.63	86.54	2.79	1.75	0.42	U.D.	4.17
Sb_2Se_3	1.50	0.83	1.86	0.92	1.08	0.00	U.D.	U.D.	1.02	0.00	U.D.	U.D.
Sb_2S_3	0.94	0.21	U.D.	4.48	0.56	0.00	U.D.	U.D.	0.57	0.00	U.D.	U.D.

Source: Abdollahramezani et al. (2020); Fang et al. (2021).

and long-wave infrared ranges. As a result, selection criteria must now prioritize materials with low optical absorption and a high refractive index change. These properties can be quantified using figures of merit (FOMs), where Δn represents the refractive index contrast, and k_a and k_c denote the extinction coefficients in the amorphous and crystalline states, respectively (Zhu et al. 2018; Ríos et al. 2015). Emerging PCMs with high FOMs across important wavelengths have been proposed. Table 6.1 compares these low-loss PCMs with GST at wavelengths such as 633 nm (nitrogen vacancy centers in diamond), 1310 nm (telecommunication O-band), and 1550 nm (telecommunication C-band).

The pronounced alteration in the complex refractive index of PCMs during phase transitions can be ascribed to a transformation in bonding patterns. In the amorphous state, atoms establish covalent bonds with adjacent atoms, adhering to the 8-N rule, which is determined by the number of valence electrons. When crystallization occurs, atoms are organized into a rigid structure that necessitates electron sharing, thus deviating from the 8-N rule. Initially, it was believed that valence electrons in crystalline PCMs resonated across neighboring bonds, creating a hybrid wave function akin to that in benzene rings, as depicted in Figure 6.1c. This phenomenon was dubbed "resonant bonding." However, recent research indicates that this bonding is of a distinct nature, falling somewhere between metallic and covalent bonds, and has been termed "metavalent bonding." The electron delocalization observed in crystalline PCMs enhances electronic polarization. As a result, there are substantial changes in the optical dielectric constants during phase transitions. Photonic switches based on PCMs operate by modulating the optical phase, absorption, or a combination of both (Feldmann et al. 2017; Ríos et al. 2019; Stegmaier et al. 2017). For example, GST at 1550 nm operates via both mechanisms due to its large refractive index ($\Delta n \sim 2.7$) and extinction coefficient ($\Delta k \sim 1$) changes. GSST, a PCM derived by substituting Te with Se in GST, operates similarly and has zero loss in the amorphous state at 1550 nm. GSST has lower absorption loss, but its change in refractive index (Δn) is reduced by over 36% compared to GST. This decrease in Δn means that devices require a larger footprint and higher switching energy due to the longer PCM and the higher crystallization temperature. As a result, GSST is more suitable for large-scale PICs where minimizing insertion loss is critical. In contrast, GST is the better choice when device size and energy consumption are more important factors. Another PCM, GeTe, shows a significant change in refractive index in the visible range, making it useful for tunable reflective displays. However, it has been less explored in integrated photonics because its FOM near 1550 nm is lower than that of GST or GSST (Figure 6.1d).

Absorption modulation proves advantageous for cavity-based switches. In these devices, alterations in the optical phase and loss contribute to an improved extinction ratio. However, it falls short when it comes to phase-only modulation applications, such as Mach-Zehnder Interferometers (MZIs) or switches based on directional couplers. To tackle this challenge, newly developed PCMs, namely Sb_2S_3 and Sb_2Se_3, have emerged. These materials are characterized by their near-zero loss in both their amorphous and crystalline states, making them a promising solution for phase-only modulation scenarios (Dong et al. 2019; Delaney et al. 2020, 2021). Sb_2Se_3 stands out as it exhibits no loss at both 1310 nm and 1550 nm wavelengths, coupled with Δn of approximately 1. This characteristic renders it highly suitable for non-volatile phase-only control in near-infrared silicon photonics. Sb_2S_3, a wide-bandgap PCM, boasts a broad transparency window in its amorphous state, spanning from 610 nm to the near-infrared region. Among the four PCMs under consideration, it is unique in enabling low-loss operation at 633 nm, where it demonstrates a high refractive index contrast of 0.94. However, Sb_2S_3 experiences a significant loss in its crystalline state near 633 nm. Consequently, device engineering is necessary to mitigate insertion loss. In summary, at 1550 nm, GST and GSST are well-suited for applications that leverage loss in one state, such as microring switches and photonic memories. On the other hand, Sb_2Se_3 is the preferred choice for phase shifter applications. At 1310 nm, GST and GSST are too lossy, even in their amorphous states, leaving Sb_2Se_3 and Sb_2S_3 as the only feasible options. Given that Sb_2Se_3 has a larger Δn compared to Sb_2S_3, it continues to be the top pick for phase-only control applications. Sb_2S_3 is the best choice for visible photonics, showing the highest FOMs among the four at 633 nm. Finally, not all chalcogenides are PCMs even if they exhibit multiple states. For instance, GeSe, with a transparency window in the amorphous state from 800 nm to near-infrared and $\Delta n \sim 0.3$ near 1550 nm when annealed at 400 °C, is attractive for low-loss operation.

Recent investigations have revealed that certain PCMs, such as GeSe, undergo irreversible transitions when subjected to optical switching with laser pulses, failing to revert to the amorphous state. Similar independent studies have also observed this unidirectional transition. In these studies, GeSe underwent thermal annealing, which transformed it into a more stable amorphous state. During this process, new bonds were formed, leading to a change in its refractive index without triggering crystallization. We also explore various tuning methods for PCM-based integrated photonics, each with distinct advantages and limitations.

Thermal annealing methods, including rapid temperature annealing or hot plate annealing in an inert atmosphere, typically lead to PCM crystallization. Although effective for large areas, this approach is slow and irreversible since it doesn't allow for re-amorphization. Moreover, it lacks the capability for individual PCM addressing, limiting system scalability. Conversely, free-space laser switching, a technique dating back to the early development of PCMs for rewritable disks, utilizes the PCM's inherent bandgap in the visible spectrum to absorb specific laser light wavelengths, such as blue or green. This absorption causes self-heating, leading to either crystallization or amorphization based on the laser's power and pulse duration. While femtosecond lasers can achieve switching speeds around GHz, this method's requirement for precise alignment and bulky optics hampers its applicability for large-scale PICs. Additionally, its effectiveness decreases with transparent PCMs like Sb_2S_3, which exhibit low optical absorption.

Another approach involves using light confined within a waveguide through evanescent coupling, which offers better scalability than free-space lasers. For instance, a 16×16 array of GST-based directional coupler switches has been demonstrated, allowing individual control over each switch. However, further increasing computational density becomes challenging as light routing through the device grows increasingly complex. Similar to free-space laser switching, on-chip optical switching heavily depends on the PCM's absorption characteristics. Large PCM patches, such as those made of GST, often fail to switch completely due to light attenuation within the PCM itself, resulting in insufficient energy at the PCM's far end to induce a phase transition.

Electrical switching presents an alternative, either by integrating PCMs into a circuit or using an external heater. When part of a circuit, the PCM undergoes phase changes due to Joule heating. This method, known as "memory switching," was originally developed for phase change memory applications. Enhancements feature the embedding of the PCM in a plasmonic nanogap formed between tapered waveguides, with metal electrodes positioned close together to minimize resistance. However, this arrangement can lead to significant ohmic losses due to the proximity of the metal to the waveguide mode. Using transparent conductors such as indium tin oxide (ITO), fluorine-doped tin oxide (FTO), and graphene can help reduce these losses. Yet, memory switching's major drawback is its limited switching volume due to filamentation, resulting in minimal changes to the optical mode—typically only about 0.5% in transmission—rendering it impractical for applications like optical computing or photonic memories. A vertically structured switch using doped silicon and ITO can achieve greater optical contrast (70%) by switching a 500 nm diameter GST patch, though the thick ITO cladding may induce substantial insertion losses.

For larger PCM patches, using external transparent heaters offers an effective solution. These heaters can be created by doping silicon, depositing transparent conductors, or transferring graphene onto waveguides. They act like miniature hotplates, providing Joule heating to the PCM. This approach is particularly beneficial for materials such as GSST, which are much more resistant than GST. As a result, GSST requires a threshold voltage that is approximately 100 times higher than the voltage needed for GST, greatly exceeding typical CMOS driving voltages. Therefore, electrical switching with an external heater is considered the most suitable option for future non-volatile silicon photonics. This method is favored for its speed, energy efficiency, low-loss, high scalability, and compatibility with various PCMs, allowing for individual pixel addressing that goes beyond the limitations of diffraction. The advantages and disadvantages of each tuning method are summarized in Table 6.2.

Table 6.2 Comparison of different tuning mechanisms in integrated photonics based on PCM.

	Switching speed	Energy efficiency	Insertion loss	Scalable?	Switch large-area PCMs?	Local switching	Multi-level operation	Compatible with transparent PCMs
Thermal annealing	slow(~mins)	low	Zero	×	✓	×	×	✓
Free-space laser	~GHz	high	Zero	×	✓	✓ (limited by diffraction)	✓	Yes; in absorptive region
On-chip optical switching	~MHz	high	Zero	○	×	✓	✓	Yes, in absorptive region
On-chip electrical memory switching	~MHz	high	Low with T.C.H	✓	×	✓	✓	Yes; if conductive
On-chip electrical heating	~MHz	high	Low with T.C.H.	✓	✓	✓	✓	✓

Source: Abdollahramezani et al. (2020); Fang et al. (2021).

GST-based PCMs boast a remarkable set of features. In their initial amorphous state, GST can be crystallized (SET) and then returned to the amorphous state (RESET), showcasing significant differences in optical constants and a dramatic variation in electrical resistivity. This dual-phase capability enables rapid switching between distinct electrical behaviors, making these PCMs ideal for potential applications in electronic FLASH memories. Already utilized as the active layer in rewritable optical data storage, their non-volatile attributes, swift read/write capabilities, scalability, and durable performance position chalcogenide PCMs as promising candidates for universal memory systems, where processing and storage occur simultaneously.

The process of reversing the amorphous to crystalline phase transition in GST involves heating it beyond its crystallization point. Reamorphization is achieved by melting the GST and then quickly cooling it, locking the atoms in a disordered state. This is typically done using nanosecond or picosecond lasers or electrical pulses to heat the GST past its melting point. A brief laser pulse excites a dense population of electron–hole pairs in the GST, raising its temperature sufficiently to melt. The alloy is then rapidly cooled at a rate of approximately 10^9 K, solidifying it into an amorphous structure. This quick quenching, facilitated by a short burst of thermal energy followed by rapid cessation, effectively freezes the molten material into a disordered state.

Such optical switching allows for swift and reversible transitions between the amorphous and crystalline states of GST, which differ markedly in their electronic structure. GST alloys are rich in Group 16 elements like tellurium (Te), selenium (Se), oxygen (O), polonium (Po), and sulfur (S), contributing to their unique optical properties. In particular, GST alloys that minimize ionic character and hybridization exhibit stark contrasts in chemical bonding between their ordered and disordered states, as depicted in Figure 6.1a.

In the crystalline form, these chalcogenide PCMs exhibit resonant bonding, characterized by electron delocalization across p–p bonding orbitals. This electron sharing across multiple crystallographic units occurs because there are not enough valence electrons per atom to complete all possible bonds in their cubic structure. Such delocalization means electrons are loosely bound to atoms, making them readily polarizable by electric fields. This phenomenon, explained by the Lorentz-Lorentz relation, leads to a significantly high refractive index due to enhanced molecular polarizability. Conversely, the amorphous state results from thermal disruption of this crystalline structure. Heat-induced atomic vibrations diminish the overlap of p–p orbitals, causing electrons to localize and reducing their polarizability, which in turn decreases the refractive index.

Chalcogenide PCMs positioned along the Sb–Te–GeTe pseudo-binary tie-line are known for their high optical absorption from the visible to near-infrared spectrum, presenting a challenge for their use in tunable photonic devices. To mitigate this, Strand et al. have successfully reduced the imaginary part of the complex dielectric constant to below 0.01 at 1550 nm by alloying chalcogenides with a small amount of other unspecified elements, enhancing their suitability for broader applications.

Developing photonic devices using GST-based PCMs requires careful consideration of several key characteristics. Previous research has confirmed that the optical behavior of chalcogenides can be modulated by transitioning between their disordered and ordered structural phases. Equally critical is the switching rate, which can achieve remarkable speeds. In electrical phase-change random access memory (PCRAM) prototypes utilizing GST225, crystallization has been observed in as little as 500 ps, whereas laser-induced amorphization completes in just 5 ps. While the ordering process is notably slower than disordering, these nanosecond-scale transitions still vastly outperform alternative optical materials like VO_2 and liquid crystals.

Recent advancements have also explored PCMs for non-volatile electronic memory applications, leveraging the stability of their metastable phases for data storage. In such systems, brief electrical pulses induce rapid switching. A major obstacle in improving data processing and transfer lies in the rate of information movement, not just across long distances but also within integrated systems. Chalcogenides and certain PCMs have been employed in optical fibers for decades, but only in the last 10 years has on-chip photonics seen substantial breakthroughs. The expansion of micro- and nanofabrication capabilities has enabled the development of integrated waveguides and diverse photonic components.

Optical information transfer is highly attractive due to its immense bandwidth and multiplexing potential. However, challenges like residual interference and photon dynamics complicate miniaturization. Both computational operations—such as arithmetic functions in processors—and data storage are fundamental for photonic computing. A pressing limitation in boosting computational speeds is the data transfer bottleneck between memory and processing units in conventional von Neumann architectures. One promising approach involves enhancing on-chip bandwidth through integrated photonic circuits. A notable drawback of current photonic systems is the lack of high-performance, non-volatile optical memory that can be seamlessly integrated at the chip level. Such memory must maintain data integrity without power, necessitating tight electronic-photonic integration. Existing solutions often involve complex designs, yet their operating speeds remain limited to the megahertz range—far slower than modern electronic memories, which achieve multigigahertz access rates.

Integrating PCMs with photonic circuits offers a compelling solution, enabling ultrafast, low-loss optical functionalities. Most materials used in integrated optics lack dynamic tunability, but PCMs have emerged as strong candidates. Recent studies demonstrate their successful evanescent coupling with nanophotonic circuits to create tunable filters and memory elements. For instance, coupling light to GST allows transmission modulation by switching its phase between amorphous and crystalline states. Here, the waveguide's optical behavior depends on GST's phase, which can be altered via high-intensity pulses—mirroring the operation of electronic phase-change memories. A key design distinction arises from GST's strong optical absorption, which would cause significant signal loss if placed directly in the light path. Instead, evanescent coupling to the waveguide minimizes attenuation by exploiting changes in the PCM's complex refractive index. Since crystalline PCMs exhibit metal-like absorption even at sub-bandgap wavelengths, phase transitions modify both the real (n) and imaginary (k) components of the refractive index, influencing evanescent interaction.

This approach has enabled data storage and retrieval directly on a photonic chip. Initial experiments achieved eight distinct transmission levels rather than simple binary storage, demonstrating the ability to transition freely between any two states—a capability surpassing traditional electronic phase-change memory. Such precise and repeatable switching is critical for reliable long-term storage, representing decades of progress in understanding PCM behavior. By selectively controlling the material's crystalline fraction—the proportion of ordered to disordered regions—the effective refractive index can be tuned, directly influencing waveguide attenuation. In essence, information is encoded in the material's phase. Since these materials were originally optimized for low-power data storage, operations require merely picowatt-level energy.

Despite its promise, the technology remains in its early stages. Key advancements are still needed, particularly in reducing energy consumption. While current power efficiency is competitive, achieving reliable multilevel storage demands greater sensitivity; so far, the lowest energy demonstrated only supports single-level writing. Nevertheless, the potential applications justify continued investment. For instance, researchers have successfully implemented wavelength multiplexing, enabling simultaneous read-write operations at different wavelengths—a feature unattainable in conventional electronic memory. This could dramatically accelerate data throughput with optimized algorithms. Additionally, photonic memories address the von Neumann bottleneck by eliminating delays between processing and storage units. Future integration with non-Von Neumann architectures could yield fully photonic processors. PCMs also hold promise for on-chip optical routing, a cornerstone of all-photonic systems. Innovative designs combining reconfigurable waveguides and plasmonic switching may further miniaturize these components. Progress in these areas will be vital for practical implementations of phase-change photonic memory and next-generation processing chips.

6.2.3 Practical Application Cases

This section examines PCM-integrated waveguide switching devices, which fall into three primary configurations: direct waveguide integration, MZIs, and directional couplers. The fundamental operating principle involves modifying waveguide transmission through controlled variations in the PCM's extinction coefficient. Initial breakthroughs by Ríos et al. established multilevel transmission (eight distinct states) in Si_3N_4 waveguides at 1550 nm using GST elements (Delaney et al. 2020). The switching mechanism employs evanescent-coupled optical pumping of GST segments (Figure 6.2a, top; Li et al. 2019). Subsequent developments introduced wavelength-selective operation through microring resonators of varying dimensions. Recent advancements have dramatically enhanced this photonic memory architecture, achieving 34 distinct transmission levels (5-bit capacity) within a single Si_3N_4 memory cell (Delaney et al. 2021). The technology has been successfully transferred to Silicon on Insulator (SOI) platforms (Ríos et al. 2015),

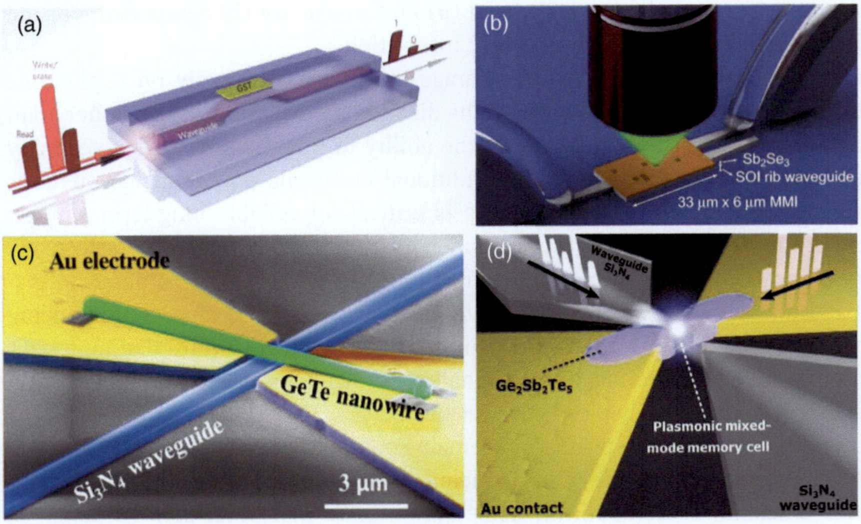

Figure 6.2 (a) Switching PCMs on Si_3N_4 waveguide via evanescent coupling. Multiwavelength write and erase using microring resonators of different radii (bottom). (b) Optical switching of low-loss Sb_2Se_3 integrated with an MZI switch. Multilevel operation and high cyclability are demonstrated with near-zero loss. (c) Mixed mode operation of GeTe nanowire on Si_3N_4 waveguide with both optical and electronic readouts. (d) Plasmonic nanogap enhanced photonic switch based on GST that can be addressed and readout both electrically and optically. *Source:* Adapted from Li et al. (2019, 2020), Lu et al. (2017), and Ríos et al. (2015).

offering reduced device dimensions and improved switching speeds. While on-chip optical addressing has largely replaced free-space laser heating, recent work demonstrates precise phase modulation in silicon photonic circuits using low-loss Sb_2Se_3 via free-space laser writing (Li et al. 2019). This approach achieved remarkable phase shifts exceeding 10π in MZI configurations without compromising extinction ratios or insertion loss, maintaining stable performance through 600 cycles with only 1.5 dB variation (Figure 6.2b). Hybrid operation modes have been explored by incorporating PCM elements like GeTe nanowires into electrical circuits (Li et al. 2020), though full electrical addressing remains unrealized (Figure 6.2c). More recently, combined electrical-optical programming and readout was demonstrated using GST in plasmonic gaps (Lu et al. 2017). However, the extremely confined active volume in this configuration results in minimal optical contrast (0.6%, Figure 6.2d), presenting significant challenges for practical implementation.

Modern data generation rates are creating unprecedented demands for high-capacity storage solutions. Traditional computing systems following the Von Neumann architecture separate memory, processing, and I/O functions into discrete components. This design creates fundamental limitations, while processors execute commands sequentially through control and arithmetic units; they cannot access memory fast enough to keep pace with modern data volumes. Optical interconnects have emerged as a potential solution, though their effectiveness is hampered by unavoidable optoelectronic conversion requirements. A more promising approach would integrate computation and storage functions directly in the optical domain, eliminating these conversion steps. PCMs provide this capability by enabling simultaneous optical processing and data storage.

Pioneering work by Pernice and Bhaskaran demonstrated this concept through an all-optical memory design using GST-coated Si_3N_4 microring resonators (Figure 6.2a). Their device features a ring resonator coupled to three waveguides (add, drop, and control), with a strategically placed GST film segment. Optical pulses delivered through the control waveguide induce reversible phase transitions in the GST between amorphous, partially crystalline, and fully crystalline states. The resonator's critical coupling condition changes dramatically with these phase transitions due to GST's substantial refractive index contrast. This creates a binary memory with 10 dB transmission contrast between amorphous ("0") and crystalline ("1") states at the through port. Pernice and Bhaskaran (2012) subsequently validated this design experimentally, characterizing state-dependent losses using balanced MZIs and ring resonators. Their implementation achieved multibit storage by exploiting intermediate crystalline states, with write operations (amorphization) using a single 100 ns pulse and erase (crystallization) employing six progressively weaker 100 ns pulses. This pulse sequencing addresses a key practical challenge-uniform crystallization requires gradual heating to prevent partial reamorphization, as different GST regions transition at different rates. The platform's capabilities were further enhanced through WDM, where microrings with distinct resonance wavelengths enabled independent addressing of multiple memory elements (Figure 6.3a) (Pernice and Bhaskaran 2012). Building on these advances, Feldmann et al. later scaled the technology to create a 256-cell photonic memory with 5-bit capacity by combining multilevel operation with WDM techniques.

An alternative approach to the decreasing-amplitude pulse train method utilizes pulse-width modulation (PWM) for state control in Figure 6.3b. This technique achieves precise intermediate states through fixed-amplitude pulses of varying durations rather than power adjustments. Experimental implementations have demonstrated 12 distinct memory levels using a PWM sequence with 30 ns pulse intervals, durations ranging from 5 to 20 ns, and constant 1.4 mW peak power in Figure 6.3b (Cheng et al. 2018). The PWM framework has been successfully adapted for basic logic operations, including OR and NAND gate functionality. Further advancements by Li et al. (2019). Pernice and Bhaskaran (2012) introduced a dual-step single-pulse programming method capable of achieving 34 distinct memory states (~5-bit resolution) in Figure 6.3c (Li et al. 2019). This approach combines both amorphization and crystallization in a single composite pulse: an initial 50 ns high-power segment resets the material, followed by a lower-power segment that determines the final crystallization degree. Compared to conventional methods operating at kHz frequencies, this technique enables MHz-range switching speeds while maintaining precise state control. The dual-pulse architecture significantly improves both programming speed and state precision, representing a major step forward in phase-change memory technology.

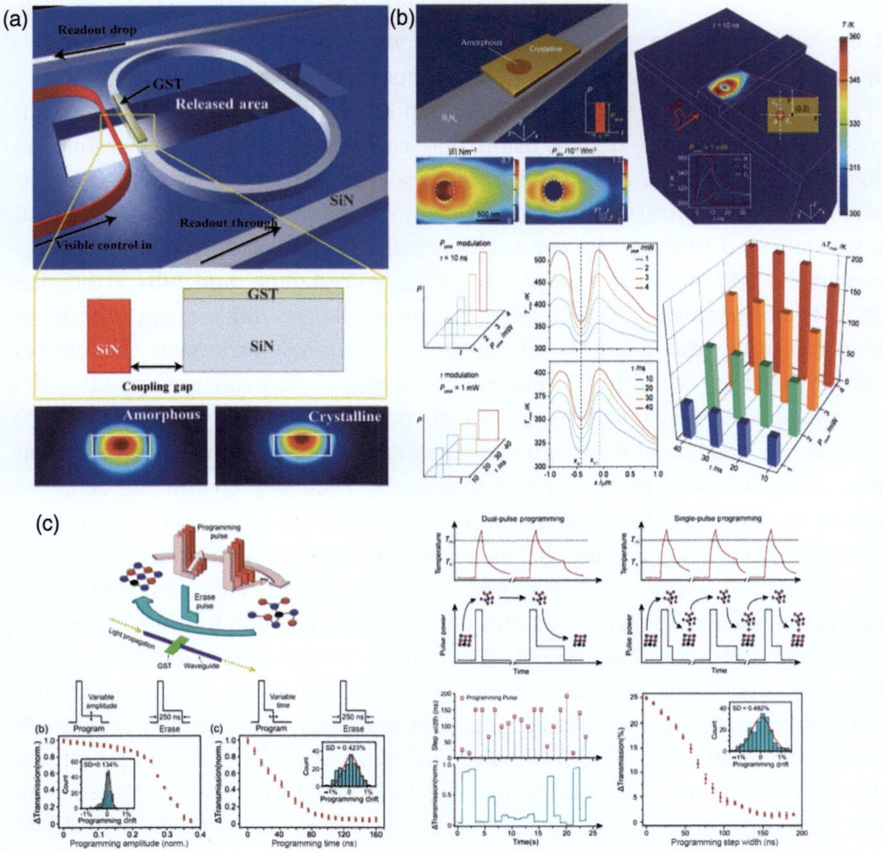

Figure 6.3 Integrated photonic memories. (a) A theoretical proposal of an all-optical memory on a Si_3N_4 platform with GST. (b) Pulse-width modulation for the access of an arbitrary memory state. (c) 5-bit photonic memory realized by the dual-step single-pulse scheme. *Source:* (a) Pernice and Bhaskaran (2012) / AIP Publishing LLC, (b) Cheng et al. (2018) / John Wiley & Sons / CC BY 4.0 and (c) Li et al. (2019) / Optical Society of America.

In this section, we explore various studies that utilize photonic memories for system-level applications, including all-optical abacuses, logic gates, photonic synapses, and optical neural networks (ONNs). We also examine topics such as optical arithmetic, vector-matrix multiplication (VMM), and convolutional neural networks (CNNs). Notably, Feldmann et al. developed a chip-scale all-optical abacus that employs an array of photonic memory cells (Feldmann et al. 2017). As depicted in Figure 6.4a, transmission levels corresponding to digits 0 (amorphous state) and 10 (fully crystalline state) are used. Incrementing by one digit involves stepwise crystallization, achieved through uniform pulses, each delivering 12 pJ over a picosecond duration. Once the PCM is fully crystallized, the device is reset with ten 19 pJ pulses. This process mimics the movement of beads in a traditional abacus, with each crystallization step representing a bead moving to the right.

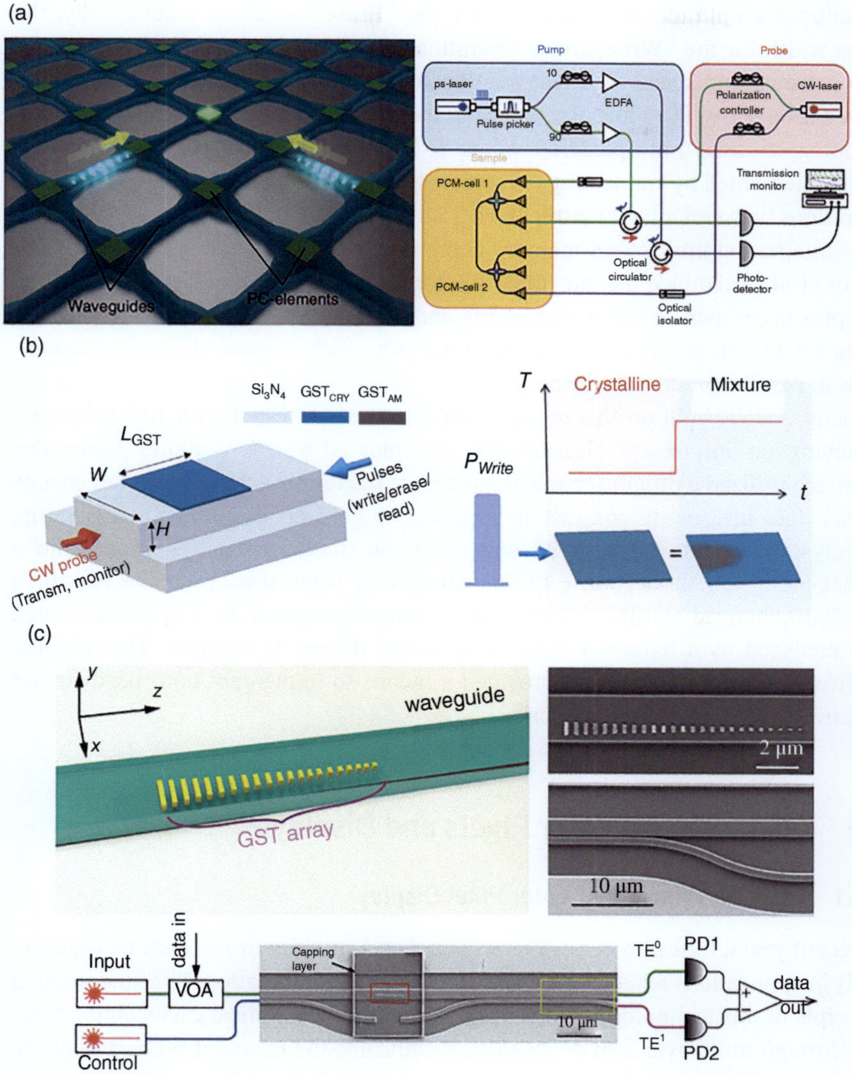

Figure 6.4 PCM-based optical VMM and neural networks. (a) A chip-scale all-optical abacus based on GST on Si_3N_4. (b) Photonic in-memory computing demonstrates optical scalar-scalar multiplication and matrix-vector multiplication. (c) The PMMC-based photonic kernel for image processing and CNN. *Source:* (a) Feldmann et al. (2017) / Springer Nature / CC BY 4.0, Ríos et al. (2019) / American Association for the Advancement of Science / CC BY 4.0, and Wu et al. (2021) / Springer Nature / CC BY 4.0.

The researchers also demonstrated precise random access within a 3×3 array by synchronizing the arrival times of dual-pulse pumps at waveguide intersections.

Furthering the concept of in-memory computing in optics, the team employed PCM-loaded waveguides for optical computation. Demonstrations included scalar operations and VMM, as illustrated in Figure 6.4b (Ríos et al. 2019). Here, double-step pulses, which are faster than the previously used consecutive pulses of reduced amplitude, were utilized for the "Erase" function. Additionally, the pulse width for the "Write/Erase" operations was optimized to balance transmission contrast and speed. A direct relationship was established between the pulse power used for writing and the resulting transmission contrast, facilitating direct mapping of scalar multiplication values to the writing and input powers. VMM was implemented by connecting two memory cells through a combiner, facilitating operations between a 1×2 matrix and a 2×1 vector.

By integrating multiple in-memory computing cells into a cross-bar array, Feldmann et al. assembled an integrated photonic tensor core capable of performing complex tasks such as image processing and digit classification via a CNN. Leveraging WDM, this photonic tensor core could potentially surpass electronic counterparts in parallel processing efficiency.

Lastly, a CNN built on this photonic tensor core was tested with 10,000 images, achieving an impressive classification accuracy of 95.3%. Another innovative approach utilized a programmable metasurface mode converter (PMMC) to encode kernel data into mode contrast in Figure 6.4c (Wu et al. 2021). In this setup, the crystalline state of the GST converted the fundamental TE_0 mode into a higher-order TE_1 mode, while in the amorphous state, the TE_0 mode remained largely unchanged. A directional coupler then out-coupled the TE_1 mode, which was analyzed by a balanced detector to assess the mode contrast. This method resolved 64 distinct levels and provided a means to implement both positive and negative weights in neural networks.

6.3 Chalcogenide Color Pixels and Displays

6.3.1 Working Principle of Color Pixel Display

In recent years, thin film optics has emerged as a prominent research focus, especially in applications related to color generation and manipulation. The fundamental principle of this technology involves precise control over optical cavity characteristics through multilayer film deposition. Specifically, the resonant properties of the cavity can be engineered by either varying the physical thickness of selected layers in the film stack or modifying the refractive indices of particular layers while maintaining their original thickness dimensions. This adjustment can modify the wavelengths reflected from the stack, thereby producing specific hues in the reflection. Structural colors typically utilize the interactions between the microstructures on the material's surface and light, such as interference, diffraction, and scattering, to achieve specific colors. Recent advances in structural engineering enable precise tuning of resonance wavelengths through careful optimization of geometric parameters. The emergence of sophisticated nanofabrication and characterization methods has significantly boosted research interest in various photonic nanostructures, including thin-film optical resonators, nano-patterned diffraction gratings, and engineered plasmonic metasurfaces. These devices exhibit unique optical properties and are highly compatible with current display and imaging technologies.

Structural colors based on thin film interference are usually composed of multilayer lossless films or dielectric-metal composite films. The surfaces of these films reflect only a portion of the incident light, while the remaining light waves undergo multiple reflections and refractions within the film layers, ultimately forming a series of reflected beams with varying optical path differences at the upper surface. Due to the phase differences between the reflected beams, light at in-phase wavelengths (phase shift of 0) experiences constructive interference, while light at out-of-phase wavelengths (phase shift of π) undergoes destructive interference, resulting in a selective color rendering effect on the surface. Researchers have designed various metal/dielectric multilayer film structures based on interference principles, including Ti/AlN, TiO_2/SiO_2, Al_2O_3/SiO_2, MgF_2/SiO_2, and Si/Si_3N_4 systems, which have shown significant application value in solar energy collectors. A typical interference structural color arises from the Fabry–Pérot (FP) resonator effect, which consists of two layers of metal films sandwiching a transparent dielectric layer. The resonator formed by the metal layers encourages multiple reflections of light waves within the cavity, allowing only those wavelengths that satisfy the resonance condition ($2nL\cos\theta = m\lambda$) to be effectively output, thus producing narrowband spectral characteristics. When the metal layers on both sides of the FP cavity are thin (<30 nm), the symmetric structure primarily exhibits transmission structural colors in specific wavelength bands; conversely, when the bottom metal layer has an optical thickness (>50 nm), the asymmetric structure suppresses transmission and creates reflection spectral valleys, resulting in reflective structural colors. The resonance wavelength can be effectively modulated through controlled variation of the dielectric layer thickness, offering distinct benefits in terms of fabrication simplicity and scalability for industrial implementations.

Nanostructure-based structural coloration demonstrates superior characteristics, including high spatial resolution, excellent durability, eco-friendliness, and compatibility with mass production techniques. Nevertheless, these systems exhibit inherent limitations as their optical resonances are permanently determined by the fixed structural geometries and material compositions established during fabrication. Since display technologies require precise control over the colors of numerous pixels, researching dynamic reconfigurable methods for structural colors is crucial. Compared to liquid crystals and electrochromic materials, PCMs possess a wide tuning spectral range and ultra-fast dynamic control capabilities. They can exist in stable amorphous and crystalline states, allowing for rapid phase transitions under optical, electrical, or thermal stimuli, accompanied by significant changes in photoelectric contrast. In 2014, Hosseini et al. utilized the reconfigurability of the chalcogenide PCM GST to create nanometer-scale, electrically driven tunable display devices, opening new avenues for research in solid-state nanometer-scale pixel displays, smart glasses, and holographic displays. This phase change display leverages the interference effects of thin film structures to achieve colored structural colors. Its simplicity and the lack of a need for high-resolution patterning processes demonstrate significant potential for applications in miniaturized display devices, enabling high resolution, high saturation, vibrant colors, and fast switching speeds in non-volatile displays.

6.3.2 Practical Application Cases

The field of thin-film optics, particularly in color rendering, has garnered significant attention recently. This interest was sparked by a seminal study by Kats et al., which demonstrated that using highly absorptive materials like metals in ultra-thin layers can generate vivid colors. The principle behind thin-film optics is relatively straightforward: it involves modulating the length of an optical cavity using a series of thin films. For instance, adjusting the thickness of one or more layers in a simple thin-film stack can alter the resonance conditions within the cavity. Alternatively, modifying the optical properties of any layer—such as its refractive index—without changing its thickness also shifts the cavity's resonance, thereby changing the wavelengths reflected and thus the color produced. The transfer matrix method is a valuable tool for calculating the reflection and transmission coefficients within multilayer thin-film structures. Incorporating PCMs as one of these layers introduces dynamic functionality, allowing for the refractive index to be altered through structural transformations triggered by electrical pulses. This capability facilitates electro-optical operations within thin-film devices.

The field of thin-film optics has been significantly advanced through the integration of PCMs. A landmark study by Hosseini et al. (2014) demonstrated an innovative nanopixel display technology utilizing electrically-tunable GST films to achieve stable color modulation. The researchers developed an advanced optoelectronic architecture incorporating a photonic cavity structure with the following key components: an active PCM layer serving as the tunable optical medium, dual transparent conductive electrodes (commonly indium tin oxide) sandwiching the PCM layer, and an optimized thin-film stack for enhanced reflectivity control (Feldmann et al. 2020). The team also showcased the functionality of cross-bar type pixels that could be switched reversibly. While various materials that exhibit changes in optical refractive index could be used, PCMs are particularly advantageous due to their significant optical contrast, electrical switchability, and well-established industrial applications, which ensure reliability and scalability (Kats et al. 2013; Heavens 1960). These achievements suggest that this technology might be scalable to larger displays and full-sized arrays, a possibility that merits further exploration.

In 2014, Hossein et al. achieved a significant breakthrough in reflective display technology by fabricating an ITO/GST/ITO/Pt multilayer system on SiO_2 substrates using magnetron sputtering (Figure 6.5a), marking the first successful application of Ge–Sb–Te alloys in this field. Their innovative design leveraged the unique optical properties of PCMS, where the controlled variation of the bottom ITO layer thickness produced distinct color outputs through thin-film interference effects, while the Pt mirror substrate significantly enhanced color saturation. The substantial contrast in refractive index (n) and extinction coefficient (k) between the amorphous and crystalline phases of GST enabled reversible color switching under electrical stimulation (Figure 6.5b). The display mechanism operated by applying precisely controlled electrical pulses to induce localized phase transitions, creating a spectrum of colored pixels that could be arranged to form complete images. This versatile optoelectronic platform not only demonstrated promising applications in reflective

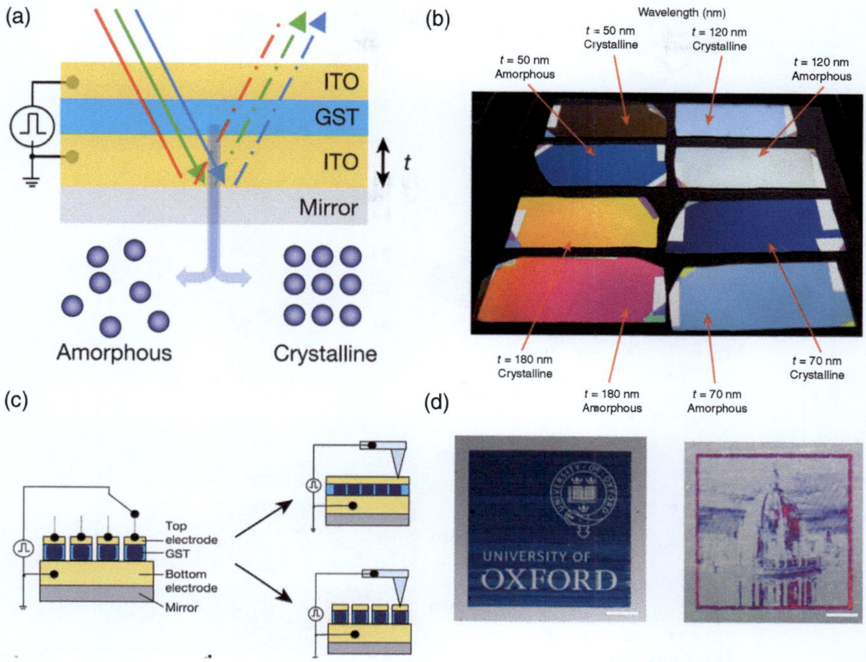

Figure 6.5 Reflective display based on Ge–Sb–Te alloy. (a) Structure of ITO/$Ge_2Sb_2Te_5$/ITO/Pt films. (b) Images of amorphous and crystalline samples with different bottom ITO thicknesses, the thicknesses of top ITO, $Ge_2Sb_2Te_5$, and Pt are 10, 7, and 100 nm, and the bottom ITO thicknesses are 50, 70, 120, and 180 nm, respectively. (c) Working principles and (d) images of reflective display. *Source:* Hosseini et al. (2014) / Springer Nature.

displays but also showed potential for adaptive optical devices, smart windows, and bio-inspired vision systems, representing a major advancement in thin-film photonics and phase-change material applications (Hosseini et al. 2014) (Figure 6.5c,d).

In addition to changing the colors of multilayer films through interference principles, researchers have cleverly utilized the in-situ phase change of PCMs to modulate their own colors. In 2019, Jafari et al. designed a photonic structure integrating a double-layer GeTe material, achieving dynamic color modulation through different phase combinations of the double-layer GeTe, as shown in Figure 6.6a (Jafari et al. 2019). By selectively thermally switching the amorphous and crystalline states of the double GeTe layers, the samples can display four colors—blue, green, red, and purple—within the same pixel (as illustrated in Figure 6.6b), and they can maintain this state with zero power, making this approach promising for low-refresh-rate passive display systems. Additionally, the study demonstrated for the first time that the phase transition temperature of GeTe can be modulated by changing the film thickness, as shown in Figure 6.6c. This is a crucial factor in achieving multiple colors in the aforementioned film structure. Notably, in a simplified GeTe (8, 15, and 20 nm)/SiO_2 (250 nm)/Pd (100 nm) film system, varying the thickness of GeTe and its crystallinity allows the samples to exhibit a range of colors, including green, blue, pink, yellow, and purple, as shown in Figure 6.6d. Therefore, the device's multicolor

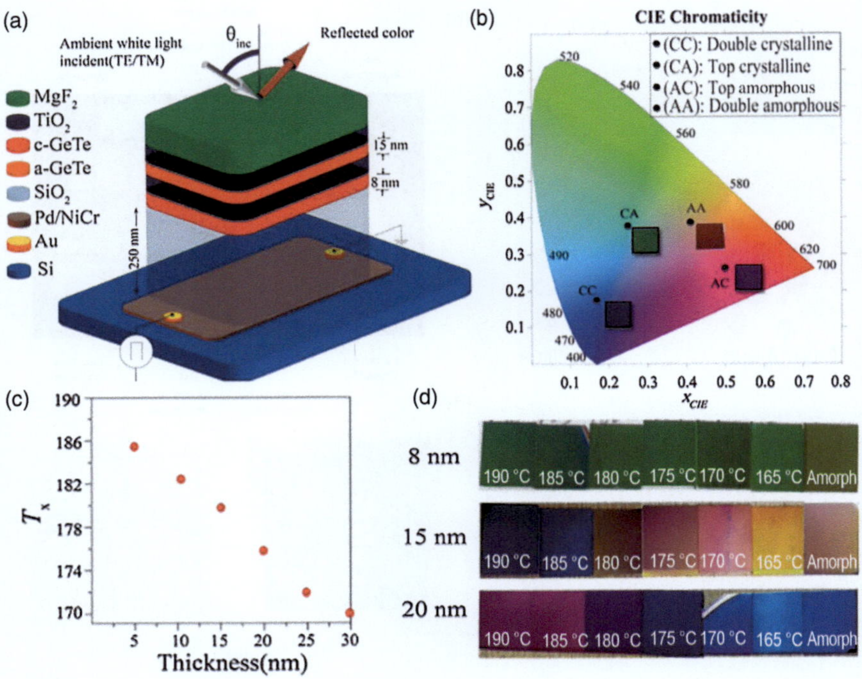

Figure 6.6 Color reflector based on double-layer GeTe material for a passive display system. (a) Photonic structure diagram of MaF_2 (60 nm)/TiO_2(85 nm)/c-GeTe (15 nm)/TiO_2 (60 nm)/a-GeTe(8 nm)/SiO2 (250 nm)/Pd-NiCr (100 nm)/Si substrate. (b) The color gamut of the samples, when the phase states of the double-layer GeTe present four states: double crystalline, top crystalline, top amorphous, and double amorphous. (d) Samples of GeTe/SiO_2/Pd films at different annealing temperatures when the thicknesses of GeTe are 8, 15, and 20 nm, respectively. *Source:* Jafari et al. (2019) / John Wiley & Sons.

capability, insensitivity to incident angles, and low polarization dependence effectively ensure its applicability in the realm of passive display systems.

Carol demonstrated the nano display function of electrically reconfigurable PCM using conductive atomic force microscopy (c-AFM) lithography technology (Ríos et al. 2016). As illustrated in Figure 6.7a, localized phase transitions were induced by applying voltage biases (0–7 V) between the AFM cantilever and the ITO substrate, leveraging Joule heating to crystallize specific regions of the initially amorphous PCM layer (Methods). This approach enabled the direct writing of $70 \times 70\,\mu m^2$ binary patterns with 500 nm feature resolution, where crystalline (white) and amorphous (black) states corresponded to 7 V and 0 V biases, respectively (see Figure 6.7b).

While the AFM-based writing system currently exhibits impedance mismatch limitations preventing ns-scale erasure pulses required for re-amorphization, previous work has established full electrical reversibility in crossbar devices through optimized multilayer configurations. Building on this foundation, they extended the characterization to sub-micron display capabilities. Figure 6.7c demonstrates a $7 \times 7\,\mu m^2$ pattern written on AIST films, where voltage modulation during

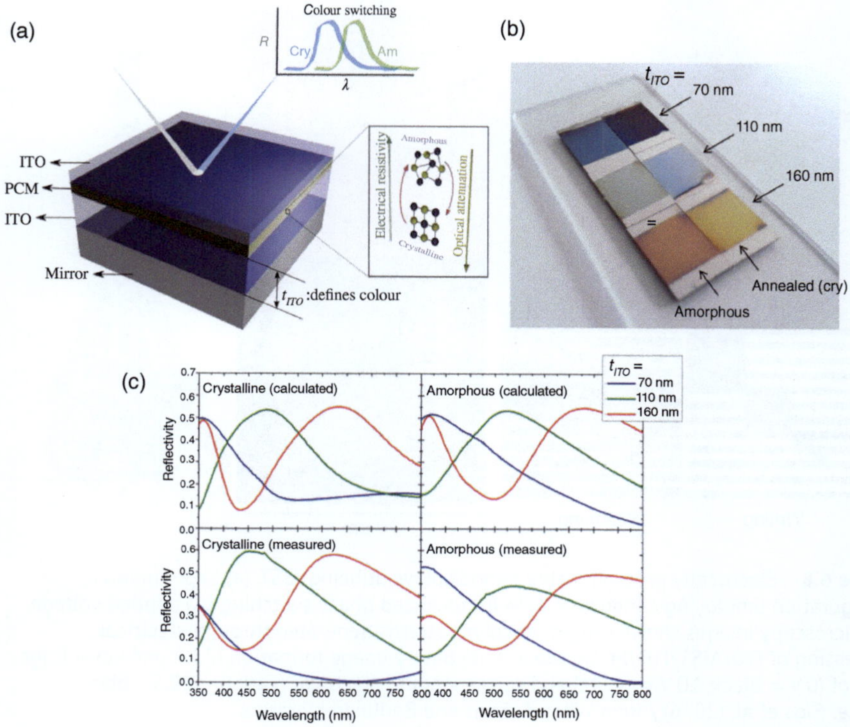

Figure 6.7 Tunable optical properties of AIST-based thin-film structures. (a) Schematic illustration of the multilayer architecture and working mechanism. (b) Visible color changes observed in sunlight for ITO/AIST/ITO stacks with varying ITO thicknesses (70, 110, and 160 nm). (c) Comparison of simulated and experimental reflectance spectra for both amorphous and annealed (crystalline) states. *Source:* Ríos et al. (2016) / John Wiley & Sons and Bodleian Libraries.

small-area scanning introduced minor crystalline nucleation anomalies (Writing image). Due to the optical diffraction limit, conventional microscopy failed to resolve these nanoscale features. Instead, they implemented a resistance-sensitive AFM mapping technique at 0.5 V bias-below the crystallization threshold but sufficient to distinguish state-dependent conductivity. The resultant conductivity map (Scanning image) successfully reconstructed the original pattern with 250–300 nm critical dimensions, despite stochastic nucleation effects.

Researchers locally switched initially amorphous PCM regions by applying voltage between a conductive AFM tip and the underlying ITO layer (Figure 6.8a,b). This approach leverages ITO's excellent conductivity to selectively crystallize individual nanopixels through Joule heating. Two addressing methods were demonstrated: Discrete point-by-point switching Continuous voltage modulation during surface scanning. Using the scanning method with 0 V (amorphous/black) and 7 V (crystalline/white) biases, the team achieved 70 × 70 μm binary patterns (Figure 6.8c) featuring remarkable 500 nm resolution elements.

PCMs have garnered significant interest in display technology due to their unique ability to dynamically modulate optical properties. PCMs such as Ge–Sb–Te and

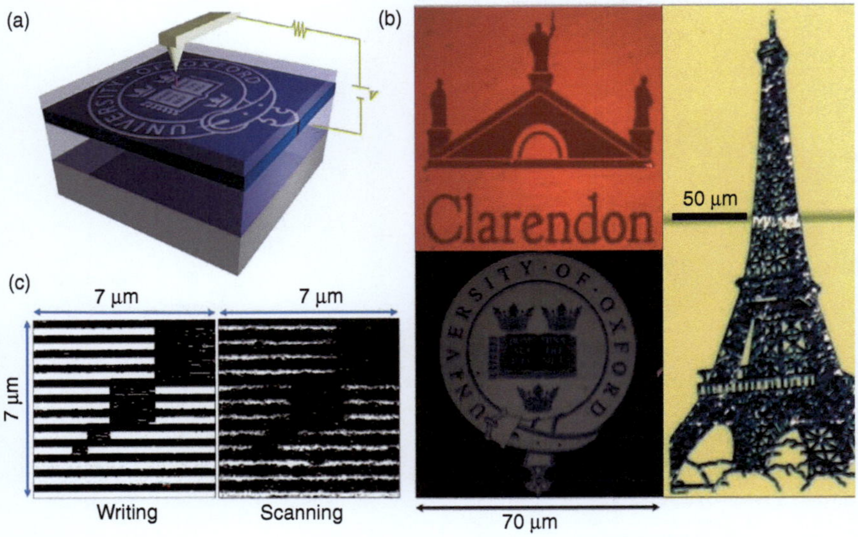

Figure 6.8 Electrically programmable nanodisplays utilizing AIST. (a) Experimental configuration employing conductive AFM for localized phase switching via applied voltage. (b) Microscopy images showcasing multicolor patterns generated through electrical addressing of ITO/AIST/ITO/Pt multilayers. (c) Binary image formation (7 × 7 μm) via voltage control (0 V = black, 10 V = white) and corresponding electrical readout at 0.5 V bias. *Source:* Ríos et al. (2016) / John Wiley & Sons and Bodleian Libraries.

GeTe can create reflective displays that achieve vibrant colors through interference effects. By utilizing multilayer structures, these displays can offer high color saturation and resolution. The capability of PCMs to maintain color states with zero power consumption makes them ideal for low-refresh-rate passive displays. This feature is particularly beneficial for applications where energy efficiency is critical, such as in wearable devices and smart glasses. Current research is focused on optimizing the optical properties of PCMs to achieve a wider color gamut and higher color fidelity. Innovations in material composition and multilayer structures are expected to result in displays that are more vivid and accurate. The integration of PCMs with smart technologies, such as sensors and adaptive optics, will facilitate the creation of intelligent displays that can adjust their properties in real-time based on environmental conditions. This could lead to applications in AR and interactive displays.

Future PCMs may be designed not only as displays but also to incorporate functionalities such as sensing, data storage, and even light modulation. This multifunctionality could revolutionize the way displays are utilized across various applications, from consumer electronics to advanced industrial systems. The future development of PCM applications in color displays will benefit from interdisciplinary collaboration, combining insights from materials science, optics, electronics, and design. The convergence of these technological domains is poised to catalyze breakthrough innovations and unlock novel applications. In essence, PCMs stand at the threshold of revolutionizing color display technologies. As material science progresses alongside manufacturing refinements and smart system integration, PCM-based solutions are emerging as fundamental enablers for future display platforms-delivering unprecedented performance metrics, remarkable design flexibility, and superior environmental sustainability.

6.4 Chalcogenide Waveguide

6.4.1 Theoretical Basis for the Characteristics of Chalcogenides and Waveguides

Chalcogenide PCMs, while well-established in electronic memories and optical storage media, represent an emerging technology for photonic applications. Their reversible phase transitions between stable structural states produce significant optical property variations ($\Delta n \sim 1$), enabling reconfigurable photonic components through simple thin-film integration on waveguide platforms. These non-volatile transitions create permanently stable optical states without requiring continuous power input. CMOS-compatible deposition techniques (sputtering/evaporation) facilitate direct integration with silicon and Si_3N_4 photonic circuits, leveraging existing manufacturing infrastructure with relatively low production risks.

This perspective highlights PCMs' potential for programmable photonics, particularly for energy-efficient applications where speed can be traded for lower power consumption. We review current advances in PCM-integrated devices and metasurfaces, while addressing critical limitations: high optical losses, binary operation modes, limited cycling endurance, and phase transition control challenges. Potential solutions and future research directions are outlined, including visible-wavelength applications, quantum photonics integration, and optogenetic interfaces. A promising approach proposes CD-inspired rewritable photonic circuits using wide-bandgap PCMs, enabling rapid prototyping with ns lasers instead of requiring advanced nanofabrication facilities or expensive fs laser systems.

Achieving reliable multistate control remains problematic with electrical addressing due to PCMs' inherent stochastic phase transition behavior. While fs lasers have demonstrated precise multilevel control, electrical methods typically only produce two stable states (amorphous or crystalline). The melt-quench amorphization process creates variable amorphous configurations, leading to inconsistent crystallization pathways during subsequent switching. Recent breakthroughs using segmented heater designs have enabled electrically controlled multilevel operation in PCM waveguide switches, achieving both phase and amplitude modulation with improved reproducibility.

6.4.2 Application of Chalcogenides in Waveguides

Achieving optimal optical contrast necessitates complete phase transitions across extended PCM regions, typically requiring external heating elements. Initial efforts employing ITO heaters on SOI waveguides demonstrated limited performance, with electrical cyclability (>10 cycles) but single-cycle optical operation. Slow crystallization kinetics (100 ms) and minimal contrast (1.2 dB) (Figure 6.9a) (Kato et al. 2017).

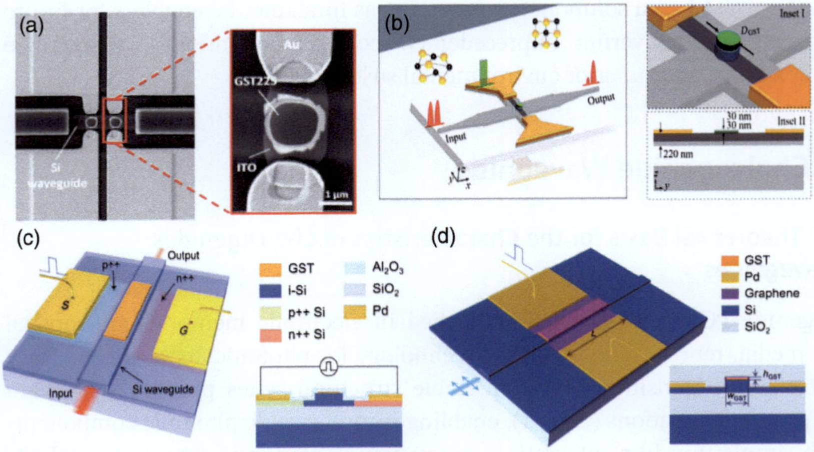

Figure 6.9 (a) Electrically reconfigurable silicon photonic switch based on GST using an ITO external heater. (b) Electrical switching of GST on a heavily doped silicon waveguide supporting multilevel operation. (c) An electrically reconfigurable silicon photonic switch based on a PIN external heater with high cyclability. (d) Simulation of a silicon photonic switch based on a graphene external heater. The graphene heater design is also compared to the silicon PIN heater and the ITO heater (bottom). *Source:* (a) Kato et al. (2017) / Optica Publishing Group / CC BY 4.0, (b) Zhang et al. (2019) / American Chemical Society, (c) Zheng et al. (2020) / arXiv, and (d) Adapted from Zheng et al. (2020).

While attributed to insufficient GST thickness, the absence of deposition details and mode simulations leaves this explanation unverified. Subsequent MZI implementations (Figure 6.9b; Zhang et al. 2019) achieved improved modulation depth (~7 dB) through ITO heating, demonstrating multilevel crystallization capability. However, these systems failed to establish reversible optical switching functionality, indicating fundamental limitations in ITO-based thermal control.

The frequent amorphization observed in these studies suggests inherent challenges with sputtered ITO, where surface irregularities and sudden resistivity jumps at 650 K generate localized heating zones (Mansha et al. 2022), resulting in partial GST phase transitions near these thermal hotspots. Material inconsistencies in both freshly deposited (Li et al. 2017) and thermally processed ITO (Lu et al. 2020) further contribute to uneven temperature distribution. Current flow through the ITO establishes preferential conductive pathways, causing selective heating in more conductive areas and consequently non-uniform PCM activation. This phenomenon accounts for the distinct responses observed-while electrical resistance shows dramatic shifts (~10^3) from minuscule crystalline regions (~nm^3) due to percolation conduction principles fundamental to phase-change memory, optical transmission exhibits minimal variation since light interacts with the complete GST volume rather than just conductive pathways. The contrast stems from differing operational mechanisms: electrical conductivity depends on continuous crystalline networks, whereas optical effects integrate responses across the entire material thickness.

Doped silicon heaters have emerged as a promising alternative, offering exceptional performance metrics including endurance exceeding 1000 cycles (Figure 6.9c,

Zheng et al. 2020), MHz-range switching speeds, and remarkable contrast ratios up to 18.6 dB ($\Delta T/T_0 = 300\%$), while maintaining full CMOS process compatibility. These heaters achieve substantially greater contrast (15× improvement) compared to ITO-based designs (Ríos et al. 2016) despite using similarly sized GST elements, primarily due to their ability to induce complete phase transitions rather than partial ones. The technology enables four distinct transmission states through controlled partial crystallization using low-power pulses, facilitated by several key characteristics: stable resistance properties up to current saturation levels exceeding 45 mA, atomically smooth crystalline silicon surfaces, and highly uniform doped region conductivity. However, this approach introduces a significant trade-off in the form of optical losses, with uniformly doped waveguides exhibiting insertion losses as high as 0.5 dB/μm, presenting a challenge for practical implementation in photonic circuits (Figure 6.9d; Zheng et al. 2020).

To mitigate the substantial losses caused by free carrier absorption, doping silicon into a PIN junction is a strategic approach. This configuration ensures that waveguiding occurs primarily in the intrinsic region, effectively reducing the insertion loss to just 0.007 dB/μm (Feldmann et al. 2017). Such doped silicon heaters are ideal for use with transparent PCMs that have wide bandgaps and exhibit minimal absorption in near-infrared wavelengths. Challenges remain in implementing on-chip optical switching due to these properties. For instance, a recently demonstrated electrically reconfigurable phase shifter utilizing a doped Si heater and low-loss PCM, specifically Sb_2Se_3, facilitates diverse applications ranging from 2 × 2 switches to optical trimming. Moreover, the operational lifetime of these devices can be substantially prolonged beyond 3000 cycles through the implementation of a protective Al_2O_3 capping layer deposited via atomic layer deposition (ALD), which forms an exceptionally dense and uniform protective barrier (Figure 6.9d).

Despite advancements with doped silicon and ITO heaters in PCM switching, issues such as slow switching speeds (~10 MHz) and high energy consumption (~nJ) persist. Recent innovations have introduced graphene as a transparent conductor for achieving high-speed (~80 MHz) and energy-efficient (19.2 aJ/nm^3 and 6.6 aJ/nm^3) non-volatile photonic switches. Graphene's minimal heat capacity and excellent thermal conductivity make it an outstanding candidate for this role. Although experimental demonstrations have shown PCM tuning using a graphene heater, integration with photonic structures remains incomplete. Current limitations predominantly arise from two key factors: the intrinsic material properties of GSST that require considerable energy to overcome phase transition thresholds, and thermal management inefficiencies in existing heater architectures. These constraints manifest as sluggish switching speeds and excessive power consumption in practical implementations. In parallel, significant progress has been made in waveguide engineering, where an advanced slot waveguide configuration has been developed to optimize optical field confinement within the GST layer. This innovative approach yields remarkable modulation characteristics, achieving a substantial effective refractive index shift ($\Delta n_{eff} \approx 1.2$) and pronounced absorption variation ($\Delta\alpha \approx 35$ dB/μm) during GST phase transitions, while simultaneously improving energy efficiency through enhanced light-matter interaction.

Photonic microcavities, including microring resonators and photonic crystal nanobeams, dramatically enhance light-matter interactions through dual spatial and temporal confinement of optical modes (Farmakidis et al. 2019). This strong confinement enables substantial modulation of both optical phase and amplitude, even with subtle refractive index perturbations in the active material. PCMs have proven particularly effective in such cavity-enhanced systems, where they enable high extinction ratios through simultaneous phase and absorption modulation. Among various PCMs, low-absorption variants such as Sb_2S_3 and Sb_2Se_3 present unique advantages due to their minimal optical loss in both amorphous and crystalline states. However, their predominant effect on the real part of the refractive index necessitates integration with sensitive photonic structures like resonant cavities or MZIs for effective detection. A notable demonstration involved the dynamic tuning of a silicon racetrack resonator through optical switching of an adjacent GST element, achieving an exceptional 12 dB extinction ratio–a marked improvement compared to conventional waveguide switches. This configuration revealed several key phenomena: (i) GST crystallization induces increased optical loss and drives the system into the undercoupling regime, manifesting as resonance linewidth broadening and spectral redshifting; (ii) time-resolved measurements using high-speed photodetectors have elucidated the thermo-optic dynamics during optical excitation, though a complete characterization of GST-induced loss mechanisms remains an important area for further investigation.

In a study by Rios et al., GST was integrated into both MZIs and microring resonators on a Si_3N_4 platform (Rudé et al. 2013), and they measured the attenuation coefficient at approximately 1550 nm wavelength. They reported that the GST in its amorphous state exhibited an attenuation of 0.1 dB/μm, while in its crystalline state, the attenuation increased to about 1.1 dB/μm. This research did not demonstrate reversible switching, primarily because the GST phase transitions were induced through thermal annealing, and the microring's coupling conditions were not ideally optimized, resulting in a modest extinction ratio of approximately 15 dB. This switch was activated by laser-induced switching, which, due to the significant optical absorption of GSST in its crystalline state, effectively disrupted the high-quality (Q) resonance of the microring. This disruption resulted in a substantial extinction ratio exceeding 40 dB, as detailed in Figure 6.10a (Zhang et al. 2019). Further advancements were noted in subsequent research, which focused on optimizing the interaction between GST and silicon microrings by adjusting the length of GST layers. This adjustment helped pinpoint the critical coupling condition necessary for achieving maximum optical contrast, as illustrated in Figure 6.10b (Zheng et al. 2018.) These findings underscore the potential of integrating PCMs like GST and GSST with photonic structures to enhance optical device performance through controlled phase transitions.

The PCM field is increasingly shifting toward on-chip actuation to enable high-density integration and scalability, overcoming challenges such as alignment difficulties and diffraction-limited resolution in laser-based switching. Among integrated photonic structures, microring resonators are highly effective for optical switching of PCMs due to their strong light confinement. While previous work

(a)

(b)

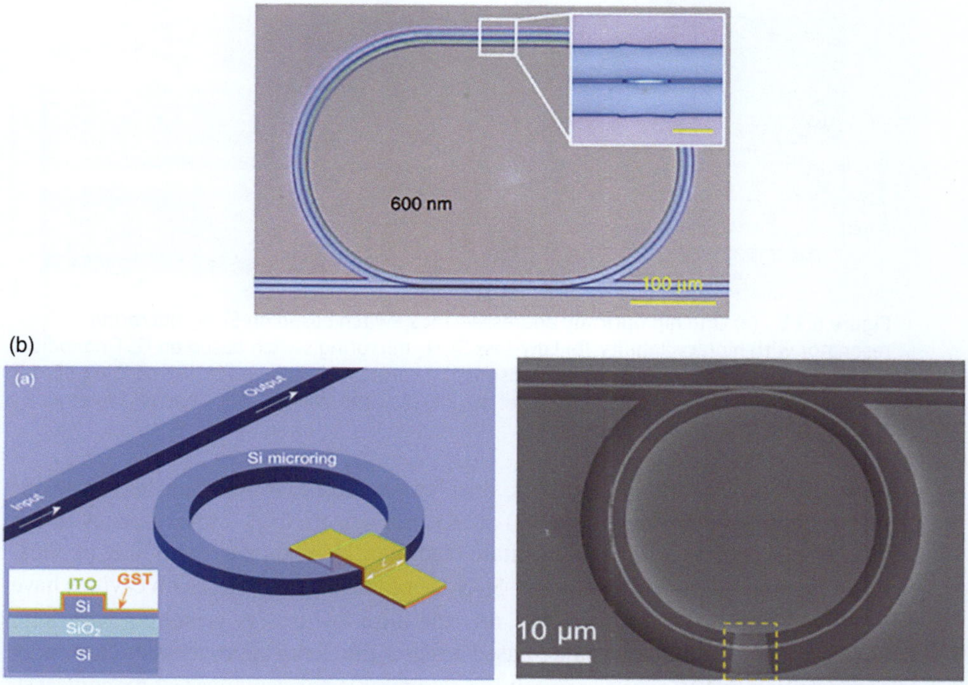

Figure 6.10 (a) Free-space optical switching of broadband low-loss GSST on Si_3N_4 racetrack resonator compared with that of GST. (b) Spectral shift and loss characterization of GST using silicon microring resonators. Multilevel optical switching by free-space laser was also demonstrated (bottom right). *Source:* Zheng et al. (2018) / Optical Society of America and Optica Publishing Group.

has demonstrated multiwavelength photonic memory using microrings, recent advances have achieved multilevel operation and endurance exceeding 1000 cycles in a 1×2 microring switch (Figure 6.11a) (Stegmaier et al. 2017). However, this design suffers from high insertion loss (~5 dB) and a modest extinction ratio (~6 dB). A more recent implementation using ALD Al_2O_3-encapsulated nanopatterned GST disks significantly improved performance, achieving <1 dB insertion loss and >20 dB extinction ratio. The conformal alumina coating prevents GST deformation during phase transitions, as illustrated in Figure 6.11b (Wu et al. 2019).

Despite its advantages over free-space optics, on-chip optical switching still faces scalability limitations due to complex routing requirements. Additionally, it cannot address wide-bandgap PCMs with negligible near-IR absorption. To resolve this, Zheng et al. introduced an electrically actuated silicon microring switch incorporating a PIN heater to thermally control GST, achieving >15 dB extinction ratio and 3000-cycle endurance (Zheng et al. 2020). Following a similar approach, Sb_2Se_3—a transparent PCM—was integrated with silicon microheaters for reconfigurable phase shifting. Another wide-bandgap material, Sb_2S_3, was paired with an ITO heater to enable near phase-only modulation without resonance broadening upon crystallization (Fang et al. 2021). The ITO heater, chosen for its transparency and

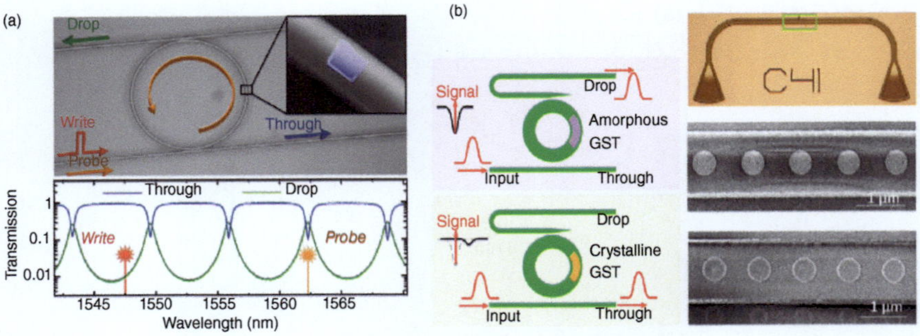

Figure 6.11 (a) On-chip optically addressed 1 × 2 switch based on Si_3N_4 microring resonator with high cyclability. (b) Low-loss Si_3N_4 microring switch based on GST nanodisks encapsulated by ALD Al_2O_3. Higher transmission contrast is achieved by the capping of ALD Al_2O_3 (top right). *Source:* (a) Stegmaier et al. (2017) / John Wiley & Sons and (b) Wu et al. (2019) / American Chemical Society.

compatibility with Si_3N_4 platforms, enabled an ultra-low insertion loss (0.48 dB) and a high extinction ratio (>30 dB) at 1550 nm. Given Sb_2S_3's transparency in the visible spectrum, it could also facilitate reconfigurable quantum photonic applications. Beyond microring resonators, Si_3N_4 photonic crystal nanobeam cavities have been explored for PCM integration, offering ultra-low energy switching with high contrast. Compared to waveguide-based designs, the cavity approach reduces energy consumption by 81% while maintaining comparable performance (~1 dB contrast).

Two fundamental challenges currently hinder broader adoption of PCM technology in photonic systems. The first concerns operational longevity-while recent developments have achieved 10^6 switching cycles, this pales in comparison to electronic memory benchmarks exceeding 10^{12} cycles. This discrepancy largely results from the substantially greater material volumes employed in photonic implementations. Scaling down device dimensions could push cycling limits toward 10^{10} operations through mitigated thermal degradation and reduced material separation.

Interestingly, existing endurance capabilities already satisfy the needs of adaptive optical systems with sporadic switching demands, such as reconfigurable computing architectures. Here, operational lifetimes could span hundreds of years before performance degradation occurs. Potential enhancements through atomic-layer deposition coatings, innovative material compositions, and subwavelength structuring could further improve reliability. The second critical obstacle lies in achieving precise phase transition control. GST's distinctive thermal characteristics-including a 650 °C melting point and low thermal conductivity (<1 W/m/K)-create delicate operational parameters where merely 0.3 V excess voltage can induce failure. This sensitivity is exacerbated by current concentration effects at structural discontinuities in silicon heating elements. Key research directions to overcome these limitations include: engineering PCM compositions with depressed transition temperatures; implementing alternative dielectric platforms with superior thermal performance; advanced thermal management through geometric optimization; and integration of two-dimensional material heating systems. Perhaps the most transformative benefit emerges in power efficiency metrics. Traditional thermo-optic components demand continuous milliwatt-level power inputs, accumulating to hundreds of joules daily just to maintain operational states. In stark contrast, PCM switching events consume mere microjoules of energy, representing a 10^7-fold improvement in static power requirements-a decisive advantage for scalable photonic networks.

The weak thermo-optic effect, while energy-efficient, induces localized heating at heater sites that can propagate as thermal crosstalk. This thermal interference fundamentally limits the integration density of photonic components and compromises operational reliability. PCM-based devices circumvent these limitations through their inherent high thermal activation threshold for state transitions. Although advanced thermal management algorithms have been developed to suppress crosstalk effects, their implementation requires sophisticated control electronics and additional software overhead. Overcoming these thermal challenges enables the realization of scalable, electrically programmable PCM-based PICs with transformative potential across multiple domains. Such next-generation PIC platforms could implement reconfigurable unitary transformations, non-blocking optical switching matrices, fault-tolerant photonic systems, multifunctional integrated photonics, neuromorphic optical computing architectures, and quantum simulation environments.

Quasi-arbitrary unitary transformations can be optically realized using reconfigurable arrays of programmable photonic units, each containing tunable beam splitters and phase shifters to perform essential matrix-vector multiplication operations for optical computing, neural networks, and quantum simulations (Figure 6.12a). While conventional implementations suffer from thermal crosstalk, high power consumption, and complex control requirements, emerging non-volatile architectures based on PCMs offer promising solutions. However, the practical realization of large-scale electrically controlled PCM systems faces significant challenges due to accumulated optical losses in arrayed configurations –although individual PCM units may exhibit acceptable losses, these compound severely in extensive arrays, making it difficult to maintain the required performance benchmarks of <0.1 dB insertion loss and >30 dB extinction ratio per unit, particularly when using traditional PCMs like GST with their inherent optical absorption. Current research efforts are therefore focused on developing low-loss PCM alternatives and novel device architectures to overcome these scaling limitations while preserving the reconfigurability needed for advanced photonic applications.

Modern PICs can achieve enhanced reliability through optical redundancy schemes incorporating non-volatile switching fabrics. The proposed architecture (Figure 6.12b, c; Chen et al. 2022) utilizes PCM-based matrices to dynamically reroute signals when failures occur in critical components like on-chip lasers, requiring zero static power while operating. Though promising, practical implementation faces challenges in achieving the essential performance metrics (<0.1 dB insertion loss, >30 dB extinction ratio) due to current material limitations and fabrication imperfections, necessitating further development of low-loss PCM formulations and precision manufacturing techniques.

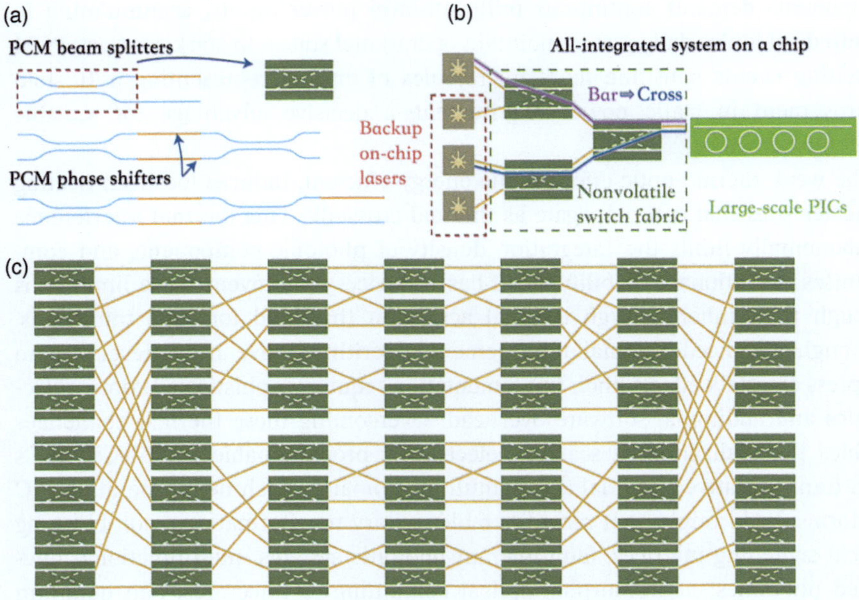

Figure 6.12 Schematic of non-volatile arbitrary unitary transformation and switching fabrics. (a) The diagram features three schematic representations of programmable units designed for arbitrary unitary transformations using PCMs. (b) The schematic suggests switching fabrics designed to incorporate optical redundancy into fully integrated chip systems. Due to the susceptibility of on-chip lasers to failures, the inclusion of multiple backup lasers is recommended. Encased in a dotted light blue box, the PCM-based switching fabric is optimally positioned to connect the laser with large-scale PICs, which may include elements like ring modulators in optical transceivers. (c) The illustration proposes a 16 × 16 Benes-type nonblocking switching fabric tailored for stable, non-volatile signal routing. The design anticipates that all included tunable beam splitters will be controlled electrically to ensure precise adjustments and signal management.

The escalating computational demands of deep neural networks (DNNs) across diverse applications has spurred interest in photonic alternatives to conventional electronic implementations. ONNs leverage the inherent advantages of light-based processing-including superior speed and energy efficiency-particularly for executing linear algebra operations fundamental to AI workloads. These systems typically encode input data in optical signals while representing network weights through programmable photonic components like MZIs or crossbar arrays.

PCMs emerge as particularly suitable for ONNs due to their non-volatile nature, providing stable weight representations post-training without constant power expenditure. Successful demonstrations include implementations of convolutional networks and neurosynaptic systems using materials such as GST. However, optically controlled configurations present practical limitations regarding alignment precision and system scalability that could hinder widespread adoption.

A more viable path forward combines electrically addressable PCM elements with electronic control circuitry, potentially through monolithic integration. This hybrid

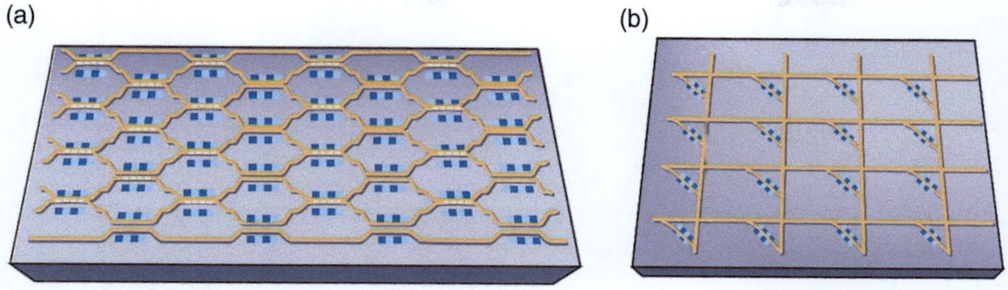

Figure 6.13 Schematic of PCM-based non-volatile platform to perform linear operations in an optical neural network. (a) A forward matrix-vector multiplier for a deep neural network based on PCM tunable beam splitters; (b) A convolutional block using a tunable attenuator.

approach (Figure 6.13) promises to deliver the energy efficiency and computational density required for next-generation optical AI accelerators while overcoming the scalability constraints of purely photonic implementations. Such architectures could enable practical, large-scale ONNs capable of handling increasingly complex machine learning tasks with significantly reduced power consumption compared to electronic counterparts.

Recent advancements in direct laser writing for photonic integration highlight a paradigm shift from conventional glass-based methods, which are constrained by permanent alterations, minimal refractive index modulation ($\Delta n \approx 0.02$), and micron-scale waveguide limitations, toward more versatile PCM platforms. Leveraging nanosecond laser technology originally developed for optical data storage, where feature sizes below 200 nm are routinely achieved, this approach enables reversible photonic circuit reconfiguration at significantly lower costs compared to femtosecond laser systems. The integration of wide-bandgap PCMs such as Sb_2S_3 with carefully engineered waveguide structures has addressed historical challenges of optical loss, as evidenced by successful demonstrations of tunable meta-optical elements. Key to this advancement is a multilayer design incorporating nanometer-scale PCM films between silicon nitride cladding and oxide substrates, which maintains optical mode confinement in low-loss regions while allowing complete, high-precision phase transitions through controlled laser pulses. By capitalizing on the refractive index contrast between amorphous and crystalline states at phase boundaries, this architecture enables the creation of lithography-free, reprogrammable photonic circuits that combine the nanoscale precision of optical storage technologies with the functional requirements of integrated photonics, opening new possibilities for adaptive optical systems.

The laser-writing technique utilizing PCMs enables unprecedented reconfigurability in PICs, extending far beyond basic waveguide routing to incorporate various functional elements like ring resonators and directional couplers (Figure 6.14; Chen et al. 2022). This innovative approach allows complete circuit reprogramming through reversible phase transitions, with functionality constrained primarily by laser precision rather than fixed architectures, essentially enabling pixel-level photonic programmability. Compared to conventional fabrication, the method offers significantly faster and more economical prototyping, though practical

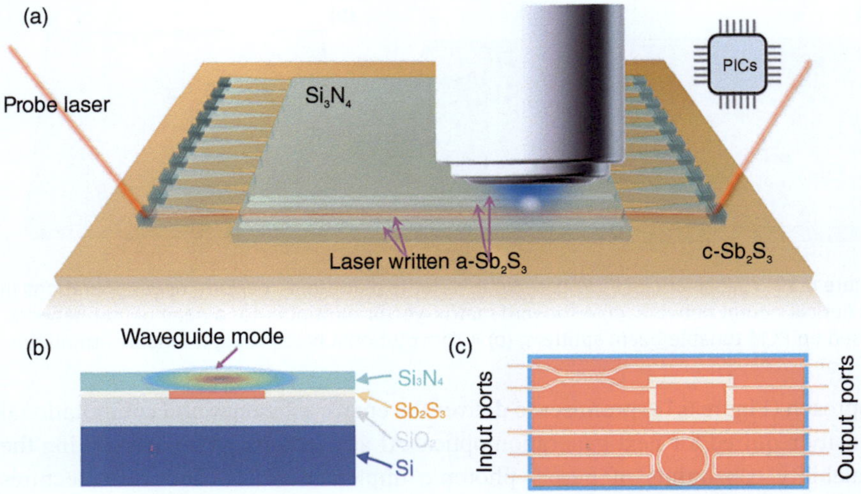

Figure 6.14 Lithography-free rewritable PICs using ns-lasers and PCMs. (a) Illustrates a proposed PIC utilizing wide bandgap PCM (Sb_2Se_3 or Sb_2S_3). Optical waveguides are created by amorphizing two Sb_2S_3 strips (highlighted in white with purple arrows) with a focused ns blue pulse laser, while keeping the central area crystalline. Prefabricated grating couplers serve as input/output interfaces. The circuit is fully reprogrammable, constrained only by material loss and the resolution of laser writing. (b) Details a material stack composed of a silicon substrate, a layer of silicon dioxide a few microns thick, a tens of nanometers thick Sb_2S_3 layer, and a Si_3N_4 cladding several hundred nanometers thick. Switching a strip of Sb_2S_3 from crystalline (dark orange) to amorphous (light orange) confines the waveguide mode predominantly in the Si_3N_4 layer to minimize optical loss. (c) Describes additional photonic components that can be fabricated on this platform, including directional couplers, multimode interferometers (MMIs), and ring resonators, with light and dark orange indicating amorphous and crystalline Sb_2S_3, respectively. *Source:* Chen et al. (2022) / American Chemical Society.

limitations in resolution and switching endurance restrict its application to larger-feature devices, positioning it as a complementary technology rather than a replacement for traditional nanofabrication. Recent developments demonstrate how this single-step laser patterning technique bypasses complex lithography processes by exploiting the refractive index contrast between PCM phases, creating fully functional circuits containing all fundamental components while enabling complete erasure and reconfiguration. This accessible platform democratizes photonics research by eliminating cleanroom dependencies, accelerating design exploration and educational opportunities through its unique combination of rapid turnaround, low cost, and rewritable capability, though it maintains distinct advantages and limitations compared to conventional manufacturing approaches.

Cao et al. (2025) present the realization of a reconfigurable Phase Change Material on Silicon (PCMoS) platform. This platform is achieved by laser direct writing of Sb_2S_3 film on a SOI substrate, enabling free-form routing between SOI waveguide ports. The PCMoS waveguide exhibits a low propagation loss of 0.7 dB/mm, strong light confinement with bend losses below 0.2 dB/90°, and reversible switching capabilities. Additionally, we successfully demonstrate the freeform routing of a 4×4 port routing network on the PCMoS platform. This method utilizes a more cost-effective laser processing system for writing and erasing, while also offering a more uniform crystallization. The research enables the implementation of free-form laser direct writing of optical waveguides to be CMOS compatible. This technology can also be extended to various platforms, including lithium niobate and III-V waveguide platforms. This expansion paves the way for broader applications such as neuromorphic computing, optical interconnects, microwave photonics, and quantum computing.

6.5 Discussion

The convergence of silicon photonics and chalcogenide PCMs heralds a transformative era in photonic engineering, enabling devices that transcend the limitations of static optical systems. This book explores the groundbreaking synergy between chalcogenide-based PCMs—notably Sb_2Se_3, Ge–Sb–Te (GST), and GST—and silicon photonic platforms, with a dual focus on non-volatile optical memory and dynamic color pixel displays. Over the past decade, chalcogenides have emerged as the linchpin for reconfigurable PICs, offering ultrafast switching, non-volatile state retention, and compatibility with CMOS fabrication. Characterized by their sulfur, selenium, or tellurium compositions, these materials exhibit reversible phase transitions between amorphous and crystalline states under thermal, electrical, or optical stimuli, inducing dramatic refractive index contrasts ($\Delta n > 1.0$) and extinction coefficient shifts that enable dynamic control over light-matter interactions at subwavelength scales. This capability addresses a critical gap in modern optics: the inability to achieve non-volatile, energy-efficient reconfiguration in on-chip devices, overcoming the high static power consumption and volatility of traditional tunable components like thermo-optic modulators.

The legacy of chalcogenides in optical data storage, rooted in rewritable discs, now evolves into photonic phase-change memory (P-PCM), where light-chalcogenide interactions enable ultrafast, non-volatile modulation of optical signals. By embedding PCMs into silicon waveguides or resonators, researchers have demonstrated devices capable of storing photonic states indefinitely, mimicking synaptic plasticity for photonic neuromorphic computing. Concurrently, chalcogenides are revolutionizing on-chip displays through structural color engineering, leveraging tunable optical constants to achieve dynamic color switching at the pixel level. Unlike pigment-based systems, these structural colors arise from interference, diffraction, or plasmonic effects in nanostructured PCMs, enabling sub-microsecond switching and zero static power consumption—critical for AR displays. However, integrating chalcogenides with silicon photonics presents challenges, including thermal crosstalk in densely packed arrays, stochastic crystallization dynamics, and endurance limits. Innovations such as backend-compatible deposition techniques, adiabatic tapers for loss mitigation, and scalable UV lithography for metasurfaces are bridging these gaps, while CMOS-compatible processes like SiO_2 trench etching and multilayer metallization enable wafer-scale integration.

This text synthesizes fundamental principles, advanced fabrication techniques, and cutting-edge applications—from non-volatile microring resonators with 25 dB extinction ratios to full-color metasurface displays achieving 92% standard Red Green Blue (sRGB) gamut coverage—providing a comprehensive roadmap for researchers. As the field advances, emerging frontiers such as machine learning-driven material discovery, monolithic 3D integration, and chalcogenide-based quantum photonics promise to redefine optical computing, sensing, and visualization. By unifying materials science, photonic design, and semiconductor manufacturing, this book charts a course toward intelligent photonic systems where light is orchestrated with unprecedented precision, marking a paradigm shift from passive transmission to active, adaptive control.

7
Chalcogenide Photonic Crystals

7.1 Introduction

Photonic crystal (PC) is an artificial material with periodic optical structure, and its design inspiration comes from the crystal structure in nature. It controls the propagation of light within a periodically arranged media, similar to the regulation of electrons by crystal structure. The unique optical property of PCs is the ability to form a photonic band gap (PBG), which prevents the propagation of light through a specific wavelength band. This unique property enables PCs to demonstrate versatile potential for use in the areas including optical communication, optical computing, and photonic integrated circuits. PCs also have the characteristics of localized field enhancement and low loss transmission. PC cavities with local field enhancement properties can greatly enhance the interaction between light and matter, making them suitable for enhancing light absorption and emission, and making PCs have potential application values in optical communication and optical computing.

Chalcogenides are a type of functional material formed by the combination of chalcogenides (sulfur, selenium, and tellurium) and non-transition metal elements (germanium, antimony, and indium). Typical representatives include phase change materials GST (Ge–Sb–Te), GSST (Ge–Sb–Se–Te), and SBS (Sb_2S_3). This type of material is known for its unique reversible phase transition between amorphous and crystalline phases, along with significant changes in optical and electrical properties, such as refractive index, reflectivity, and resistivity, during the phase transition process, making it important for optical storage (such as DVDs and Blu ray discs) and new storage devices (such as phase change memory and PCM). In addition, chalcogenides exhibit excellent optical nonlinear properties, wide spectral transparency windows, and adjustable band gaps, demonstrating great potential in fields such as photonics, infrared optics, and holographic displays. Its unique phase transition mechanism and multifunctional properties make it a hot research direction in information storage, optoelectronics, and smart materials.

Sulfide-based PCs are a type of photonic device designed based on sulfur-based materials, such as Ge–Sb–Te and As_2S_3. By utilizing their high refractive index, wide-infrared transparency window, and tunable optical properties, precise control of light propagation can be achieved within the PC structure. Owing to their distinctive optical properties, these materials have shown extensive potential for

Chalcogenide Nanophotonics, First Edition. Tun Cao, Ying Su, Yinghan Li, and Hongze Gao.

applications in fields including photonic crystal fibers (PCFs), PC cavities, and photonic crystal waveguides (PCWs). Sulfide-based PCFs achieve low loss and high nonlinearity optical transmission by introducing periodic air pore structures into the substrate material to form photonic bandgap effects. The material demonstrates excellence in optical communication, super-continuum generation, and mid-infrared (mid-IR) band gas sensing due to its high nonlinear coefficient and wide-infrared transparency. For example, PCFs based on As_2S_3 and Ge–As–Se materials have been implemented in the laboratory and demonstrated good performance. The PC cavity achieves high localization and enhancement of the optical field by designing high-Q microcavity structures in chalcogenides, making it suitable for low threshold lasers, nonlinear optical devices, and quantum optical experiments. Especially the reconfigurable PC cavity based on the Ge–Sb–Te material, which can dynamically control the cavity performance by utilizing its phase transition characteristics, has become a research hotspot. PCWs utilze the significant refractive index differential and minimal loss characteristics of chalcogenides to achieve efficient optical transmission and manipulation in periodic structures, and have important applications in optical signal processing and integrated photonics in the mid-IR band. For example, As_2S_3-based PCWs have been used to achieve efficient four-wave mixing (FWM) and nonlinear optical effects, providing an important platform for next-generation optical communication and computing technologies. Although sulfur-based PCs have shown promising prospects in fields such as optical fibers, cavities, and waveguides, further technological challenges such as material processing, integration, and stability need to be addressed to promote their large-scale application in practical optical systems and devices.

7.2 Chalcogenide PC Platform

Chalogenide PCs have become a key target in the scope of photonics owing to their distinctive photonic attributes, including high refractive index, wide-infrared transmission range, and high nonlinear optical coefficients. In recent years, research has mainly focused on optimizing material properties, device fabrication, and expanding applications.

Sulfide glasses, such as sulfides, selenides, and tellurides, have drawn a great deal of attention owing to their high transmittance and nonlinear optical coefficients in the infrared band. For example, materials such as As_2S_3 and As_2Se_3 exhibit excellent optical properties in the mid-IR band (3–5 μm and 8–12 μm) (Troles et al. 2007). By regulating the crystal structure, researchers can improve the optical properties of materials. For example, introducing alkali metal or alkaline earth metal cations can regulate the band gap, while introducing transition metal tetrahedra containing d^{10} configurations can enhance nonlinear optical activity (Wu et al. 2023). Sulfide-based PCs are widely used in the manufacturing of fiber Bragg gratings (FBGs) and PCFs. For example, As_2S_3 fiber has been used to manufacture high-sensitivity biosensors (Guo et al. 2022). Significant progress has been made in the design and manufacturing of multimode and single-mode photonic PCFs. For example, researchers have developed microstructured optical fibers for mid-IR transmission, which have a large area of the mode field and minimal bending loss (Troles et al. 2007).

Sulfur-based compound optical fibers have been used to manufacture highly sensitive biosensors. For example, Bragg gratings in As_2S_3 fibers can be used to detect the interactions of biomolecules (Guo et al. 2022). Through femtosecond laser direct writing technology, researchers have fabricated high-precision gratings in As_2S_3 glass for use in photonic devices. Significant progress has been made in the design and manufacturing of multimode and single-mode PCFs. For example, researchers have developed microstructured optical fibers for mid-IR transmission, which have a large-mode field area and low bending loss. In 2024, researchers proposed a Tb-doped chalcogenide multi-core fiber for high-power mid-IR laser applications (Wu et al. 2025). Sulfide-based PCs have important applications in infrared imaging, sensing, and communication fields. For example, sulfide glass is used to manufacture night vision devices and thermal infrared cameras. These materials are also widely used in the branch of nonlinear optics, such as in frequency converters and optical limiters. Sulfide-based PCs can be used to manufacture optical components for optical coherence tomography (OCT) and fluorescence imaging.

7.2.1 Introduction to PCs

The concept of PCs was independently proposed in 1987 by S. John (John 1987) and E. Yablonovitch (Yablonovitch 1987). They found that by periodically arranging media with different refractive indices, a photonic bandgap can be formed to regulate the transmission of light. This discovery laid the foundation for the study of PCs. In 1989, Yablonovitch and Gmitter were the first to experimentally verify the presence of three-dimensional photonic band structures, and subsequently, theoretical research on PCs rapidly developed (Yablonovitch et al. 1991). The theoretical research on PCs is mainly based on electromagnetic theory and band theory in solid-state physics. Common methods include plane wave expansion (PWM), transfer matrix method, and finite-difference time-domain method (FDTD). These methods are employed to compute the band structure and photonic bandgap characteristics of PCs, helping to design and optimize PC devices. In recent years, research on PCs has been continuously active worldwide. The research areas include the preparation methods, optical properties, and applications in different wavelength bands (such as microwave, infrared, and visible light) of PCs. For example, the Institute of Chemistry has made a series of advances in the field of PC preparation and application research, developed convenient methods for preparing large-area PCs, and prepared a series of high-performance optical devices.

7.2.1.1 Classification of PCs

PCs can be categorized according to their dimensionality, material composition, and functional properties, each category offering unique advantages for optical applications in waveguides, sensors, lasers, and quantum photonics. The most common way to classify PCs is based on their periodic structure in space, which

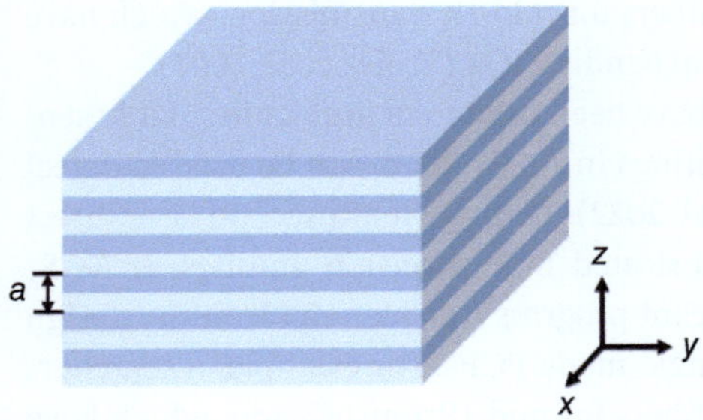

Figure 7.1 The multilayer film regarded as a one-dimensional PC. This structure is referred to as a one-dimensional PC, which is characterized by the fact that its dielectric function fluctuates periodically just along the z-axis. It is formed by two types of materials with varying dielectric constants arranged in alternating layers (represented in blue and light blue), arranged with a spatial period a. Each layer is supposed to be homogeneous and extends endlessly along the x-axis and y-axis. The periodicity along the z-direction is also considered to extend infinitely.

affects how they control light propagation. Based on the spatial dimension of their periodic structures, PCs can be classified into the following types:

One-Dimensional PCs A one-dimensional PC refers to a material with a periodic structure in one direction, usually made up of alternating layers of media with high and low refractive indices. Its structure is simple and easy to prepare, and common forms include multilayer dielectric films and Bragg reflectors.

As shown in Figure 7.1, the simplest PC is composed of alternating layers of materials with different dielectric constants, which is a multilayer film. This arrangement is not an innovative concept. The periodic change in the refractive index leads to the formation of photonic bandgaps—frequency ranges in which the propagation of light is prohibited. These band gaps can be engineered by modifying the thickness and the difference in refractive index of the constituent layers, allowing precise control over the optical reaction of the structure. Due to their strong wavelength selectivity and ease of integration, one-dimensional PCs have been discovered to applications not only in traditional optical coatings but also in emerging areas such as photonic integrated circuits, biosensing, and low-threshold lasers. Moreover, their compatibility with a diverse variety of materials, such as polymers, semiconductors, and even perovskites, makes them highly versatile for both passive and active photonic functions. As can be observed later, this kind of PC can act as a mirror (a Bragg mirror) for light with a frequency within a specified range, and it can localize light modes if there are any defects in its structure.

Two-Dimensional PCs A two-dimensional PC exhibits periodicity in two spatial directions—typically the x- and y-axes—while remaining uniform along the third direction, usually the z-axis. A representative example is shown in Figure 7.2, consisting of a square lattice of high-index dielectric rods that are inserted in a lower-index background material. In theoretical treatments, these rods are often assumed to be infinitely tall, ensuring perfect translational symmetry along the

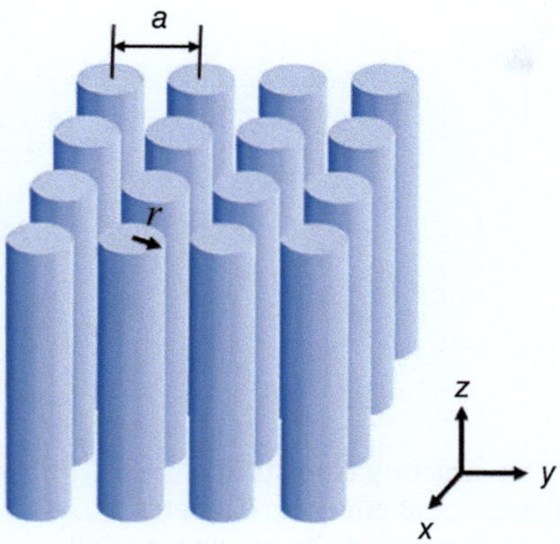

Figure 7.2 Two-dimensional PC. The structure is composed of a square-shaped lattice made up of dielectric rods. These rods have a radius of *r*, and the lattice has a constant of a. It maintains uniformity along the *z*-axis (assuming the rods have infinite height), and it shows periodicity in both the *x* and *y* directions. This configuration creates a two-dimensional PC, as the dielectric function varies periodically within the *xy* plane while remaining constant along the vertical direction. The symmetry and periodicity of the lattice play a crucial role in shaping the photonic band structure and determining the existence of in-plane bandgaps.

z-axis. For specific values of the lattice constant and dielectric contrast, such a structure can support a PBG that occurs in the *xy* plane. Within this frequency range, electromagnetic waves incident from any in-plane direction cannot propagate through the crystal; instead, they are strongly reflected due to Bragg scattering. This is in contrast to one-dimensional PCs, such as multilayer films, which only block light propagation along a single direction.

The two-dimensional PC's ability to suppress light transmission in all directions within the plane makes it particularly valuable for controlling in-plane light propagation. This property allows for the creation of advanced photonic devices, including waveguides, resonant cavities, beam splitters, and optical logic circuits. Furthermore, by introducing controlled defects into the lattice, it is feasible to generate localized states or guided modes, allowing for highly compact and integrated photonic components.

Three-Dimensional PCs Three-dimensional PCs exhibit cyclical variation of the dielectric function along all three spatial dimensions (*x*, *y*, and *z*). Unlike their one- and two-dimensional counterparts, 3D PCs are capable of manipulating light propagation in all directions, enabling the realization of a complete PBG—i.e., a frequency range in which electromagnetic waves are forbidden to propagate in any direction or polarization. This unique property opens the door to unparalleled manipulation of light–matter interactions, including the suppression of spontaneous emission, photon localization, and highly efficient confinement of light.

A variety of 3D PC geometries have been proposed and realized, including diamond-like lattices, woodpile structures, and inverse opals. These architectures typically require high symmetry and a large refractive index contrast to open

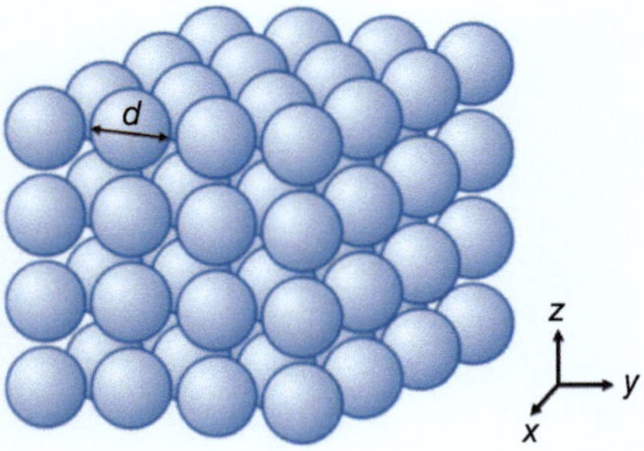

Figure 7.3 Three-dimensional opal-structured PC. This structure consists of a face-centered cubic (FCC) lattice of monodisperse dielectric spheres with diameter *d*, arranged in a periodic three-dimensional configuration. The periodicity occurs along all three spatial directions, forming a close-packed structure with a lattice constant determined by the sphere size. The crystal exhibits long-range order and symmetry characteristic of opal lattices, making it a versatile platform for controlling light propagation in three dimensions.

a complete bandgap. For example, in the case of inverse opals, a template of close-packed spheres is infiltrated with a high-index material and then selectively removed to create a periodic air-sphere lattice. Figure 7.3 depicts a schematic illustration of a basic opal-structured three-dimensional PC. Fabrication of such structures remains challenging, often involving advanced techniques such as self-assembly, direct laser writing, or layer-by-layer stacking. Despite these challenges, 3D PCs hold great promise for next-generation photonic devices, such as thresholdless lasers, zero-threshold nanocavities, and ultracompact optical circuits.

7.2.1.2 Basic Principle of PCs

A fundamental feature of PCs is their ability to prohibit the propagation of light within specific frequency ranges known as photonic bandgaps. Within these bandgaps, electromagnetic modes are forbidden, and as a result, light at those frequencies cannot propagate through the structure. This concept was first clearly articulated by Yablonovitch, who proposed that PCs can be used to suppress spontaneous emission by manipulating the photonic density of states (DOS) (Yablonovitch 1987).

Spontaneous emission, a fundamental quantum process, depends critically on the availability of optical states at the emission frequency. Specifically, the probability of spontaneous emission is directly proportional to the photonic DOS at the frequency of the emitted photon. When an atom or quantum emitter is embedded in a PC and its emission frequency falls within the photonic bandgap, the number of available photonic states at that frequency drops to zero. Consequently, the emission process is effectively suppressed, leading to strong inhibition of spontaneous emission.

On the other hand, PCs can be designed and fabricated to boost spontaneous emission as well. This takes place when the structure is engineered to accommodate many optical states at frequencies, such as at the band edges or in defect-induced localized modes. By introducing controlled impurities or structural defects into the PCs, localized states can be created within the bandgap. These impurity states often exhibit high quality (Q) factors and strong field confinement, leading to significantly

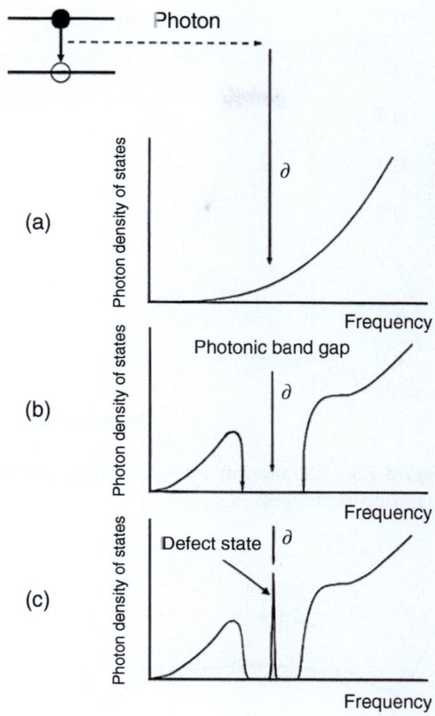

Figure 7.4 The influence of photon bandgap on atomic spontaneous emission. (a) In free space; (b) in PCs (where spontaneous emission is suppressed); (c) in defective PCs (where spontaneous emission is enhanced).

enhanced light–matter interactions and increased spontaneous emission rates at those frequencies. This effect is illustrated in Figure 7.4, where an impurity mode appears within the bandgap, enabling emission at a frequency that would otherwise be forbidden.

The appearance of photonic bandgap depends on the structure and dielectric constant ratio of PCs. Generally speaking, when the dielectric constant ratio between the two media in a PC gets larger, the incident light will be scattered more strongly, and the probability of the emergence of a photonic bandgap will increase. There is also an important factor that affects the existence of the bandgap: the geometric structure of the crystal. In 1990, Ho and Chan from the US Soukoulis group were the first to successfully predict the existence of a complete photonic bandgap in a three-dimensional PC featuring a diamond structure, occurring between the second and third energy bands (as shown in Figure 7.5) (Ho et al. 1990). Yablonovitch artificially manufactured the first three-dimensional PC with a full bandgap in the laboratory in 1991. A triangular pattern of air holes was distributed on the plane of a high dielectric material substrate, with each hole drilled three times in a direction perpendicular to the central axis (35°–26° from the substrate). These three drilling directions were clamped at a 120° angle to each other (as shown in Figure 7.6) From a three-dimensional perspective, this is a face centered cubic structure (Yablonovitch et al. 1991). However, later research has shown that this structure does not have an absolute bandgap. In nature, natural three-dimensional PCs that possess a complete bandgap are extremely scarce. To acquire PCs that

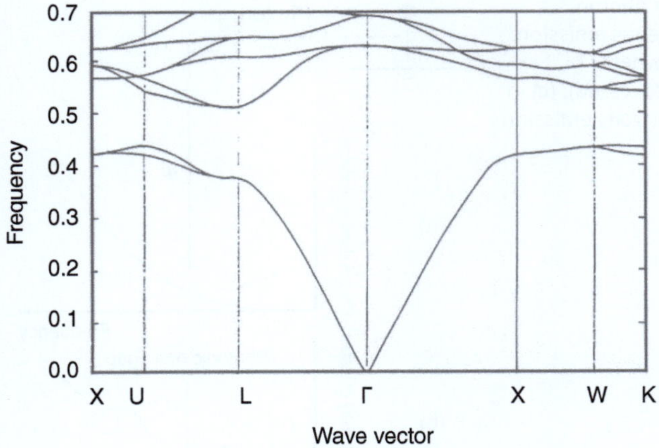

Figure 7.5 Calculation results of photonic bandgap in three-dimensional diamond-structured PCs.

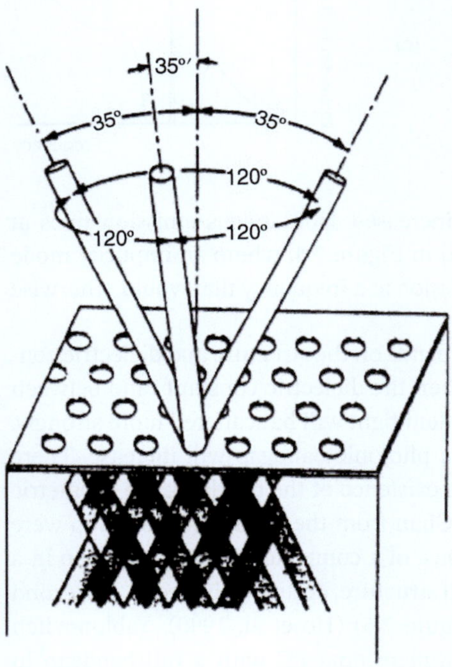

Figure 7.6 Three-dimensional PCs produced by Yablonovitch. *Source:* Ho et al. (1990) / American Physical Society.

meet our requirements, it should be a feasible and practical effective method to artificially modify the symmetry structure of some crystals.

Another main feature of PCs is photon localization. John proposed it in 1987: photons exhibit strong Anderson localization in a superlattice composed of carefully designed disordered dielectric materials (similar to what we currently refer to as PCs) (John 1987). If a certain amount of defect is added to the PC and photons with frequencies of the defect state are in harmony, they might be trapped at the position

Table 7.1 Comparison of characteristics between PCs and semiconductors.

	PC	Semiconductor
Structure	Periodic arrangement of materials featuring distinct dielectric constants	Periodic potential field
Research object	Propagation of electromagnetic waves (light) in crystals: bosons	Transport behavior of electrons in fermions
Eigen	$\left[\Delta x \left[\frac{1}{\xi(x)}\Delta x\right]\right] \boldsymbol{H}(x) = \frac{\omega^2}{c^2} H(x)$	$\left[-\frac{\hbar^2}{2m}\Delta^2 + V(\boldsymbol{x})\right] \cdot \psi(\boldsymbol{x}) = E\psi(\boldsymbol{x})$
Eigenvector	Electric field strength, magnetic field strength: vector	Wave function: scalar
Characteristic	PBG	Electronic bandgap
	Local patterns at the defect location	Defect state
	Surface state	Surface state
Scale	Electromagnetic wave (light) wavelength	Atomic size

of the defect. Once the light strays from the defect, it will quickly lose its intensity. In an ideal situation where the PC has no defects, there is no light–attenuation mode as per the periodic demands of its boundary conditions. However, once the original symmetry of the crystal is destroyed, a defect state with extremely narrow bandwidth may appear in the center of the PC bandgap (as shown in Figure 7.4c). Photon crystals have both point imperfections and linear imperfections. Light is localized in the plane perpendicular to the line defects and can only propagate along the direction of the line defects. Point defects seem to be completely enveloped by a total reflection wall. By using point defects, light can be "trapped" in a specific position and cannot propagate outward from any direction, which is equivalent to a microcavity. Table 7.1 compares the characteristics of PCs and semiconductors. It is not difficult to see from the table that PCs and semiconductors have astonishing similarities in their physical concepts. We can transplant the research methods of semiconductors to PCs. Of course, it must be noted that photons are bosons and electrons are fermions.

Before the 1980s, people believed that spontaneous radiation was a random phenomenon that could not be artificially controlled. In 1946, Purcell proposed that spontaneous radiation could be artificially altered, but it did not receive much attention. It was not until the concept of PCs was proposed that people changed their views (Yablonovitch 1987). The emission of photons arises not as an intrinsic characteristic of isolated particles but emerges from dynamic coupling between quantum systems and their surrounding field environment. The likelihood of spontaneous emission can be determined using Fermi's golden rule:

$$W = \frac{2\pi}{n}|V|^2\rho(w)$$

In the formula, $|V|$ is called the zero-point short element and is the state density of the optical field. Based on the fact that the probability of spontaneous emission is proportional to the state density, PCs are capable of suppressing spontaneous emission. If an atom with spontaneous emission is positioned within a PC and its spontaneous emission frequency happens to fall inside the bandgap, the probability of spontaneous emission is also zero as the state density of $\rho(w)$ at that particular frequency inside the bandgap is zero, which suppresses spontaneous emission. Conversely, when impurities are introduced into the PC, defect states featuring high-quality factors and a high DOS will emerge within the photonic bandgap. This can boost spontaneous emission, as depicted in Figure 7.4c. This phenomenon of regulating spontaneous emission is referred to as the Purcell effect.

As discussed earlier, PCs refer to materials featuring periodic structures, which are capable of regulating the propagation of light, similar to how semiconductor crystals control electrons. Based on their dimensionality, PCs are typically classified into one-dimensional (1D), two-dimensional (2D), and three-dimensional (3D) types. Each type of PC exhibits different characteristics in light propagation control, PBGs, and photon localization effects.

A one-dimensional PC typically is composed of alternating layers of two materials with different refractive indices, creating a periodic structure. In this structure, the propagation of light is suppressed, resulting in a PBG where certain frequencies of light cannot propagate through the crystal. This band gap effect is widely used in devices such as optical waveguides and filters. Since the one-dimensional PC has periodicity in only one direction, light primarily propagates along that direction, limiting its range of applications.

In contrast, a two-dimensional PC has a periodic structure along two directions. This structure not only forms a PBG in one direction but also restricts light propagation within the plane, allowing for more precise control over the light's propagation path. Furthermore, the structure of a 2D PC can induce photon localization effects, a feature that has important applications in optical waveguides, photonic integrated circuits, and other technologies.

A three-dimensional PC possesses a fully periodic arrangement in three-dimensional space, capable of forming a PBG throughout all directions, thereby completely prohibiting the propagation of specific frequencies of light. The PBG phenomenon in 3D PCs is the most pronounced, enabling complete control over light. Moreover, photon localization effects are particularly significant in 3D PCs, making them highly valuable in advanced optical, communication, and quantum information processing applications. Despite the complexity of their structure, 3D PCs offer precise regulation of light beams, making them an essential material for next-generation optical devices.

7.2.2 Characteristics of Chalcogenide PCs

7.2.2.1 Characteristics of Chalcogenide Materials

Owing to their distinctive physical and chemical properties, chalogenide materials possess extensive application prospects in the domains of optics and optoelectronics. The following is a description of the characteristics based on high refractive index, wide-band transparency, high nonlinearity, and tunability:

High Refractive Index Sulfur-based materials have a high refractive index, usually ranging from 2 to 3. This high refractive index characteristic enables sulfur-based materials to achieve more significant light bending effects in optical design, thereby optimizing the performance of optical components. For example, the high refractive index of sulfur-based glass gives it significant advantages in the design of infrared optical components. In addition, by modifying the composition of the material, its refractive index can be further controlled to meet different application requirements.

Wide-Band and Transparency Sulfur-based materials exhibit excellent transparency over a wide wavelength range, particularly in the infrared spectrum. Sulfur-based glass has excellent transmission performance in the mid-IR (3–5 μm) and far-infrared (8–12 μm) regions, and the transmission range of some materials can even extend to 20 μm. This wide-band transparency makes it an important material in fields such as infrared imaging, optical communication, and optical storage. In addition, the optical bandgap of sulfur-based materials can be optimized by adjusting their composition to meet specific application requirements.

High Nonlinearity Sulfur-based materials exhibit significant nonlinear optical properties, making them of significant application value in nonlinear optical devices. For example, the high nonlinearity coefficient of sulfur-based materials makes them excellent in domains such as optical switches, optical computing, and optical signal processing. In addition, sulfur-based materials also exhibit high efficiency in nonlinear frequency conversion, including second-harmonic generation and difference frequency generation. This high nonlinearity makes it a highly promising material platform in integrated photonics.

Tunability The optical and physical properties of sulfur-based materials can be adjusted through multiple methods. For instance, by changing the composition ratio of the material, the refractive index and transmittance range of sulfur-based glass can be flexibly controlled. In addition, the phase transition characteristics of sulfur-based materials also make them have broad application prospects in tunable optical devices. For example, antimony-based sulfide compounds exhibit significant refractive index changes under thermal, optical, and electrical stimulation and can be used for multibit operations and phase-change memory. This tunability makes sulfur-based materials have important potential applications in fields such as optical switches, optical storage, and optical sensors.

In summary, sulfur-based materials have shown great potential for applications in the fields of optics and optoelectronics because of their high refractive index, wide-band transparency, high nonlinearity, and tunability.

7.2.2.2 Advantages of Chalcogenide PCs

Chalcogenide PCs offer unique advantages for controlling and manipulating light at the nanoscale, making them extremely appealing for use in optical communication applications, sensing, nonlinear optics, and infrared technologies. Their exceptional optical properties, including a high refractive index, wide-band transparency, strong nonlinearity, and tunability, enable superior performance compared to traditional dielectric PCs.

Advantages of High Refractive Index A prominent merit of chalcogenide PCs lies in their remarkably high refractive index, which typically ranges from 2.0 to 3.5, depending on composition. The substantial disparity between the chalcogenide material and the surrounding medium boosts light confinement and slow-light effects, leading to improved photonic bandgap formation and better control over wave propagation. High refractive index materials can achieve more compact photonic device designs, reduce device size, and thus achieve miniaturization. Compared to conventional silicon-based PCs, chalcogenide PCs can achieve stronger optical localization, enabling miniaturized, high-efficiency optical devices, such as waveguides, filters, and resonators, with reduced footprint and enhanced performance.

Advantages of Wide-Band Transparency Chalcogenide materials exhibit excellent transparency over a broad spectral range, particularly in the mid-wave infrared (3–5 μm) and long-wave infrared (8–12 μm) regions. This wide-band transparency makes chalcogenide PCs ideal for infrared photonic applications, such as thermal imaging, remote sensing, and free-space optical communication. Unlike traditional semiconductors that suffer from strong absorption in the infrared, chalcogenide PCs enable efficient light manipulation in these longer-wavelength regions, facilitating the development of advanced mid-IR photonic circuits and biochemical sensors.

High Nonlinearity for All-Optical Signal Processing Another significant advantage of chalcogenide PCs is their high third-order optical nonlinearity, which is usually two to three orders of magnitude greater than that of silica. This strong Kerr nonlinearity enables ultrafast nonlinear optical effects, such as self-phase modulation, FWM, and super-continuum generation, making them well-suited for all-optical signal processing, ultrafast switching, and optical frequency conversion. Additionally, their high nonlinear refractive index enhances nonlinear photonic bandgap engineering, allowing for dynamic control over light propagation within the PC structure. This capability is essential for developing next-generation optical computing and high-speed telecommunication technologies.

Tunable Optical Properties for Reconfigurable Photonics Chalcogenide PCs also offer excellent tunability, allowing their optical properties to be dynamically modified through various external stimuli, including thermal, electrical, and optical excitation. Many chalcogenide materials, like $Ge_2Sb_2Te_5$ (GST) and GeSe, exhibit phase-change properties, enabling reversible switching between the amorphous and crystalline phases. This transition of phase induces significant changes in refractive index and optical absorption, allowing for reconfigurable photonic devices, programmable metasurfaces, and tunable filters. Additionally, chalcogenides can be engineered for electro-optic and all-optical tuning, enabling adaptive beam steering, variable optical attenuators, and dynamically adjustable photonic circuits.

Chalcogenide PCs combine high refractive index, broad-infrared transparency, strong nonlinearity, and tunability, making them highly advantageous for cutting-edge photonic applications. Their ability to provide strong optical confinement, efficient mid-IR manipulation, ultrafast nonlinear response, and dynamic tunability positions them as key materials for next-generation optical communication, sensing, and signal processing technologies. As research continues to advance fabrication techniques and material engineering, chalcogenide PCs are expected to play a critical role in the future of photonic integration and infrared optics.

7.2.3 Preparation Process of the Chalcogenide PC Platform

The preparation process of sulfide-based PC platforms covers various structures from one-dimensional to three-dimensional, involving various technologies such as magnetron sputtering, electron beam lithography, holographic lithography, and flexible substrate integration. These processes not only enable high-performance optical devices but also provide important technological foundations for the future development of integrated optics and photonics.

7.2.3.1 Preparation Process of One-Dimensional Chalcogenide PC Platforms

The fabrication process of one-dimensional photonic crystals (1D PhC) typically involves the deposition of multilayer thin films, with common techniques including radio frequency magnetron sputtering (RF sputtering), ion-assisted sputtering, molecular beam epitaxy, and plasma-enhanced chemical vapor deposition. In the preparation process, it is essential to choose appropriate materials as high refractive index layers (such as $Ge_2Sb_2Te_5$, SiO_2, TiO_2, and ZnO) and low refractive index layers (such as SiO_2 and Si_3N_4) and accurately control the thickness of each layer according to design requirements, usually a quarter wavelength thickness. For example, when using $Ge_2S_2Te_5$ (GST), GST and SiO_2 are deposited layer by layer on the substrate through sputtering technology to form periodic structures. The structural example is shown in Figure 7.7 (Gadhwal et al. 2022). In some cases, defect layers (such as SiO_2 doped with rare earth metals, ZnO and GaAs) may also be introduced to enhance nonlinear optical response. After preparation, induce phase transition or structural optimization of the material through annealing or photochemical processes to achieve the desired optical properties.

7.2.3.2 Preparation Process of Two-Dimensional Chalcogenide PC Platforms

In the preparation process of two-dimensional chalcogenide PC platforms, the preparation of PCWs is an important step. Its material selection is chalcogenides, such as $Ge_2Sb_2Te_5$. The preparation approach involves focused ion beam (FIB) etching technology, which serves to carve periodic structures onto chalcogenide thin films. Meanwhile, electron beam exposure technology can also be used to prepare PCWs on insulating silicon on insulator (SOI) substrates, as illustrated in Figure 7.8 (Reprinted from [Wang et al. 2013]). In terms of structural design, two-dimensional PCWs typically adopt a triangular lattice structure with apertures and periods of approximately 200 nm and 550 nm, respectively. This type of waveguide has broad

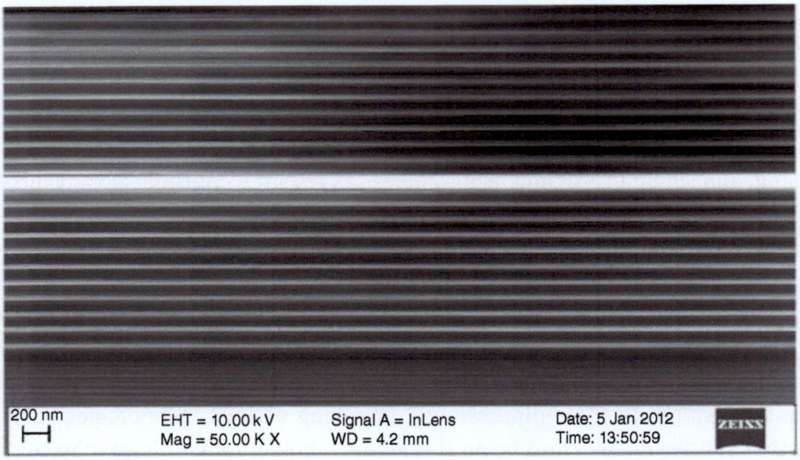

Figure 7.7 SEM image of the cross-section of a 1D microcavity with a ZnO defect layer in the middle; the dark and bright areas correspond to the SiO_2 and TiO_2 layers, respectively. *Source:* Reproduced from Gadhwal et al. (2022) / Informa UK Limited.

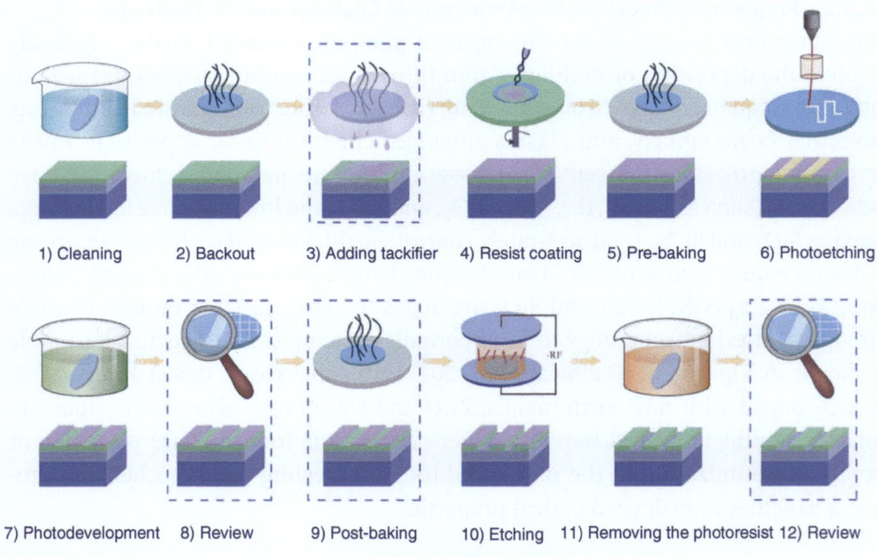

Figure 7.8 PCWs are manufactured through FIB etching.

application prospects in photonic devices like optical splitters, optical logic NOT gates, and sensors.

7.2.3.3 Preparation Process of Three-Dimensional Chalcogenide PC Platforms

The manufacturing of three-dimensional PCs typically involves several advanced methods, including self-assembly, direct laser writing, and nanolithography. A common approach for creating three-dimensional photonic structures is through self-assembly, where colloidal nanoparticles, such as silica or polystyrene spheres, spontaneously arrange into highly ordered structures. These nanoparticles are placed on a substrate in a regulated way, typically forming FCC or hexagonal close-packed lattices, which are the basic frameworks of many PC structures. To form the PC, a high-index material is infiltrated into the voids of the ordered lattice using a process like chemical vapor deposition or atomic layer deposition. After infiltration, the template can be removed through methods such as selective etching or dissolution, leaving behind a periodic PC with well-defined microstructures.

Alternatively, direct laser writing, which uses a focused laser beam to pattern a photoresist material, can be employed to directly fabricate three-dimensional PCs while precisely controlling the lattice structure. With this method, it is possible to generate complex and personalized designs at the nanoscale, providing flexibility in tailoring the bandgap properties of the crystal.

Another widely used technique is two-photon polymerization, which allows for the construction of three-dimensional structures in a layer-by-layer fashion. Using this approach, complex PC structures can be produced with submicron resolution, making it particularly suitable for high-performance photonic devices. The first type is the "Woodpile" structure, which is made of chalcogenides such as As-Se-Te. The preparation process involves depositing As-Se-Te thin films on a substrate using vacuum evaporation technology, writing periodic patterns on the film using dual-beam interference holographic lithography technology, and then selectively etching the unexposed portion to form a three-dimensional PC structure. The final structure is shown in Figure 7.9 (Feigel et al. 2001). In this structural design, each layer is a diffraction grating, and the layers are perpendicular to each other, making it of great application value in optical communication and integrated optical devices.

After fabrication, these three-dimensional PCs display distinctive optical characteristics, such as PBGs, which can be manipulated by adjusting the lattice constant, material properties, and structural geometry. These crystals find use in numerous applications, including optical filters, waveguides, and photonic cavities. In these applications, their capacity to manage light propagation is extremely crucial.

Figure 7.9 Schematic diagram of three-dimensional PC structure.

7.3 Applications

7.3.1 Chalcogenide PCFs

Chalcogenide photonic crystal fiber (Ch-PCF) is a specialized kind of optical fiber made from chalcogenide glass, which contains elements including sulfur (S), selenium (Se), and tellurium (Te). These fibers are constructed to have air holes arranged periodically in their cross-section, creating a PC structure that allows for tailored light propagation. Chalcogenide glass is a type of non-oxide glass that consists of elements like sulfur, selenium, and tellurium, which has high nonlinearity, a wide-infrared window, and adjustable refractive index. These properties make chalcogenide PCFs have significant advantages in photonic applications in the mid-IR wavelength range. Chalcogenide glass has a nonlinear refractive index that is several orders of magnitude greater than that of quartz glass, which is suitable for nonlinear optical applications. In the mid-IR wavelength band ranging from 2 to 20 μm, chalcogenide glass exhibits high transparency, which is suitable for infrared optical applications. The refractive index of chalcogenide glass can be adjusted by modifying its composition, which helps to optimize the performance of PCFs. Chalcogenide PCFs usually have periodic arrangement of air hole structure, and by designing the arrangement and size of these air holes, one can control the propagation characteristics of light. Combined with the high nonarity of chalcogenide glass and the low loss characteristics of the PC structure, chalcogenide PCFs show excellent performance in the mid-IR wavelength. Chalcogenide PCFs generate mid-IR super-continuum as broadband light sources for spectroscopy. They can be used for chemical and environmental sensing, such as detecting gases such as methane (CH_4) and carbon dioxide (CO_2). They can also assist in advanced imaging techniques such as OCT in biomedical imaging. They can also be used in nonlinear optical devices to achieve optical switching and signal processing in communications.

In order to achieve higher power laser transmission, increasing the mode area is the most straightforward and efficient approach. Meanwhile, to guarantee the quality of the beam, the laser must also meet the single-mode transmission conditions in the optical fiber. In conventional step-index optical fibers, it is extremely challenging to fulfill the requirements of a large-mode area and single-mode transmission at the same time. The advent of PCFs offers a novel approach to resolving this issue. In 2006, Person et al. prepared a photonic-crystal optical fiber with a mode field area of 108 μm using a glass composed of $Ge_{20}Ga_5Sb_{10}S_{65}$ and determined the single-mode transmission performance of the fiber through near-field spectroscopy (Le Person et al. 2006). As the structural design and preparation process of sulfur-based PCFs become more mature, in 2007, Troles et al. prepared sulfur-based PCFs with a large-mode field area using $Ge_{20}Ga_5Sb_{10}S_{65}$ as the base material (Troles et al. 2007). Figure 7.10 shows sulfur-based PCFs of different structural sizes. When the outer diameter of the fiber is 240 μm, the ratio of the aperture (d) to the hole spacing ($Λ$) is 0.31, and when the hole spacing is 14 μm, the mode field area of the fiber is 150 μm^2. When the outer diameter of the fiber becomes 400 μm and the hole spacing is 28 μm, the mode field area is 1000 μm^2. In theory, through reasonable

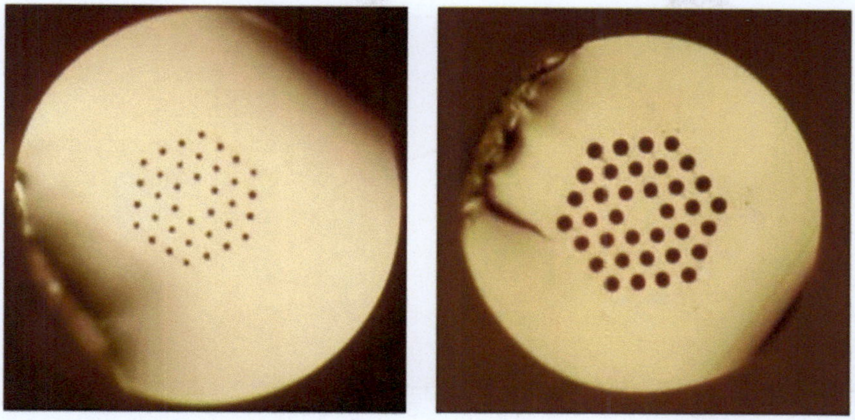

Figure 7.10 Chalcogenide large-mode field PCFs with different structural dimensions. *Source:* Adapted from Fatome et al. (2009).

structural design, the mode field area of sulfur-based photonic-structured optical fibers can even be increased to tens of thousands of square micrometers and still maintain single-mode transmission, which provides ideas for realizing high-power infrared laser transmission ranging from hundreds of watts to even kilowatts.

During the progress of large-mode-area PCF development, researchers proposed an all-solid-state design. This is because compared with the porous microstructure cladding, the all-solid-state design can prevent defects such as pore deformation that occur during the fiber-drawing process, eliminate the impact of environmental factors on fiber transmission, and prevent water contamination in the air.

In 2019, Ren et al. prepared a fully solid-state large-mode-field chalcogenide PCF with a mode field area of 5200 μm^2 and a fiber loss of 5.2 dB/m at a wavelength of 4 μm (Ren et al. 2019). In 2020, Feng et al. demonstrated a few-mode PCF for mid-IR high-power laser transmission (Feng et al. 2020). The actual measured fiber loss was 7.8 dB/m at a wavelength of 2 μm, and numerical simulation showed that its mode field area reached 10,500 μm^2. From the overall situation, the lowest loss of sulfur-based large-mode field PCF is several dB/m, which is far from the calculated limit loss of 10^{-2} dB/m. This is mainly because the total loss of PCF is composed of the material loss, defect loss, and bending loss. Due to the high overlap of mode field and glass material, the fiber loss of PCFs is very sensitive to material absorption, structural defects (capillary deformation collapse, scattering loss caused by the interface effect), and bending; at the same time, due to the special cladding structure, the beam quality of PCF transmission is more easily affected by the thermal lens effect. These problems limit the ability of sulfur-based PCF to transmit high-power lasers.

Rare-earth-doped chalcogenide glass fiber lasers and amplifiers offer significant potential for applications across various fields, particularly due to their crucial role in the near and mid-IR wavelength range. Recent advancements in the development of rare-earth-doped optical fibers made from chalcogenide glasses have significantly accelerated the progress of active devices designed to operate at wavelengths exceeding 3 μm (Mescia et al. 2012). These glasses are renowned for their chemical

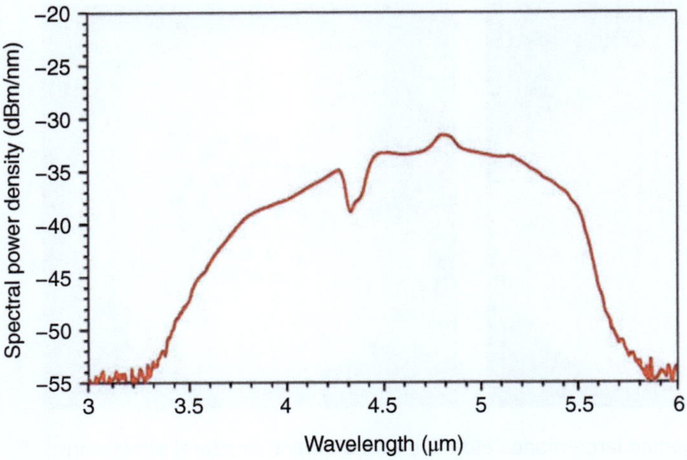

Figure 7.11 Chalcogenide large-mode field PCFs with different structural dimensions. *Source:* Adapted from Fatome et al. (2009).

and mechanical robustness, low toxicity, and substantial glass-forming capabilities, which enable the production of fibers with low loss. Moreover, their higher refractive index and lower phonon energy boost the rates of radiative decay, optimize the absorption and emission cross-sections of radiative electronic transitions, and reduce nonradiative multiphonon relaxation. All these factors lead to an improvement in quantum efficiency.

The electronic energy levels of rare-earth ions enable a multitude of advantageous transitions over a wide spectrum, spanning from 2 to 12 μm. Nevertheless, just a small number of glass hosts can efficiently support transitions at longer wavelengths. The comparatively low phonon energy of chalcogenide glasses makes it possible for efficient laser transitions to occur between closely spaced electronic states, thereby facilitating a diverse range of infrared transitions. Furthermore, the significant solubility of rare-earth elements in specific chalcogenide glasses further promotes the production of efficient doped lasers and amplifiers. This is achieved by minimizing ion clustering and lessening the concentration quenching effects. For instance, Figure 7.11 illustrates a diode-pumped Pr^{3+}-doped selenide glass fiber source, which exhibits broadband emission in the 3–5 μm wavelength range (Fatome et al. 2009).

The latest developments in the creation of rare-earth-doped optical fibers crafted from chalcogenide glasses have substantially driven forward the advancement of mid-IR laser technologies. Notably, the outstanding transparency of these fibers within the mid-IR spectrum enables thorough scanning throughout the whole spectral span of biomolecules. Additionally, the inherent chemical resistance of chalcogenide glass contributes to its strong biocompatibility with biological systems. Prudenzano et al. proved that the new Er^{3+}-doped chalcogenide PCF laser is feasible at 4.5 μm wavelength (Prudenzano et al. 2010). To enhance the pumping efficiency, the new cascade laser source operating at 4.5 m and 2.7 μm was theoretically studied with an erbium ion concentration of $N_{Er} = 7 \times 10^{26}$ ions/m^3. The concentration of

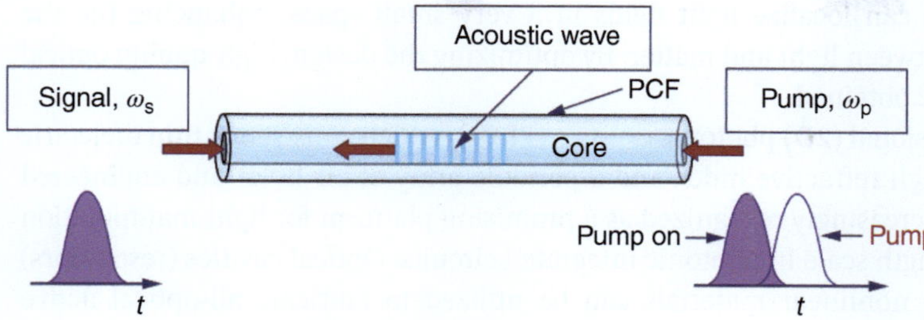

Figure 7.12 Schematic of the principle of SBS in PCF.

erbium and the length of the fiber were fine-tuned to achieve the highest possible output power for both signal wavelengths. The numerical outcomes indicate that the laser features a slope efficiency approaching the theoretical maximum, a low pump-power threshold, a short fiber length, and broad tunability within the mid-IR wavelength range.

The researchers also explored the slow light phenomena and potential applications of chalcogenide PCFs. In a theoretical research, Sinha and his colleagues explored the application of a single-mode As_2Se_3 chalcogenide PCF for generating tunable slow light, as depicted in Figure 7.12 (Sinha et al. 2015). Their simulations analyzed crucial elements like the highest allowable pump power to ensure undistorted output pulses, Brillouin gain, and the time lag that pulses encounter while traveling through the designed PCFs. The authors discovered that a Brillouin gain of approximately 88 dB/m and a time delay of around 137 ns could be achieved in a 1-m-long PCF when pumped with 100 mW of power. The findings additionally showed that both the time lag and Brillouin gain could be regulated by altering the pump power and the structural characteristics of the As_2Se_3 chalcogenide PCF. Diouf et al. introduced a stimulated Brillouin scattering slow light As_2S_3 PCF for temperature sensing (Diouf et al. n.d.). Under an input of 50 mW, the temperature measurement range is 23 °C and the resolution is 0.7 °C.

Recent advancements in chalcogenide PCFs have led to significant improvements in nonlinear optical performance, mid-IR super-continuum generation, and high-power applications. These fibers are becoming essential for next-generation optical communication, sensing, and biomedical imaging technologies.

7.3.2 Chalcogenide PC Cavity

Chalcogenide PC cavities are microcavity structures based on chalcogenide glasses (such as compounds of sulfur and tellurium), which exhibit high nonlinearity, a wide-infrared transparency window, and tunable refractive index. The core feature is to achieve a strong optical field and high-quality factor (Q value) through the periodic structure of the PC, which has significant advantages in mid-IR photonic applications. The nonlinear refractive index of chalcogenide glasses is considerably higher than that of conventional glasses, making them suitable for nonlinear optical applications. They have high transparency in mid-IR wavelength range of 2–20 μm.

PC structures can localize light fields in a very small space, enhancing the the interaction between light and matter. By optimizing the design, high-quality optical cavities can be obtained.

Two-dimensional (2D) photonic crystals (PhCs), which consist of a thin dielectric slab with a high refractive index and a periodic array of air holes and engineered defects, are increasingly recognized as a promising platform for light manipulation at the wavelength scale in photonic integrated circuits. Optical cavities (resonators) that integrate nonlinear materials can be utilized to fabricate all-optical active devices. By making use of highly nonlinear materials and high-Q microcavities, which boost the intensity of light, the power demands can be substantially decreased. It is widely recognized that the required input power is directly proportional to V/Q^2, in which V represents the mode volume. These devices, which have the potential to operate at multiple wavelengths, can serve as all-optical logic gates (or "photonic transistors") and lay the groundwork for more sophisticated all-optical circuits. Chalcogenide glasses also exhibit extensive photostructural effects. It is commonly acknowledged that photosensitivity stems from structural reconfigurations triggered by the absorption of photon energy by light near the electronic band edge of the material, which leads to alterations in refractive index and density. The potential to adjust cavity resonances via photosensitive chalcogenide films integrated into PhC devices seems quite promising.

Double heterostructure cavities trap light through the "mode gap" effect. These cavities are formed by altering a small section of the "W_1" waveguide, which locally increases the waveguide's effective index. This is typically accomplished by making geometric adjustments to the PC lattice, such as expanding the lattice period in the short segment of the waveguide. Alternatively, local refractive index modifications using photosensitivity have also been suggested as a potential approach.

Grillet et al. offer an inspiring initial showcase of photoinduced defects in chalcogenide planar PCs (Grillet et al. 2010). The authors successfully created a defect cavity in a PCW by utilizing photosensitivity. By employing in situ monitoring techniques, they could monitor the emergence and evolution of the cavity resonance as the exposure advanced. As a result, they could exert better control over the ultimate properties of the defect structure. Figure 7.13 illustrates the refractive index profile needed to fabricate a double heterostructure-type cavity via local modification of the refractive index. This structure is based on a "W_1" waveguide, which includes a short segment with an increased refractive index—achieved by removing a single row of holes from the hexagonal lattice. Ideally, this refractive index variation should follow a Gaussian distribution, with a contrast Δn of approximately 0.04 to optimize the quality factor (Q-factor). The samples were fabricated on a 300-nm-thick AMTIR-1 ($Ge_{33}As_{12}Se_{55}$) chalcogenide glass layer, and the hole array was patterned using FIB milling. The period of the structure is 520 nm, with the hole radius being 0.3 times the period. The photosensitivity of the samples was evaluated using the technique described in the paper, revealing that exposure to 633 nm light caused a decrease in the refractive index, and the maximum change that was observed was approximately 0.13.

Lee et al. conducted experiments using the setup shown in Figure 7.14, which is composed of two components, an exposure system and a measurement system

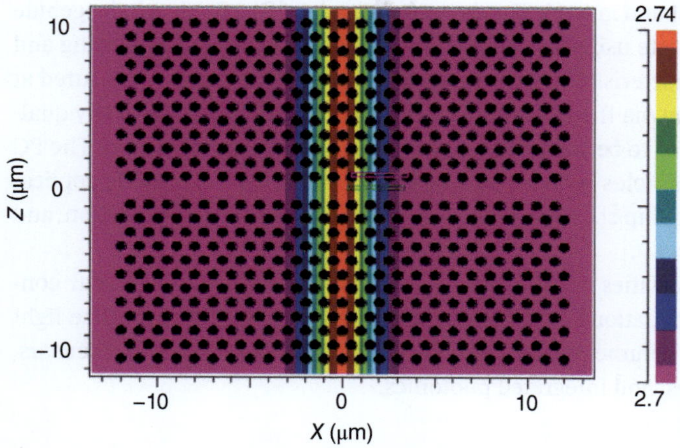

Figure 7.13 Depicts a diagram showing the index profile of a double-heterostructure cavity formed by local refractive index modification. *Source:* Gr llet et al. (2010) / SPIE.

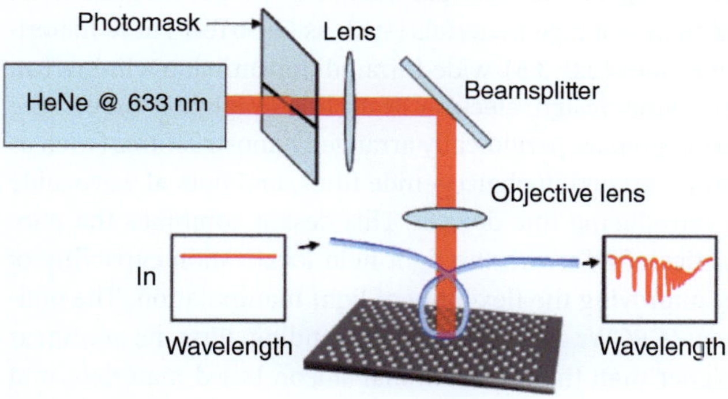

Figure 7.14 The schematic diagram of the experimental setup is shown. Selective exposure is realized by projecting a photomask onto the sample. Meanwhile, measurements are carried out by monitoring the transmission spectrum of a fiber taper that is in contact with the PC. *Source:* Figure is reprinted from Lee et al. (2008).

(Lee *et al.*, 2008). The exposure system projects the image of the photomask onto the sample, allowing for selective exposure. A 633-nm HeNe laser with a power of 6 mW served as the light source for the exposure. The measurement setup includes a tapered fiber, a broadband light source, and an optical spectrum analyzer (OSA). To perform the measurement, the taper is brought into contact with the PC sample, and the transmission spectrum of the taper is recorded. A reduction in the transmission spectrum signals coupling to the modes of the PC. The authors employed an annular taper with a waist diameter of 1.5 μm and approximated the contact length with the waveguide to be 12 μm. The spectrum was recorded at intervals of 30 seconds during the exposure process. In this experiment, a dark stripe with a nominal width of 8 μm was projected onto the sample, resulting in an index profile similar to the one shown in Figure 7.13.

The quality factor Q serves as a crucial parameter for evaluating the performance of PC cavities. Lin et al. was the first to manufacture a waveguide PC cavity that worked in the mid-IR band in 2013. The device was made of $Ge_{23}Sb_7S_{70}$ chalcogenide glass on a CaF_2 substrate using a combination of photolithographic patterning and FIB milling. The characteristics of the waveguide-coupled cavity were measured at a wavelength of 5.2 μm via the fiber end-fire coupling method, and the cavity quality factor was measured to be about ~2000 near the critical coupling region. The PC cavity device platform holds promise as a valuable component for various applications, for example, on-chip chemical detection, optical wireless communication, and thermal imaging.

Chalcogenide PC cavities offer a promising platform for high-Q optical confinement, mid-IR applications, and nonlinear optics. Their ability to localize light strongly within small volumes makes them ideal for next-generation optical sensors, quantum technologies, and integrated photonics.

7.3.3 Chalcogenide PCWs

Chalcogenide PCW is an advanced photonic device based on chalcogen elements (sulfur, selenium, and tellurium). Its core materials include chalcogenide glass (such as As_2S_3 and As_2Se_3) and phase-change materials (such as GeSbTe). These materials possess a high refractive index (2.0–3.5), wide-infrared transmission window (up to 20 μm). In terms of structural design, electron beam lithography or ion etching technology is usually used to prepare periodically arranged nanostructures (such as hexagonal holes or columnar arrays) in chalcogenide films, and optical waveguide channels are formed by introducing line defects. This design combines the photonic bandgap characteristics of PCs with the light-field localization capability of waveguides, significantly improving the flexibility of light manipulation. The optical properties of chalcogenide PCWs are extremely outstanding. First, its nonlinear effect is 10^3–10^4 times higher than that of traditional silicon-based materials, and the enhancement of the third-order nonlinear polarizability ($\chi^{(3)}$) makes it an ideal platform for ultrafast all-optical signal manipulation (for example, wavelength conversion and super-continuum generation). Second, high-refractive-index contrast (such as the combination of As_2S_3 and silicon substrate) can form a wide photonic bandgap, which effectively limits the propagation of light in defective waveguides and reduces energy loss. In addition, chalcogenides have excellent optical properties in the mid-IR band (1.5–20). The excellent light transmittance of the waveguide (200 μm) makes up for the deficiency of silicon materials in this band, providing a new way for gas sensing and biomedical imaging. Dynamic adjustability is another highlight: for example, GeSbTe (GST) phase-change material can achieve rapid regulation of the refractive index up to $\Delta n = 2.0$ by inducing the transformation between the crystalline and amorphous states via laser or electrical pulse, thereby giving the waveguide optical switch, reconfigurable filtering, and other functions. Combined with thermo-optical or electro-optical effects, external excitation (such as temperature and electric field) can further adjust the transmission characteristics of the waveguide, laying the foundation for intelligent photonic networks.

In practical applications, this type of waveguide has shown broad prospects. In the realm of nonlinear optics, its pronounced nonlinear properties are utilized. It

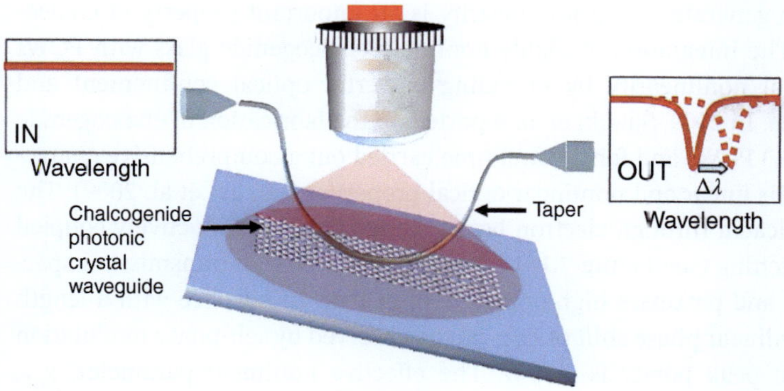

Figure 7.15 A diagram that shows the principle underlying the photosensitive post-tuning of a chalcogenide PCW. *Source:* Lee et al. (2007) / Optical Society of America.

is used to design all-optical switches and super-continuum light sources, such as realizing picosecond signal processing in the communication band or providing a wide-spectrum light source for mid-IR spectral analysis. Dynamic adjustability has given rise to reconfigurable photonic devices, such as programmable optical path filters, modulators, and optical memories based on phase-change states, which are of great value in optical computing and photonic neural networks. In the field of mid-IR photonics, chalcogenide waveguides can accurately detect the characteristic absorption spectra of gases such as CO_2 and CH_4, promoting the development of environmental monitoring technology; their infrared light penetration ability is also suitable for label-free imaging of biological tissues, assisting pathological diagnosis. In addition, thanks to the low intrinsic loss (<0.1 dB/cm) and compact micron-scale structure of chalcogenide glass, chalcogenide PCWs can be heterogeneously integrated with other materials (such as silicon-based optoelectronic devices) to build multifunctional photonic chips that support complex functions such as FWM and optical delay lines.

Like chalcogenide PC cavities, chalcogenide PCWs can also be photosensitively post-tuned. Lee et al. used the photosensitivity of $Ge_{33}As_{12}Se_{55}$ chalcogenide glass materials to near-bandgap light to perform post-tuning experiments regarding the dispersion of two-dimensional PCWs (Lee et al. 2007). When the material was exposed to 633 nm light, both its refractive index and volume changed. As a result, the resonant coupling wavelength between the tapered fiber and the W_1 waveguide mode shifted by over 5 nm, as shown in Figure 7.15. This is the first demonstration of the principle of photosensitive post-tuning in planar PC devices. The observed wavelength shift followed a saturating exponential growth pattern as the exposure increased, suggesting the potential for precise control over the photosensitive tuning process. This study highlights the promising nature of this post-processing technique for applications that require fine-tuning of existing structures, such as in single-photon source technologies or it can be used to directly create resonant cavity structures by means of light.

The ability to generate strong nonlinearity is an important property of chalcogenide PCWs. The integration of highly nonlinear chalcogenide glass with PCWs enhances optical nonlinearity by providing powerful optical confinement and slow-light effects. In 2009, Suzuki et al. reported on the fabrication of chalcogenide glass (Ag-As_2Se_3) PCWs and for the first time carried out a comprehensive characterization of their linear and nonlinear optical properties (Suzuki et al. 2009). The waveguide fabricated through electron beam lithography and inductively coupled plasma (ICP) etching (see Figure 7.16) exhibits a characteristic transmission spectrum of a PCW and possesses high optical nonlinearity. In a device with a length of 400 μm, a nonlinear phase shift of 1.5π can be observed by self-phase modulation when the input peak power is 0.78 W. The effective nonlinear parameter, γ_{eff}, which is estimated according to these observations, reaches 2.6×10^4 W/m. FWM is additionally detected inside the waveguide, while two-photon absorption at the optical communication wavelength is sufficiently small, resulting in an advantage value greater than 11. Later, Suzuki et al. reported enhanced nonlinearity due to low-dispersion slow light in Ag-As_2Se_3 chalcogenide PCWs (Suzuki and Baba 2010). They devised and manufactured a third-row lattice-shifted waveguide. Then, they verified the existence of slow light by inspecting its group refractive index spectrum and analyzing the transmission of picosecond pulses. Even with slow light enhancement, two-photon absorption remains negligible. Regarding self-phase modulation, symmetrical spectral broadening and central tilt were detected, which suggested a phase shift of 1.5π. The slow light enhancement reduced the required input power by half, aligning well with theoretical predictions. This leads to a notable effective nonlinear parameter of 6.3×104 W (1 m), which stands as the highest value ever reported for any nonlinear waveguide so far. Owing to the slow light effect, the conversion efficiency of FWM improved from −22 dB to −14 dB, as shown by the red circles in Figure 7.16. This research attains a well-proportioned optical nonlinearity by integrating lower operating power and shorter device length. Although there remains room for enhancement in optical coupling and spectral flatness, these results evidently show the potential of high-efficiency, compact, on-chip nonlinear devices featuring ultrafast response times.

In a similar vein, Monat et al. demonstrated triple-harmonic generation (THG) within a dispersion-engineered slow-light PCW fabricated from AMTIR-1 chalcogenide glass (Monat et al. 2011). Owing to the comparatively low loss and dispersion within the slow-light region (c/30), coupled with the high nonlinearity of the material (~2), they achieved a conversion efficiency of $1.4 \times 10^{-8}/W^2$, which is 30 times higher than that of similar silicon waveguides. Additionally, uniform visible light modes were detected along the waveguide. These findings expand the possible applications of THG on slow-light platforms, for example, the direct observation of the spatial development of propagating modes.

PCWs can produce slow light effects, and the researchers studied slow light in chalcogenide waveguides. In 2015, Han et al. studied the broadband slow-light properties of $Ge_{20}Sb_{15}Se_{65}$ sulfur-based PC slab waveguide by altering the radii of the air holes in the two innermost rows (Han et al. 2015). Using the three-dimensional PWM method, the band structure, group refractive index, and dispersion of the

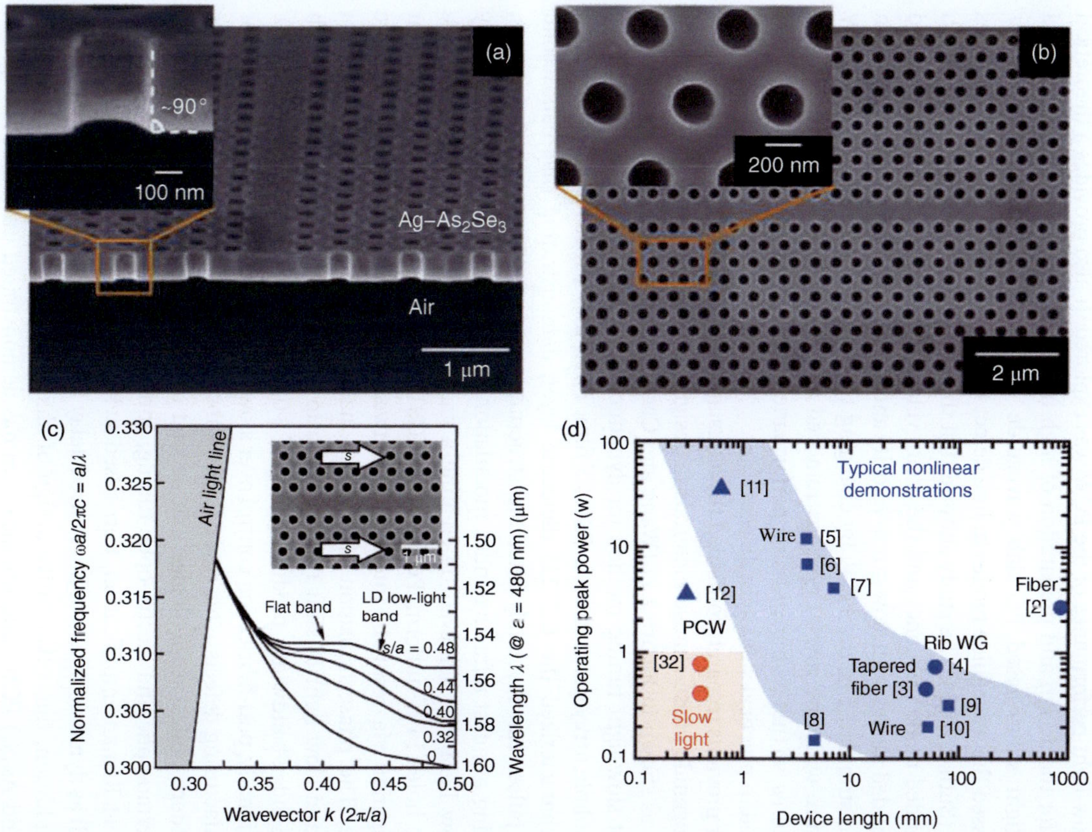

Figure 7.16 The SEM image of a manufactured Ag-As_2Se_3 PCW is presented. (a) Cross-sectional view; (b) top view. (c) Photonic band diagram of the Ag-As_2Se_3 LS-PCW. (d) Correlation between the operating power and the device length in the context of optical nonlinear phenomena, like self-phase modulation (with a 1.5π nonlinear phase shift) and FWM across various material waveguides. *Source:* (a,b) Suzuki et al. (2009) / Optica Publishing Group and (c,d) Suzuki and Baba (2010) / Optical Society of America.

waveguide were calculated, and their relationship with the radius of the inner air holes was analyzed. At the same time, the radius of the first and second air holes was optimized to obtain a symmetrical sulfur-based PCW structure with zero group velocity dispersion, and the group refractive index of 125, 40, and 18 was obtained within a 20% variation range, corresponding to the bandwidth of 1.7, 5.6, and 9.7 nm at the communication wavelength, and the impact of the refractive index on the slow-light performance of the PCW was explored. It also provides a theoretical basis for the design and preparation of high nonlinear, low-dispersion broadband slow-light sulfur-based PC slab waveguide devices. In the same year, Suthar discussed the waveguide modes in line defect PCWs with As_2S_3 square-rod lattices in air (Suthar 2015). The study showed that the size of the defect rods affects both the guided mode frequency and the group velocity. The author also demonstrated that guided modes with slow group velocity and very large group velocity dispersion can be stimulated in the PCW. By changing the size of the defect rods, the PCW structure with one row of gratings removed can be adjusted compared to other structures, so that its transmission performance can be fine-tuned. Such PCWs can be used in practical slow-light devices.

Our previous research combined the adjustable refractive index characteristics of chalcogenide phase-change materials to design the first mid-IR tunable slow-light PCW, as shown in Figure 7.17 (Wang and Cao, 2020). The researchers proposed a robust slow-light tuning method for dynamically reconfigurable PCWs based on a nanohole array penetrating a GST225 film. First, they experimentally fabricated and characterized the GST225 dielectric layer, demonstrating that nanosecond laser pulses can reversibly alter the structural state of GST225 within ~150 ns, enabling ultrafast refractive index modulation. Theoretically, the reconfigurability of slow light was achieved by creating a waveguide via removing a single line of air holes in a uniform PC based on GST225. This design allows slow-light propagation via resonant interference between guided modes and the periodic lattice. The phase transition of GST225 enables the operational wavelength of flat-band slow light to shift from 3575 nm to 4905 nm while maintaining a high group-index-bandwidth product (GBP) of ~0.24 and low dispersion with $\beta_2 \sim 0$. Additionally, post-fabrication control over slow-light properties can compensate for manufacturing defects. This research has substantial implications for the progress of modern optical technologies that depend on strengthened linear effects (such as thermo-optic and electro-optic tuning, and optical gain) and nonlinear effects (such as Raman amplification, Kerr-based switching, and wavelength conversion), as well as dynamic time-slot interleaved delay lines.

In addition, the collaborative design of chalcogenide compound (such as GeSbTe and As_2S_3) PCWs and microcavities has demonstrated unique advantages in the fields of all-optical information storage, nonlinear optics, and dynamic photonic devices and has become a research hotspot in integrated photonics in recent years. Grillet et al. demonstrated that integrating PCs with a novel category of highly nonlinear materials, chalcogenide glasses, is a highly promising method for achieving all-optical chips (Grillet et al. n.d.). The authors investigated various kinds of PC resonances. The coupling technique to chalcogenide PCWs and cavities via tapered

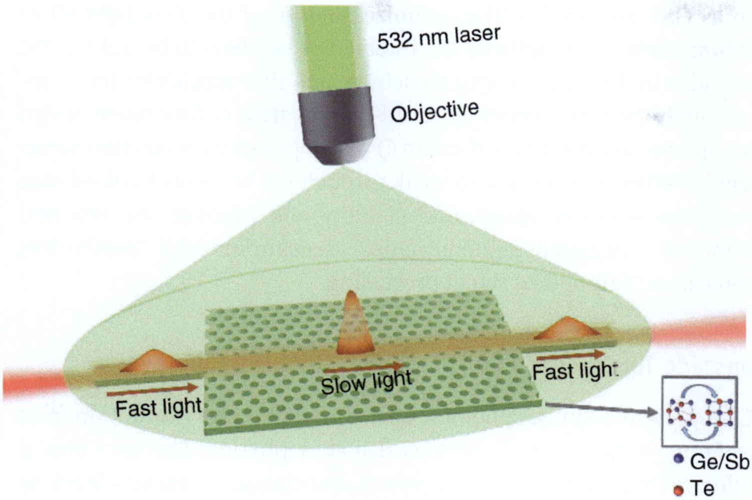

Figure 7.17 Schematic of a reconfigurable slow-light PCW with an array of air nanoholes passing through a 200-nm-thick GST225 film suspended in air. *Source:* Reprinted from Wang and Cao (2020).

nanowires is detailed. Resonant guiding in chalcogenide glass PC films using the Fano-probe technique was demonstrated. Strong resonances were observed in the transmission spectrum of normally incident light, and these resonances are associated with the Fano coupling between free-space and guided modes. Findings that are in good agreement with the outcomes of –3D-FDTD simulations were achieved. Moreover, it was found that the guided mode close to the center of the first Brillouin zone accounts for the main spectral characteristics. Singh et al. proposed a channel drop filter by positioning a defect cavity adjacent to the line defect waveguide in an As_2S_3 two-dimensional PC (Singh and Bhargava, 2022). The filter structure can realize multi-drop channel transmission under larger cavity size, and obtain a good quality factor for the research. This filter based on two-dimensional sulfide PhC can be used for filtering applications and can also be used for externally controlled optical switching applications.

Chalcogenide PCWs have shown great application potential in the field of mid-IR photonics due to their unique optical properties and flexible structural design. Their core advantages are first reflected in the wide-spectrum transparency and low transmission loss in the mid-IR band (2–20 μm) (for example, the light transmission range of $Ge_{23}Sb_{12}S_{65}$ materials covers 1–12 μm, and the intrinsic loss is <0.1 dB/cm), providing an efficient platform for gas sensing, thermal imaging, and free-space communication. Combined with high nonlinear optical response (the nonlinear refractive index n_2 of As_2Se_3 can reach 10^{-17} m^2/W) and slow-light effect (group refractive index $n_g > 50$), chalcogenide waveguides have achieved significant efficiency improvements in nonlinear applications such as FWM and optical parametric oscillation. In terms of dynamic tunability, wavelength-selective dispersion control and high-speed optical switching are achieved through nanosecond-level refractive index modulation ($\Delta n > 1.5$) or photothermal effect of phase change materials (such as GST and Sb_2S_3). The optimized design of the slow-light effect (flat band structure) further strengthens the interaction between light and matter, showing unique value in the fields of optical caching and high-sensitivity detection. In the future, through the introduction of multi-physics field collaborative design (such as electro-optical and mechanical control) and topological protection structures, chalcogenide PCWs are expected to break through the bottlenecks of existing devices in power consumption, speed, and stability and promote the practical application of mid-IR integrated photonic chips in environmental monitoring, quantum photonics, and full-optical signal processing.

7.3.4 Chalcogenide Topological PCs

Topological PCs represent a kind of photonic structure that draws inspiration from topological insulators in the field of condensed matter physics. The core idea is to achieve topological boundary states with robust transmission characteristics by designing the band topological properties of PCs. In comparison to conventional PCs, the boundary states of topological PCs are topologically protected and naturally immune to structural defects, disordered perturbations, and backscattering. Therefore, they show unique advantages in the fields of low-loss optical transmission and high-fault-tolerant photonic devices. In recent years, the combination of chalcogenides (such as GeSbTe and Sb_2S_3) with topological PCs has given rise to dynamically reconfigurable and nonlinearly enhanced topological photonic devices.

In 2019, our group was the first to put forward the concept of robust control of topological edge states in dynamically reconfigurable topological PCs composed of GST225 rod arrays (Cao et al. 2019). The topological characteristics of PCs are realized through the optical equivalent of the spin Hall effect. This gives rise to topological edge states at the boundary between the non-trivial and trivial regions of the crystal. In these states, light is capable of traveling along a path of any shape to a specific point without experiencing back reflection. Theoretically, we obtain the reconfigurability of topological edge states by building a two-dimensional PC with a deformed honeycomb lattice composed of GST225 rods. Thermoelectric models show that a bias voltage can reversibly change the phase of GST225 within 150 nanoseconds, resulting in an extremely rapid alteration of the refractive index of GST225, which then modifies the spectral position of the topological edge states. As depicted in Figure 7.18, we show that the PC is capable of swiftly switching the propagation of topological edge states at a wavelength of 2174 nm. This showcases the flexibility of the proposed reconfigurable platform to achieve distinctive device functionalities. These discoveries are crucial for expanding the design possibilities of next-generation optical communication devices that depend on the spin-orbit angular momentum of light, spin-dependent transport, and high-intensity optical currents.

Kang-Hyok et al. theoretically probed into the stable optical bistability and multistability within a topological cavity-coupled waveguide system. They made use of both single-color and two-color pumping combined with Kerr nonlinearity. They cited a chalcogenide valley-Hall photonic system that can support

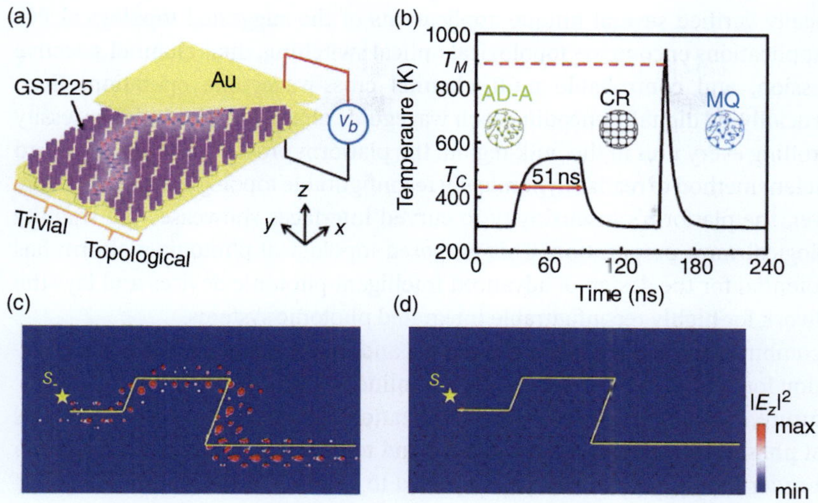

Figure 7.18 Two-dimensional topological PCs constructed with GST225 and their ability to "turn on/off" light propagation. (a) Schematic diagram of a topological two-dimensional PC made of GST225 pillars. (b) Left: Schematic diagram of a unit cell consisting of six GST225 pillars. Right: A magnified view of the hexagonal unit cell. At a wavelength of 2174 nm, the propagation of light along the diamond path can be "switched on/off" by toggling between the (c) amorphous state and the (d) crystalline state. *Source:* Reprinted from Cao et al. (2019).

second-order topological corner states as an instance. The proposed system exhibits a high Q-factor resonance of 1.86×10^4, a substantial intensity enhancement of approximately 8000 within the communication range, and a low-threshold optical bistability of 6 kW/m for a single pump. The system's bistability remains resilient to structural defects, owing to its topological properties. Notably, under two-color pumping, the system demonstrates multiple bistabilities and multistabilities, enabling optical switching across several stable states. The findings from this study have promising applications in photonics, such as all-optical transistors, multi-level optical switches, logic gates, and memory devices, all of which remain unaffected by structural imperfections that typically arise during the manufacturing process.

Valley twist states and valley-polarized chiral edge states are distinguished by topologically protected one-way propagation, presenting a hopeful method for manipulating light waves. In recent years, these states have attracted considerable attention in the realm of topological photonics. However, achieving flexible and tunable dispersion between two distinct topological states, along with the development of a dynamically controllable photonic platform to switch topological wave routing, continues to present a major challenge. Han et al. put forward a reconfigurable topological wave routing structure within the telecommunication frequency range. In this structure, a cylinder made of the phase-change material Sb_2S_3, which has a tunable refractive index, is inserted into each topological channel to dynamically modify the dispersion of the topological edge state (Han et al. 2024). By switching between the amorphous and crystalline phases of Sb_2S_3, the researchers numerically verified several unique applications of the suggested topological PC. These applications encompass topological optical switching, dual-channel selective transmission, and controllable multi-channel cross-waveguide operation. Even more crucially, by digitally encoding each waveguide channel without the necessity of controlling every unit in the bulk region, the platform provides a straightforward and efficient method to realize dynamically reconfigurable topological wave routing. Moreover, the platform's insensitivity to curved interfaces showcases the stability of topological wave propagation. This proposed topological photonic platform has great potential for the design of advanced intelligent photonic devices and lays the groundwork for highly reconfigurable integrated photonic systems.

The combination of chalcogenide materials and topological PCs opens up a new dimension for dynamically reconfigurable, nonlinearly enhanced photonic systems. In the future, through the coordinated optimization of material innovation (such as ultrafast phase-change chalcogenide alloys) and topological structure design (such as high-order topological states), it is expected to achieve "intelligent topology" of photonic devices and promote integrated photonics into a new era of self-adaptation and anti-interference.

7.4 Discussions

Chalcogenide PCFs, chalcogenide PC cavities, and chalcogenide PCWs are all photonic devices based on chalcogenide glass materials (such as compounds of sulfur, selenium, and tellurium), which have the characteristics of high nonarity, wide-infrared transparency window, and tunable refractive index. They are different in structure, function, and application but also have connections.

Name	Chalcogenide PCF	Chalcogenide PC cavity	Chalcogenide PCW
Structure	A fiber optic structure composed of periodically arranged air holes	A microcavity made of periodic structures, such as air holes or high-refractive-index-contrast structures	Waves made of periodic structure Guides, used to guide the light field along a specific path
Features	Primarily employed for light transmission, particularly for long-distance and low-loss transmission in the mid-IR band	Localized and enhanced light field, achieving high-Q optical resonance	Efficient transmission and positioning of light fields supporting nonlinear optical effects and integrated photonics applications
Applications	Infrared optical communications, nonlinear, super-continuum lasers	Nonlinear, optical switching, sensing, on-chip integration	Nonlinear, reconfigurable devices, on-chip integration, filters, all-optical signal processing

All three are based on chalcogenide glass materials, which have high nonlinearity, a wide-infrared transparency window, and adjustable active index. The periodic structure of PCs serves to realize the localization, transmission, or enhancement of the optical field. All three have significant advantages in mid-IR photonic applications, especially in nonlinear optics, sensing, and integrated photonics. In practical applications, the three can be used in combination. For example, chalcogenide PCFs are applicable for optical transmission. Meanwhile, chalcogenide PCWs can be utilized for optical signal processing, and chalcogenide PC cavities are employed for optical field enhancement and filtering. Chalcogenide PCFs, chalcogenide PC cavities, and chalcogenide PCWs each have their own advantages in mid-IR photonics, but there is also potential for synergistic applications. In the future, with breakthroughs in materials, manufacturing technologies, and application scenarios, these devices will have a significant impact on nonlinear optics, quantum optics, integrated photonics, and infrared sensing promoting the rapid development of mid-IR photonic technology.

8

Chalcogenide Metamaterials

8.1 Introduction

In materials science research, "metamaterials" are typically defined as materials with artificially designed properties capable of manipulating electromagnetic waves. Chalcogenide metamaterials are an emerging category of synthetic materials exhibiting remarkable optical, electronic, mechanical, and thermal properties. These materials are usually compounds of chalcogenides (such as sulfur, selenium, and tellurium) and other metal or nonmetal elements, which can achieve precise modulation of electromagnetic waves by designing their microstructure. Due to their unique crystal structure and chemical composition, chalcogenide metamaterials exhibit excellent optical absorption properties, electrical conductivity, and superconductivity in the optical field. In particular, they are capable of attaining nearly perfect optical absorption within the spectral range and have garnered significant interest in new optoelectronic devices and superconducting materials. By accurately controlling the geometry, size, and arrangement of the materials, the absorption of light can be effectively increased, showcasing exceptional potential in applications such as optoelectronic devices, sensors, and optical filters.

Chalcogenide metamaterials are not only tunable in terms of their microstructure, but their physical properties can also be regulated by altering the synthesis conditions. For example, varying the ratio of different alloys or applying external pressure can effectively influence the material's conductivity and optical response. Additionally, chalcogenides achieve high absorption and scattering of light through mechanisms such as surface plasmon resonance and quantum confinement. These characteristics provide a strong theoretical foundation for developing high-efficiency energy conversion devices, sensors, and optoelectronic devices.

Chalcogenide metamaterials also excel in the study of perfect absorbers. Absorbers constructed with these materials can achieve nearly 100% light absorption efficiency over a wide wavelength band. This characteristic endows these metamaterials with broad applicability across multiple frontier domains of science and technology, such as photovoltaic power generation, sensors, and thermoelectric devices. For instance, combining chalcogenides with gold nanoparticles to form composite structures can achieve tunable light absorption properties, providing new ideas for the development of sensors as well as photoelectric conversion devices.

Chalcogenide Nanophotonics, First Edition. Tun Cao, Ying Su, Yinghan Li, and Hongze Gao.

Chalcogenide metamaterials also show a unique promise for thermal management and thermoelectric applications. For example, they are an ideal candidate for designing efficient thermoelectric materials because of their high thermal conductivity and excellent thermoelectric performance. By introducing phase change materials, such as chalcogenides, it is possible to modulate the thermal conductivity properties of the material and realize intelligent regulation of heat flow. This property holds promise for implementation in critical areas including thermal management and energy harvesting, to meet future sustainability challenges.

This chapter provides an introduction to chalcogenide metamaterials, systematically categorizing the different types and exploring the structural features and performance traits of each kind. Additionally, it highlights the applications of chalcogenide metamaterials in areas such as optical switching, perfect absorbers, and beam steering, analyzing their advantages in broadband absorption, photonic devices, and sensors. By integrating recent research findings and practical case studies, the chapter reveals the potential applications of chalcogenide metamaterials. Finally, it sums up the advancements and difficulties encountered in present-day research, with the goal of providing a theoretical basis and practical perspective for the subsequent development of chalcogenide metamaterials.

8.2 Chalcogenide Metamaterials Platform

Owing to their distinctive electronic structure, optical response, and dynamic modulation abilities, chalcogenides have emerged as a significant platform within the realm of metamaterials. In recent years, with breakthroughs in two-dimensional materials, nanostructure engineering, and phase transition technology, chalcogenide metamaterials have shown revolutionary potential in photonics, energy, flexible electronics, and biomedicine.

The material properties and design principles of chalcogenide metamaterials can be summarized in the following three points. Among them, one aspect is the electronic and optical characteristics. The bandgap coverage of the chalcogenides ranges from zero bandgap to broad forbidden bands, realizing a full spectral response from the infrared to the visible. The monolayers of the transition metal tungsten/molybdenum chalcogenides are all S–M–S sandwich structures, i.e., one metal atomic face separates the sulfur atoms in two hexagonal faces by covalent bonding, while the layers are held together by feeble van der Waals forces. These materials possess not only structures and properties comparable to those of graphene but also bandgap widths that change according to the number of layers. Take MoS_2 as an example, the bulk material of it shows an indirect bandgap of approximately 1.2 eV, while when thinned to a single layer thickness, it shows a direct bandgap of approximately 1.8 eV (Kuc et al. 2011). This type of bandgap and the variation in bandgap width make them shine in the fields of optoelectronic detection, nanoelectronic devices, and gas detection (Figure 8.1).

The second property of chalcogenide metamaterials is phase transition modulation and dynamic response. Chalcogenide metamaterials (e.g., $Ge_2Sb_2Te_5$,

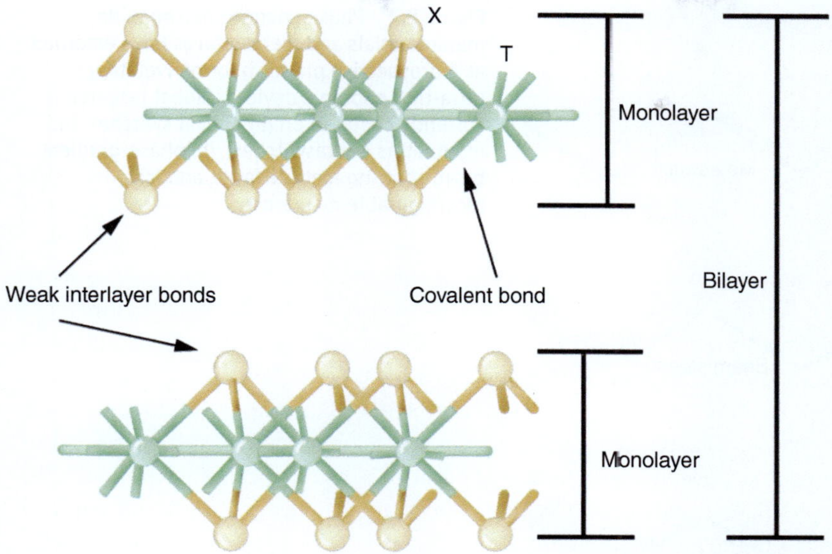

Figure 8.1 The TX_2-type layered transition metal dichalcogenides (T: transition metal; X: chalcogenide) feature a characteristic atomic architecture. Each monolayer consists of a three-atomic-layer X–T–X sandwich structure with covalent bonding between constituent atoms, while adjacent layers are interconnected through weak interlayer bonds.

chalcogenide glasses) have become the core materials for reconfigurable photonic devices by virtue of their unique phase transition dynamics and tunability of optical response (Li et al. 2024). Chalcogenides (e.g., $Ge_2Sb_2Te_5$, GST) can be reversibly transformed in the crystalline/amorphous state by light pulses or electrothermal stimulation. For example, GST is induced by femtosecond lasers, when phase transition speeds up to nanoseconds and refractive index changes in the range $\Delta n > 2$ ($\lambda = 1550$ nm). This property stems from the dynamic reorganization of covalent bonds of the sulfur genus elements (Te, Se, etc.) in the chalcogenides: weak bonds are easily broken in the amorphous state, whereas a network of strong bonds is formed in the crystalline state, resulting in sudden alterations in optical and electrical characteristics. The weak coupling properties of the interlayer van der Waals bonds during phase transitions in polymerized chalcogenide superlattices (e.g., GeTe–Sb_2Te_3) allow the chalcogenide element layers to slide and form intermediate suitable structures. This self-organization effect gives materials the ability to "remember" external stimuli, such as Te-doped chalcogenide glasses that achieve ultra-10^5 cycle lifetimes through dynamic reorganization of Se–Te bonds. In terms of dynamic response, the crystallization activation energy of GST is as low as 2.3 eV, and the phase transition energy consumption is only 0.1 nJ/bit, which is much lower than that of conventional phase transition materials. Chalcogenide glasses are triggered by femtosecond lasers, reducing the response time to picoseconds and meeting the demand for high-speed optical switching. The all-chalcogenide composite structure realizes over 1000 reversible phase transition cycles within the near-infrared wavelength range, while the optical loss remains below 0.5 dB.

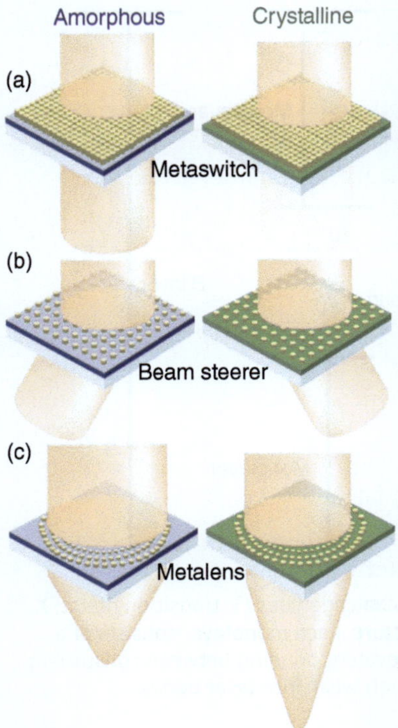

Figure 8.2 Phase-change chalcogenide metamaterials and metasurfaces have emerged as a prospective platform for nonvolatile, ultra-thin photonic devices. Global research is currently centered on (a) optical switches and modulators (metaswitches), (b) phase-gradient beam-steering metasurfaces, and (c) reconfigurable metalenses.

In terms of multidimensional tunability, the phase transition temperature window can be extended by component design to fit different environmental requirements. The 2H/1T phase transition of 2D transition metal chalcogenides (e.g., MoS_2, WS_2) is tuned by electrostatic doping or intercalation, realizing the dynamic adjustment of the light absorption coefficient from $10^5\ m^{-1}$ to $10^7\ m^{-1}$ (Figure 8.2).

The third point, chalcogenide metamaterials have significantly improved interfacial charge transport efficiency by vertically stacking two-dimensional transition metal chalcogenides to construct van der Waals heterojunctions, which has become a core research direction in the field of optoelectronics and integrated circuits. Different two-dimensional transition metal chalcogenides are vertically stacked by chemical vapor deposition or mechanical transfer techniques to form atomically flat interfaces. The layers are solely held together by van der Waals forces, thus circumventing the lattice mismatch issue that conventional heterojunctions face. For example, the MoS_2/WSe_2 heterojunction has a layer spacing of about 0.7 nm, and the formation of type II energy band pairs at the interface promotes the separation of photogenerated electron–hole pairs (Pi et al. 2019). In addition, the introduction of sulfur vacancies or Te doping at the heterojunction interface can optimize the carrier concentration distribution. For example, the interfacial defect density of Se-doped MoS_2/WSe_2 heterojunctions is reduced to $10^{10}\ cm^{-2}$, with a 30% increase in the charge transport efficiency. Through strain engineering, the lattice constants of MoS_2 can be modulated, which allows for the continuous adjustment of the bandgap ranging from 1.2 eV to 1.9 eV.

From the classification perspective, the chalcogenide metamaterials platform can be unfolded based on a three-dimensional framework of dimensional–functional–dynamical properties.

When categorized by dimension and structure, chalcogenide metamaterials can be divided into two-dimensional layered materials, quasi-one-dimensional nanostructured materials, and three-dimensional bulk and thin-film materials.

Two-dimensional layered materials include three types of transition metal chalcogenides, noble metal chalcogenides, as well as heterojunctions and superlattice structures of chalcogenides.

- **Transition Metal Chalcogenides:** MoS_2, WS_2, etc., are prepared by chemical vapor deposition or mechanical stripping, and their monolayer structures have a direct bandgap and high quantum efficiency, which are suitable for photodetectors and light-emitting diodes.
- **Noble Metal Chalcogenides:** Such materials excel in high-speed field effect transistors and quantum devices by virtue of their high carrier mobility and air stability (e.g., PtS_2, $PdSe_2$).
- **Heterojunctions and Superlattice Structures of Chalcogenides:** The MoS_2/WSe_2 heterojunction constructed by van der Waals stacking can form a type II energy band arrangement to enhance the photogenerated carrier separation efficiency, while the Moore superlattice modulates the exciton behavior through a periodic potential field.

Quasi-one-dimensional nanostructures can be categorized into nanotubes and Janus structures, quantum dots, and nanocrystals.

- **Nanotubes with Janus Structures:** WS_2 nanotubes are synthesized by the template method, and their bent structure leads to energy band splitting and enhanced bulk photovoltaic effect for self-supplied photodetectors.
- **Quantum Dots and Nanocrystals:** PbS quantum dots achieve narrow-band light absorption through size modulation for near-infrared imaging and solar cells.

Furthermore, three-dimensional bulk and thin-film materials include chalcogenide glasses and copper-based chalcogenides. For example, the Ge–As–Se–Te system, with a wide infrared transparent window and ultrafast nonlinear optical response, is used in photonic integrated devices; Cu_2Se, based on the "phononic liquid–electronic crystal" mechanism, is used to optimize the thermoelectric properties.

According to the classification of functional properties, chalcogenide metamaterials are mainly used in the fields of optoelectronics and photonics, energy conversion and storage, sensing, and biomedicine. The specific applications of chalcogenide metamaterials will be introduced in the next subsection.

Categorized by the dynamic response mechanism, chalcogenide metamaterials can also be classified into phase transition-driven materials and external field modulated materials. As mentioned before, the phase-change-driven material $Ge_2Sb_2Te_5$ supersurface realizes dynamic tuning of the reflection spectrum through laser-induced crystalline-amorphous transition, which can be used for dynamic holographic displays. The electric field tuning of the energy band shift of MoS_2/WSe_2 heterojunction can dynamically switch the photoconductivity mode of the device, which is suitable for reconfigurable logic circuits.

In summary, the platform of chalcogenide metamaterials can be categorized by the dimensions of material dimensions, functional properties, dynamic response mechanisms, etc., and its multidisciplinary intersections provide abundant design possibilities for the fields of optoelectronics, energy, and biomedicine.

8.3 Applications

8.3.1 Chalcogenide Optical Switches

In modern telecommunications, imaging, and data storage industries, the demand for rapid data processing and advanced computing capabilities has driven significant technological advancements. Optical networks, which utilize light pulses rather than electrons as information carriers, have emerged as a superior alternative to traditional electronic networks, offering enhanced speed and efficiency. Nevertheless, a crucial obstacle in the deployment of optical networks is the creation of rapid and dependable switching mechanisms for light signals at different network nodes. This switching process necessitates the precise isolation of specific signals from incoming channels at input ports and their subsequent routing to designated output ports (Gustafson et al. 1987).

Current switching solutions predominantly employ optoelectronic circuits that require the conversion of optical signals to electronic form for modulation and processing, followed by reconversion to optical signals for transmission. While functional, this dual conversion process introduces significant latency, ultimately limiting data processing and transfer speeds within the network. Consequently, the development of all-optical switching alternatives has become a paramount objective in optical network technology, with key design requirements including minimal latency, compact physical dimensions, and low power consumption.

Recent advancements in chalcogenide phase-change materials (ChG–PCM) have enabled the realization of various optical switches that meet these criteria, facilitating the creation of reconfigurable photonic networks with optimized power efficiency and spatial requirements (Zhang et al. 2020). However, the successful integration of such technologies into existing infrastructure necessitates the consideration of backward compatibility with current commercial systems. Therefore, an ideal solution must not only provide all-optical operation capabilities to address current bottlenecks and latency issues but also maintain compatibility with electrical addressing systems to ensure seamless large-scale deployment.

In this context, optoelectronic switches made of metamaterials (metaswitches) that take advantage of the reconfigurable characteristics of ChG-PCM have emerged as an especially promising platform. These devices combine the benefits of all-optical operation with electrical compatibility, potentially serving as a transformative technology for next-generation telecommunications networks while maintaining integration capabilities with existing infrastructure.

In recent years, research efforts have focused on two primary categories of metaswitches: hybrid plasmonic and all-dielectric configurations (Meng et al. 2020). Hybrid plasmonic metaswitches are engineered through the integration of nanostructured metamaterials or metasurfaces composed of noble plasmonic metals with chalcogenide phase-change materials (ChG-PCM), which serve as a dynamically switchable dielectric medium. This combination makes use of the distinctive optical features of the two materials to attain effective switching functionality.

On the contrary, all-dielectric metaswitches are either built completely from ChG-PCM or formed as hybridized all-dielectric metasurfaces. These structures usually comprise single-layer or multilayer dielectric stacks that are nanostructured with features smaller than the wavelength, which allows for accurate control of light–matter interactions. Both approaches offer distinct advantages in terms of performance, scalability, and compatibility with existing photonic technologies, making them promising candidates for next-generation optical switching applications.

Hybrid plasmonic metaswitches leverage the phenomenon of localized surface plasmon resonances (LSPRs), which are generated by the collective oscillations of conduction electrons, which are induced by light at the interface between metallic structures and dielectric materials on subwavelength scales. These resonances exhibit a strong dependence on the refractive index of the surrounding medium. Consequently, by dynamically modulating the refractive index of the chalcogenide phase-change material (ChG-PCM), a significant alteration in the optical response of the metamaterial device can be achieved. Despite their potential, hybrid plasmonic metaswitches are hindered by substantial ohmic losses intrinsic to metallic components, particularly in higher frequency ranges such as the UV–Vis spectrum (Khurgin 2015). These losses are exacerbated by the inherent properties of metals and the challenges associated with their nanofabrication.

When depositing ultra-thin metal films, which are usually deposited by standard methods such as evaporation or sputtering, discontinuous growth often occurs on commonly used substrates like silicon, glass, or quartz. This lack of continuity causes the percolation threshold in film thickness to become evident (Abeles et al. 1984; Kazmerski and Racine 1975; Heavens 1991), which can be mitigated by employing a wetting layer (Park et al. 1997; Denier et al. 1993; Logeeswaran et al. 2007; Chen et al. 2010) or using lattice-matched substrates (Palmberg et al. 1967). However, these solutions introduce additional fabrication steps, complicating the process and limiting scalability. Furthermore, ultra-thin metallic films exhibit a granular structure characterized by smaller grains compared to thicker films. The increased grain boundary scattering of free electrons in these fine-grained structures contributes significantly to ohmic losses, with grain boundary scattering being the dominant loss mechanism in ultra-thin films (Naik et al. 2013). Additionally, employing lithographic techniques for nanopatterning metallic films frequently leads to rough surfaces and edges, which in turn further intensify the losses caused by the enhanced scattering of free electrons (Kretschmann 1972).

A further crucial challenge is the noncompatibility between noble metals and standard complementary metal-oxide semiconductor (CMOS) fabrication processes, posing significant barriers to the integration of plasmonic metamaterials with conventional electronic components. This lack of compatibility arises from the propensity of metals to spread into silicon, creating deep traps that significantly impair the performance of nanoelectronic devices (Yau and Sah 1972). While diffusion barriers can be implemented to prevent metal diffusion—as demonstrated with copper in silicon-based fabrication processes—metals like gold and silver remain incompatible with current CMOS technologies. These limitations underscore the need for alternative approaches that can achieve strong electric and magnetic resonances with minimal losses while maintaining compatibility with established fabrication processes. Such advancements would pave the way for more efficient and scalable integration of plasmonic metaswitches into next-generation photonic and electronic systems.

Reconfigurable metaswitches derive their functionality from the unique properties of chalcogenide phase-change materials (ChG-PCM), which can trigger reversible phase transitions between the amorphous and crystalline states via optical or electrical stimuli. This ability to change phases offers outstanding compatibility and flexibility in the field of optoelectronics, making ChG-PCM a cornerstone of modern reconfigurable photonic devices. The switching mechanisms in these metaswitches are primarily achieved through two distinct tuning approaches: global and local tuning, each offering tailored control over the amplitude, polarization, and phase response of the device. In the global tuning scheme, the phase transition of the entire metasurface is induced, resulting in a collective modulation of the device's optical response. This approach is relatively straightforward and effective for applications requiring uniform changes across the entire metasurface. However, for more sophisticated applications, local tuning is often preferred. In this scheme, individual subwavelength structures, or meta-atoms, are designed to allow independent control over their optical responses. This is typically accomplished by incorporating a layer of ChG-PCM into the meta-atom design or by directly nanostructuring the ChG-PCM itself, which functions as a dielectric medium with a high refractive index. By selectively switching the phase of the ChG-PCM at specific locations, precise local tuning of each meta-atom's optical response can be achieved. This degree of control allows for unprecedented manipulation of the phase and polarization characteristics of the device, opening new possibilities for advanced photonic applications.

In 2010, Samson et al. presented a plasmonic metamaterial with tunable properties by integrating asymmetric split rings (ASR) etched into a 70-nm gold film with a 200-nm layer of phase-change material GLS (Samson et al. 2010). As depicted in Figure 8.3a, there is a metamaterial structure. The GLS layer was deposited on a silicon nitride membrane. When it crystallized through electrical bias, it led to a 150-nm blue shift in plasmonic resonance. This phase transition allowed the device to function as an electro-optic switch, achieving a 4:1 optical contrast in the 950–1800 nm range (Figure 8.3b), demonstrating the possibility of phase-change materials being used in reconfigurable optical devices.

While early demonstrations enabled switching from the amorphous to crystalline phase of chalcogenide glass (ChG) films via electrical bias, achieving all-optical switching and reversible transitions remained a challenge. In 2013, Gholipour et al. addressed this limitation by utilizing $Ge_2Sb_2Te_5$ (GST), a phase-change material widely employed in rewriteable optical discs and electronic phase-change

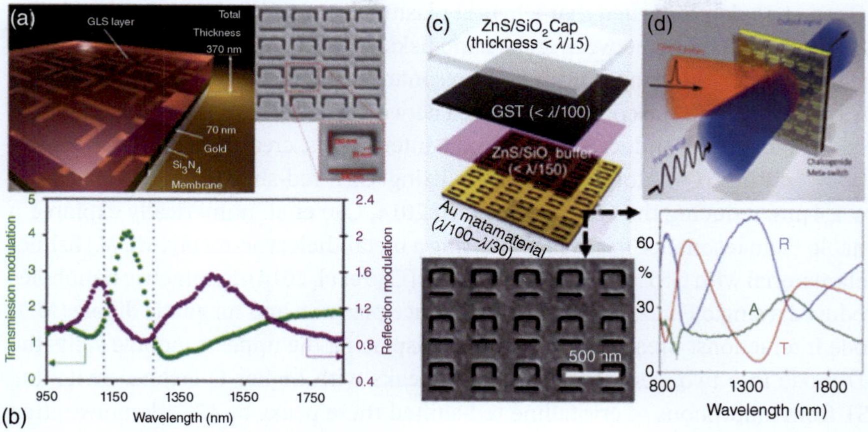

Figure 8.3 (a), (b) Plasmonic electro-optic switch based on ChG-PCM. (a) A schematic diagram of the hybrid metadevice structure (on the left) and a scanning electron microscope (SEM) picture of an array of asymmetric split rings (ASR) (on the right). (b) The experimentally measured contrast of transmission and reflection modulation caused by the phase transition of the Ge-Ga-S (GLS) layer. (c) An artistic illustration of the metamaterial structure (at the top) and an SEM photo of an array of ASR (at the bottom). (d) A schematic showing the operation of the device, emphasizing its all-optical functionality (at the top). The reflection, transmission, and absorption spectra are presented (at the bottom). *Source:* S'amson et al. (2010) / AIP Publishing, Gholipour et al. (2010) / Europe PMC.

random-access memory (PCRAM) technologies, to realize a fully reconfigurable metaswitch (Figure 8.1c–d) (Gholipour et al. 2013). The GST layer was reversibly switched between amorphous and crystalline states using a pulsed laser source at $\lambda = 660$ nm, with pulse intensity and duration optimized for each transition.

The metamaterial consisted of an array of $50 \times 50\,\mu m$ asymmetric split rings (ASR). These ASRs were created through focused ion beam (FIB) milling on a 50-nm-thick gold film. The unit cell sizes were 400 nm and 900 nm, which were respectively designed for the near-infrared and mid-infrared ranges. A 15- to 40-nm amorphous GST film was sputtered onto the gold layer, followed by a 100-nm capping layer of ZnS/SiO_2 or pure SiO_2 to prevent GST degradation. A 10- to 30-nm ZnS/SiO_2 buffer layer was inserted between the gold and GST to inhibit gold diffusion during the melt/quench process required for the crystalline-to-amorphous transition. The crystallization of the GST layer induced by the laser led to an increase in the real part of its refractive index, causing a redshift of about 200 nm in the resonant features. This all-optical switching mechanism achieved a contrast ratio of up to 4:1, exceeding the 2.5:1 threshold required for optical interconnects in short-range data processing applications (Reed et al. 2010; Crow 1989; Sanders 2008). More recently, Zhu et al. tunable metaswitch in operation was demonstrated in the near-infrared (NIR) range using a similar design, featuring a gold ASR array atop a thin GST layer (Zhu et al. 2020). By employing in situ baking to control the phase of the GST, they achieved continuous tuning of the plasmonic resonances, further advancing the field of reconfigurable metaswitches.

Chen et al. demonstrated that a hybrid plasmonic metaswitch using a $Ge_2Sb_2Te_5$ (GST) layer is placed between gold nanodisks and a quartz substrate in a sandwiched manner, enabling transmission resonance tuning over 500 nm (Chen et al. 2013). Without GST, no coupling occurred between nanodisk dipoles and diffractive modes. Adding GST induced strong dipole interactions, creating a sharp resonance at λ = 1.89 μm (Q-factor = 17). Crystallizing GST red-shifted the resonance to λ = 2.4 μm, reducing the Q-factor to 13. In 2014, Cao et al. numerically explored a tunable Fano resonance in the mid-IR using a metal-dielectric-metal (MDM) fishnet metamaterial with GST as the dielectric layer (Cao et al. 2014). Symmetric nanoholes produced a single Fano resonance from surface plasmon and magnetic dipole (MD) mode interactions. Breaking symmetry by displacing the upper nanohole splits the MD mode into two, generating two Fano peaks with higher Q-factors. Switching GST from amorphous to crystalline red-shifted these peaks by 42%, demonstrating broad tunability for multiwavelength applications, as shown in the plot.

Despite their promising performance, ChG–PCM-based optical switches face challenges such as phase transition speed, durability, and compatibility with existing CMOS fabrication processes. Future research must focus on optimizing material properties, developing low-loss structures, and exploring integration strategies with current optoelectronic platforms. Overall, ChG–PCM optical switches offer a promising pathway toward high-speed, low-power, and reconfigurable photonic devices, with the potential to take on a crucial part in the optical communication, imaging, and data storage systems of the next generation.

8.3.2 Chalcogenide Perfect Absorbers

Chalcogenide metamaterials demonstrate revolutionary potential for perfect absorption applications owing to their exceptional optical characteristics and dynamic tunability. The perfect absorber refers to a photonic material that achieves near-unity (≈100%) absorption efficiency for incident electromagnetic radiation. The basic principle can be attributed to the combination of phenomena such as interference, diffraction, and plasma resonance of light. When light waves strike the surface of an ideal absorber, part of the light waves are confined by free carriers inside the material and converted into thermal energy, and part of the light waves interact with the structural features of the material through reflection, transmission, etc. This coherent interference establishes standing wave patterns that enable near-total optical absorption.

Specifically, metamaterial absorbers typically consist of a three-layer structure known as a metal/insulator/metal structure. The top layer is composed of periodically aligned metal nanopatterns, which are usually realized in practical processing using an electron beam or DUV lithography. Chalcogenides are used as dielectric layers to achieve precise tuning of light absorption bands using their tunable bandgaps. The bottom layer is a continuous metal layer that reflects unabsorbed light, creating a resonant cavity effect that localizes and converts light energy into heat or electricity. Through meticulous optimization of layer thickness, refractive index, and material composition, resonant effects can be selectively induced within target spectral bands. Such resonance enhancement significantly strengthens light–matter interactions, thereby promoting highly efficient photon capture. For instance, by designing micro- and nanostructures with specific geometries, the surface plasmon resonance effect can be activated, resulting in localized photons and consequently enhancing the material's light absorption capacity.

Chalcogenide metamaterials in perfect absorbers work on light absorption enhancement mechanisms, and these materials achieve ultra-broadband, high-efficiency absorption through the cooperative effect of surface plasmon resonance, photon localization, and exciton effects.

Surface plasmon resonance is an excitation phenomenon resulting from the interplay between light and free electrons on the surface of a metal. When incident light irradiates a metal surface at a specific wavelength and angle, the electric field of the light interacts with the free electrons on the metal surface to form electromagnetic waves (plasmons) along the metal surface. These plasmons can effectively enhance the localized electric field on metal surfaces, significantly improving the absorption and scattering properties of light. Surface plasmon excitation critically depends on three key parameters: the metal's optoelectronic properties, incident light angle, and excitation wavelength. Resonance occurs when the photon frequency coincides with the plasmon mode frequency, producing dramatic local field enhancement through coherent energy concentration. Chalcogenides are employed as the dielectric layer to match the wavelength of the surface plasmon resonance by virtue of their tunable bandgap characteristic, which significantly enhances the light absorption efficiency through the synergistic effect of exciton resonance and surface plasmon excitations of the metal nanostructures.

The photon localization effect occurs due to the unique propagation properties of photons in metamaterials or photonic crystals, enabling photons to be effectively trapped within specific subwavelength structures. This phenomenon typically manifests through interactions with the material's microstructure or defects. In particular, the low-symmetry crystal structure of chalcogenides results in an anisotropic distribution of electronic states. When integrated with the surface plasmon resonance stimulated by metal nano-arrays, this results in the creation of a powerful localized electromagnetic field at the interface, remarkably boosting the optical energy density. The realization of the localization effect depends not only on the refractive index and optical bandgap of the material but also on its thickness, shape, and arrangement. When a light wave enters a metamaterial, photons of certain frequencies create localized modes within the material due to reflections and refractions at the interfaces. This interaction results in photons being confined to specific regions. For example, in the case of metal-sulfur compounds, the localization phenomenon can be significantly enhanced by modulating the surrounding environment to improve light absorption efficiency. This process leads to non-radiative attenuation of photons, effectively reducing reflections and transmissions by allowing photons to transition between different electronic energy states.

The exciton effect of chalcogenide metamaterials in perfect absorbers stems from their unique strong exciton binding energies, which are significantly higher than those of conventional semiconductors, allowing for exciton stabilization and suppression of carrier non-radiative complexation at room temperature. For example, the exciton energy of monolayer MoS_2 couples with the plasmonic exciton resonance mode on the surface of metal nanostructures to form an exciton–plasmonic exciton hybridization state, which localizes the light field at the sub-nanometer scale and enhances the light absorption efficiency to more than 98%.

Here are some examples of the applications of chalcogenide metamaterials in perfect absorbers. In terms of broadband near-perfect absorbers, large-area broadband near-perfect absorption can be realized by coupling thin chalcogenide films with gold nanoparticles, which provides a new idea for broadband absorber design. The proposed architecture harnesses the unique photonic characteristics of chalcogenide materials synergistically with nanoparticle-induced localized surface plasmon resonance, significantly boosting photon capture efficiency. Cao et al. developed a broadband near-unity absorber by integrating $Ge_2Sb_2Te_5$ chalcogenide thin films with Au nanoparticles, achieving high-efficiency photon capture spanning visible to near-infrared spectra. Through a facile self-assembly deposition approach, the team realized 92% near-unity absorption across the 400–1000 nm band, with performance stability maintained under varying illumination angles and polarization states. (Cao et al. 2019) The absorption enhancement mechanism is revealed by time-domain finite-difference method (FDTD) simulations: the plasmonic resonance in Au nanoparticles cooperates with the dielectric loss properties of GST films, collectively enhancing both photon confinement and energy conversion efficiency. In addition, the absorber maintains stable absorption performance at 85 °C. This chalcogenide-based metamaterial demonstrates the promising potential for broadband near-perfect absorption, offering novel approaches for photonic applications including solar energy conversion, photodetection, and stealth technologies (Figure 8.4).

In another part of the team's work, a reconfigurable perfect absorber based on the coupling of $Ge_2Sb_2Te_5$ with gold nanoparticles was proposed (Cao et al. 2019). The structure achieves near-perfect absorption of incident light (up to 98% absorption) through the interaction of a gold/chalcogenide bilayer with gold nanoparticles. The properties of phase transition in GST (switching between amorphous and crystalline states) enable the absorber to achieve dynamically tunable absorption properties in different wavelength ranges. During the phase transition from the amorphous state to the crystalline state, the GST absorption peak exhibited a significant redshift from 980 nm to 1520 nm, while retaining high absorption efficiency. This reconfigurability makes the absorber applicable in optical switches, thermal imaging, and optical modulators. Furthermore, the structure is created using a lithography-free self-assembly process, which offers advantages such as low cost, easy mass production, and a new approach for realizing reconfigurable photonics devices. Devika et al. similarly investigated the possible application of chalcogenide metamaterials in the domain of reconfigurable perfect absorbers. Ternary chalcogenides, especially type I–III–VI_2 ternary chalcogenides with chalcopyrite structure, such as $CuInSe_2$ and Cu_2SnS_3, are ideal materials for realizing high efficiency of light absorption due to high absorption coefficients and adjustable bandgaps. Dynamic modulation

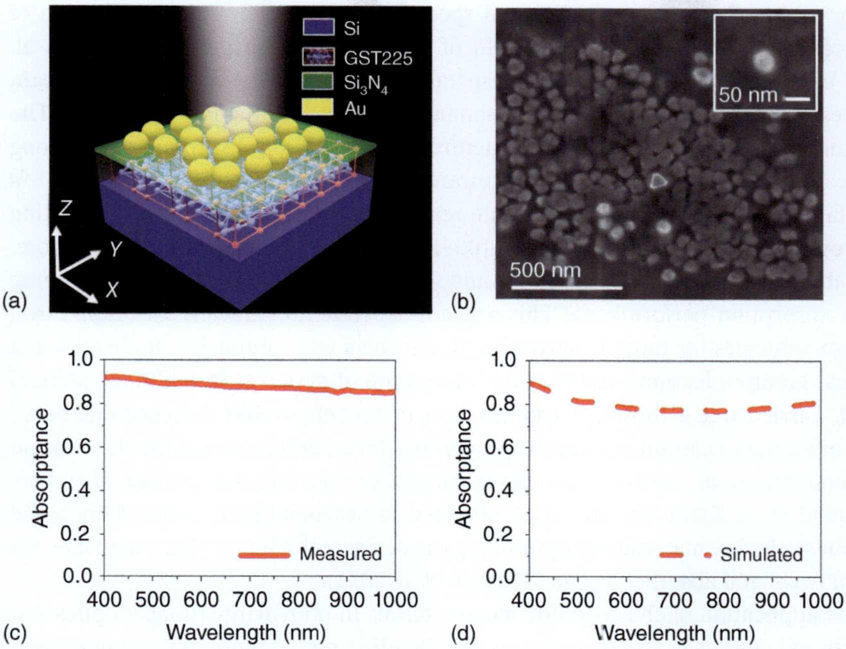

Figure 8.4 (a) Illustrates the plasmonic absorber architecture, featuring an Au NP array coupled with a 35-nm-thick amorphous GST225 layer and a Si_3N_4 protection layer (48–52 nm NP diameters), all deposited on a 200 μm silicon substrate. (b) Presents a cross-sectional SEM image normal to the light propagation direction. (c, d) Compare experimental and simulated absorption spectra at normal incidence. *Source:* Reprinted from Cao et al. (2019).

of light absorption can be realized by designing rational metamaterial structures, such as combining gold nanoparticles with thin films of chalcogenides. The phase transition property of chalcogenides further enhances the reconfigurability of the absorber, enabling near-perfect absorption at different wavelengths. This dynamically tunable perfect absorber demonstrates significant potential across multiple domains including optoelectronics, sensing technologies, and thermoelectric systems, offering novel design strategies for advanced photonic components.

The design of narrow-band perfect absorbers using phase change materials, specifically chalcogenides, is currently a major area of research. Phase transitions in chalcogenide metamaterials induce substantial modifications in key optical parameters including refractive index and conductivity. These changes provide an opportunity to create narrow-band perfect absorbers. Through engineered phase transition control, these materials can attain targeted wavelength absorption with optimized efficiency. An ultrathin perfect absorber constructed using chalcogenide metamaterials realizes efficient absorption of infrared light by adjusting the temperature and phase transition state of the material. These absorbers demonstrate significant utility for photothermal energy conversion and optical detection applications (Yang and Ramanathan 2015). Oliveira et al. devised a composite narrow-band plasmonic absorber relying on chalcogenide phase-change materials with unique performance in infrared spectroscopy, expanding the application of chalcogenides (e.g., GeTe) in the field of narrowband absorption (Oliveira et al. 2021). By exploiting the reversible phase transition characteristics of PCMs, the team successfully demonstrated tunable resonance peak positioning in the absorber. The engineered 3D hybrid plasmonic structure achieves >95% absorption with strong optical modulation capability across infrared wavelengths. Precise tuning of GeTe's crystallization degree enables accurate resonance wavelength control, facilitating narrowband absorption within the 1100–1900 nm spectral window. Furthermore, the absorber demonstrates excellent angular stability (0°–45°) while maintaining high absorption performance. This angle-independent characteristic offers novel design strategies for tunable narrowband absorbers with promising applications in optical communications and thermal management systems. In addition, Mandal et al. carried out a thorough examination of reconfigurable chalcogenide-based metamaterials, examining material properties, device architectures, and fabrication methodologies to advance the development of narrowband perfect absorbers (Mandal et al. 2021). Tripathi et al. obtained narrowband perfect absorbers based on phase change materials by optimizing the design, which provides a new method for narrowband absorber design (Tripathi et al. 2023).

The application of chalcogenide metamaterials in nonvolatile tunable optics and functional devices is also of great interest. By utilizing the phase transition properties of chalcogenides, it is possible to design optical devices with nonvolatile tunable optical properties, such as optical switches and optical memories. Such devices combine rapid operation, low energy requirements, and nonvolatile information storage capabilities, offering novel pathways for advancing optical information processing and storage technologies (Xiao et al. 2020).

8.3.3 Chalcogenide Beam Steering

Chalcogenide-based beam steering is an advanced optical technique that leverages the unique properties of chalcogenide materials to control the direction of light without mechanical movement. Chalcogenides are made up of elements like sulfur, selenium, and tellurium in combination with germanium, arsenic, or antimony, exhibit high refractive indices, strong nonlinear optical properties, and excellent transparency in the mid-wave infrared (MWIR) and long-wave infrared (LWIR) regions. These characteristics make them particularly well-suited for applications in LiDAR, infrared imaging, optical communication, and augmented reality. Unlike traditional beam steering methods that rely on mechanical mirrors or MEMS devices, chalcogenide-based approaches utilize solid-state mechanisms such as phase-change materials (PCMs), electro-optic tuning, and metasurface engineering. Phase-change chalcogenides, for example, $Ge_2Sb_2Te_5$ (GST), are capable of reversibly transitioning between amorphous and crystalline states. This transition modifies their optical characteristics, enabling them to dynamically steer light. This enables rapid, energy-efficient beam steering with improved reliability and durability, as there are no moving parts involved. Additionally, chalcogenide metasurfaces and holographic elements allow for ultra-thin, flat optical components

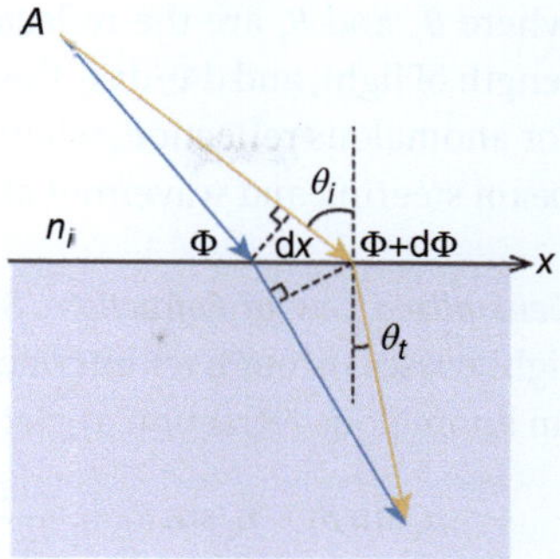

Figure 8.5 Schematic diagram employed to deduce the generalized Snell's law of refraction.

that can further miniaturize beam steering systems. As research continues to refine these materials and integrate them with silicon photonics, beam steering based on chalcogenides is ready to take on a vital part in the advancement of next-generation optical and photonic technologies. This section mainly introduces the principle, classification, and research status of chalcogenide metasurface beam steering.

8.3.3.1 Generalized Laws of Reflection and Refraction

By introducing an abrupt phase shift, also known as phase discontinuity, at the boundary between two media, it becomes possible to reexamine the laws of reflection and refraction using Fermat's principle. When a plane wave is incident at an angle θ_i, assuming that the two paths are extremely close to the actual path of light (as shown in Figure 8.5), there is no phase difference between them (Yu et al. 2011):

$$\left[k_0 n_{\mathrm{i}} \sin\left(\theta_{\mathrm{i}}\right) \mathrm{d}x + (\Phi + \mathrm{d}\Phi)\right] - \left[k_0 n_t \sin(\theta_t)\mathrm{d}x + \Phi\right] = 0$$

Here, θ_t represents the angle of refraction. Meanwhile, Φ and $\Phi + \mathrm{d}\Phi$ represent the phase discontinuities at the points where the two paths intersect the interface. The separation between these points of intersection is given by $\mathrm{d}x$. n_i and n_t are the refractive indices of the two media, and $k_o = 2\pi/\lambda_o$, where λ_o is the wavelength in vacuum.

The general laws of reflection and refraction extend the classical laws of light interaction at an interface to account for more complex optical phenomena, such as gradient-index materials, structured surfaces (e.g., metasurfaces), and nonlocal effects. Traditional laws state that the angle of reflection equals the angle of incidence, and that refraction follows Snell's law, which relates the incident and refracted angles to the refractive indices of the two media. However, in structured or inhomogeneous materials, these laws must be modified to include additional phase shifts, momentum changes, and polarization-dependent effects.

Generalized Law of Reflection Unlike classical reflection, where the reflected wave follows a simple geometric path, the generalized law incorporates an additional phase gradient imposed by surface structures, such as metasurfaces. This leads to a modified reflection angle given by

$$\sin\theta_r = \sin\theta_i + \frac{\lambda}{2\pi}\frac{\mathrm{d}\Phi}{\mathrm{d}x}$$

where θ_r and θ_i are the reflected and incident angles, respectively, λ is the wavelength of light, and $d\Phi/dx$ is the phase gradient imposed by the interface. This allows for anomalous reflection, where light is reflected at nontraditional angles, enabling beam steering and wavefront shaping.

Generalized Law of Refraction Similarly, the classical Snell's law is modified when light passes through an interface with a spatially varying phase response, leading to an anomalous refraction angle:

$$n_t \sin\theta_t - n_i \sin\theta_i = \frac{\lambda}{2\pi}\frac{d\Phi}{dx}$$

where n_i and n_t are the refractive indices of the two media, and θ_t is the transmitted (refracted) angle. This modification allows for engineered wavefront control, including negative refraction and asymmetric light propagation.

These generalized laws are crucial for modern photonic applications, including beam steering, flat optics, and metasurface-based optical devices, enabling unprecedented control over light propagation beyond traditional optics.

8.3.3.2 Classification of Chalcogenide Beam Control

Chalcogenide-based beam control can be broadly classified into reflection-based and transmission-based approaches, depending on how the material manipulates the incident light. These methods leverage the unique optical properties of chalcogenides, including their high refractive index, strong nonlinear effects, and phase-change capabilities, to dynamically steer and shape light for applications in LiDAR, optical communication, and infrared imaging.

Reflection-Based Chalcogenide Beam Control In reflection-based beam control, chalcogenide materials are engineered to modify the direction or properties of reflected light. Chalcogenide phase-change materials (PCMs) like $Ge_2Sb_2Te_5$ (GST) and Sb_2Se_3 can undergo reversible structural transitions, inducing tunable phase shifts that enable dynamic control of the reflected wavefront. By programming spatial phase gradients, metasurfaces can achieve anomalous reflection, steering beams at unconventional angles. C. David Wright et al. put forward a novel approach for dynamic beam steering by utilizing an integrated plasmonic metasurface. with a phase change material (PCM) $Ge_2Sb_2Te_5$ (GST), operating in the near-infrared (NIR) range (1530–1570 nm) (de Galarreta et al. 2018). The authors devise, manufacture, and characterize a metal–insulator–metal (MIM) structure. In this structure, the GST layer can achieve nonvolatile and reconfigurable beam control by transitioning between its crystalline and amorphous states, as depicted in Figure 8.6. When it is in the crystalline phase, the metadevice reflects light in a mirror-like fashion, whereas in the amorphous state, it reflects light at predesigned anomalous angles, allowing for mechanically-free beam steering. A key advantage of this approach is that the PCM layer is protected from oxidation using ITO capping layers, ensuring durability, while its minimal thickness ($\approx\lambda/60$) reduces optical losses and facilitates efficient phase transitions. Unlike conventional beam steering techniques that rely on mechanical movement or free-carrier injection, this

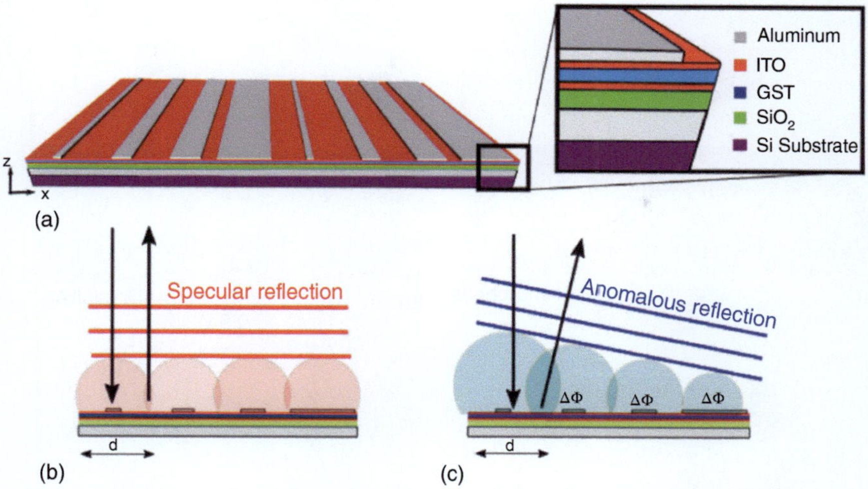

Figure 8.6 Design of a reconfigurable phase–change beam steering apparatus: (a) General view of the device's structure and the materials used. Utilization of Huygens' principle to illustrate wavefront reconstruction under normal incidence when the GST layer is in the crystalline (b) and amorphous (c) phases. *Source:* de Galarreta et al. (2018) / John Wiley & Sons / CC BY 4.0.

metadevice offers fast switching speeds, low power consumption, and enhanced efficiency, with demonstrated dynamic switching using optical excitation and potential for electrical and thermal control. The metasurface can be pixelated to enable multi-angle beam steering, and by modifying the antenna design, it could achieve polarization-independent operation and full 2π phase control, making it suitable for LiDAR, free-space optical communication, imaging, and dynamic holography. The study highlights the feasibility of phase-change metadevices as a scalable and compact solution for next-generation photonic applications, with the potential to operate in both NIR and MIR spectral regions where GST offers low optical losses and high refractive index contrast.

Pu Mingbo et al. presented a quasi-continuous metasurface featuring dynamically reconfigurable phase modulation through the integration of catenary meta-atoms. and the phase-change material (PCM) $Ge_2Sb_2Te_5$ (GST) into a metal–insulator–metal (MIM) architecture (Song et al. 2021). Two key applications of this design are presented. First, by arranging catenary atoms in a one-dimensional configuration, a switchable beam deflection metasurface is fabricated. When the GST changes from its amorphous state to the crystalline state, the reflection of the incident light switches from anomalous reflection at a predetermined angle to normal specular reflection, as depicted in Figure 8.7. Next, a switchable meta-axicon for generating Bessel beams is presented by arranging catenary atoms in a helical pattern. When in the amorphous state, the meta-axicon is triggered to generate a highly focused Bessel beam with enhanced forward propagation. Conversely, when in the crystalline state, the Bessel beam is fully deactivated, and there is no forward propagation. Moreover, through meticulous design of the meta-axicon in its amorphous

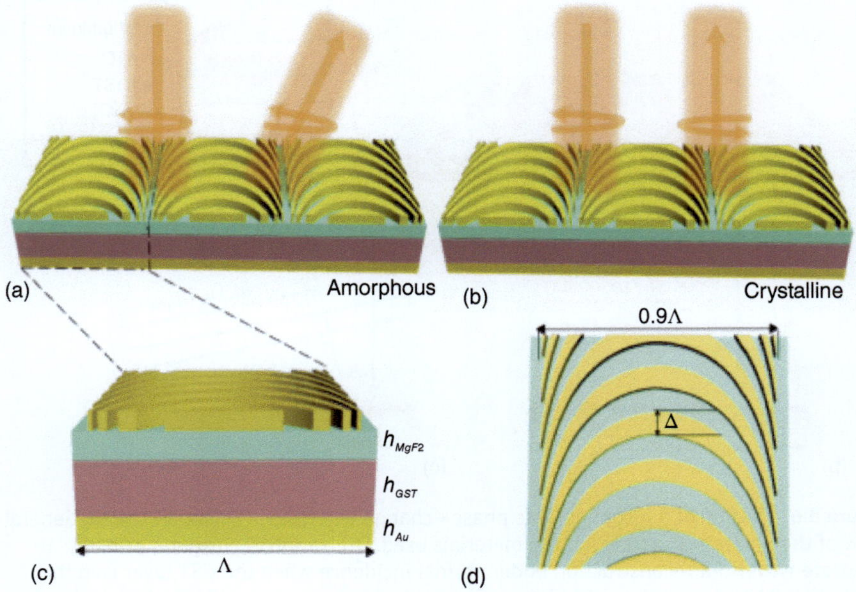

Figure 8.7 Quasi-continuous metasurfaces are fabricated with catenary–$Ge_2Sb_2Te_5$ hybrid meta-atoms for dynamic beam steering. Figures (a) and (b) artistically illustrate the switchable beam deflection in the (a) amorphous and (b) crystalline states caused by the phase change. Figure (c) displays the cross-sectional view, while Figure (d) shows the front view of a single unit cell with catenary structures. *Source:* de Galarreta et al. (2018) / John Wiley & Sons / CC BY 4.0.

state, a diffraction-limited Bessel beam with a reduced full-width at half-maximum (FWHM) is obtained. This research emphasizes a hopeful method for active photonic devices capable of switchable phase adjustment, offering high switching contrast and dynamic beam manipulation, which could be highly beneficial for applications requiring large switching ratios and precise optical control.

Juejun Hu et al. demonstrated an on-chip electrically switchable platform that enables both binary switching and quasi-continuous tuning of phase-change material (PCM)-based active metasurfaces (Zhang et al. 2021). In contrast to traditional thermal or optical switching approaches, electrical actuation represents a substantial progress in the direction of fully integrated, chip-scale reconfigurable optics. By utilizing the recently developed PCM $Ge_2Sb_2Se_4Te_1$ (GSST), the study achieves key benefits, including reduced optical losses and enhanced reversible switching volume, leading to improved performance. Exploiting strong light–PCM interactions, the metasurface showcases ultra-broadband adjustment of meta-atom resonances that cover half an octave, which is a vital ability for dynamic phase and amplitude control. As an illustration of the concept, the researchers present the active switching of a Huygens' metasurface beam deflector, showcasing precise wavefront manipulation, as shown in Figure 8.8. Furthermore, the design of electrical switching and the strategy of geometric optimization can be adapted to other advanced heating materials like silicon and graphene, paving the way for a wide array of next-generation

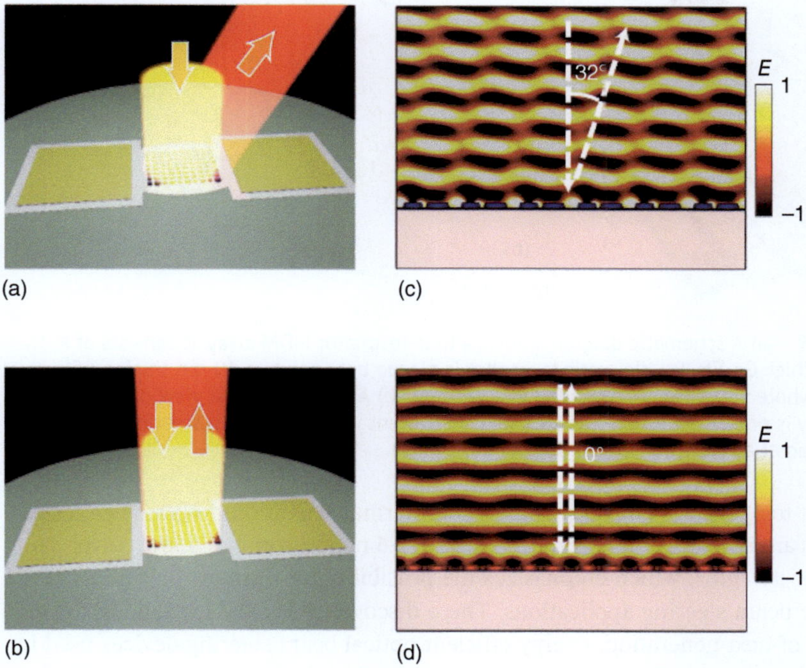

Figure 8.8 Schematic rendering showing the reconfigurable metasurface beam deflector in its amorphous (a) and crystalline states (b). Simulated electric field amplitude profiles at a wavelength of 1550 nm are presented to demonstrate the beam steering features of the device in its amorphous state (c) and crystalline state (d). *Source:* Zhang et al. (2021).

reconfigurable metasurface devices. These devices hold potential for applications in optical communication, beam steering, and dynamic holography.

Transmission-Based Chalcogenide Beam Control Transmission-based approaches modify the properties of transmitted light as it propagates through or exits a chalcogenide medium. Similar to reflection-based methods, phase-change chalcogenides enable tunable transmission-based optical elements. When incorporated into metasurfaces or waveguides, these materials allow for dynamic control of transmitted light direction and phase. Chalcogenide-based metasurfaces can impose a spatially varying phase profile on transmitted light, enabling effects like negative refraction, beam deflection without traditional optical components.

Our group demonstrated a low-power, ultrafast mid-infrared (MIR) beam steering device using an array of Au/$Ge_2Sb_2Te_5$ (GST)/Au multilayer strips, as shown in Figure 8.9 (Cao et al. 2015). The phase-change material (PCM) GST facilitates dynamic beam control, with varying strip widths in the amorphous state producing different phase shifts. As a result, the transmitted light has a deflection angle of 259°, while the reflected light has a deflection angle of 74°. Furthermore, by transitioning GST from its amorphous to crystalline state, the device achieves a wide beam steering range. For transmitted light, it is 11°, and for reflected light, it is 22°. To further analyze the switching dynamics, a thermal model is constructed to monitor the

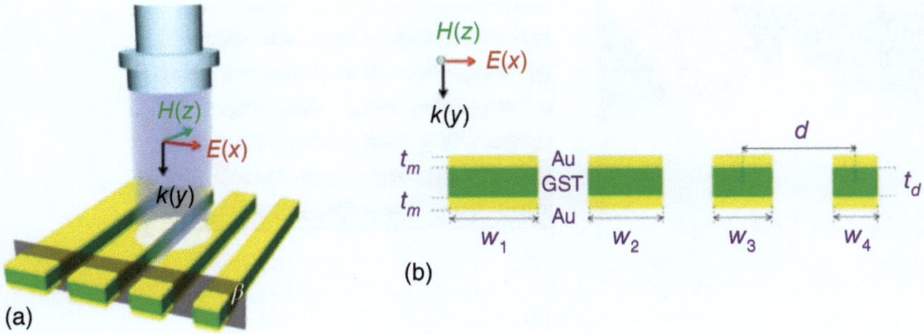

Figure 8.9 (a) A schematic diagram shows a four-resonator MDM array. It consists of a 250-nm-thick $Ge_2Sb_2Te_5$ dielectric layer that is placed between two 40-nm-thick Au films, and the whole structure is suspended in a vacuum. (b) A cross-sectional view of the MDM strip array is presented, where the strips have different widths for beam steering. *Source:* Cao et al. (2015) / Optical Society of America.

transient temperature changes in the photothermal activation process. The model forecasts an incredibly rapid response time of 3.6 nanoseconds at a low power density of 2.6 μW/μm^2, which emphasizes the possibility for ultralow-power, ultrafast photonic beam steering applications. These discoveries lay the foundation for the creation of next-generation, energy-efficient optical beam steering devices used in LiDAR, free-space optical communication, and infrared imaging.

Parikshit Moitra et al. explored the potential of Sb_2S_3 as a phase-change material (PCM) for all-dielectric metasurfaces that can be programmed and operated within the visible spectrum (Moitra et al. 2023). Owing to its low optical loss, high refractive index, and a substantial refractive index contrast (~0.5) between amorphous and crystalline states, Sb_2S_3 enables high-efficiency, nonvolatile, and reversible wavefront manipulation. The researchers developed a high-resolution nanofabrication process to create nanoscale Sb_2S_3 patterns, which were then used to design Huygens' metasurfaces capable of programmable optical phase control. Unlike conventional designs where PCMs only serve as thin-film overlays, this work fully integrates Sb_2S_3 into the metasurface nanoantennas, resulting in a large resonance shift (~60 nm) and a transmission phase shift exceeding 1.5π upon phase transition, as shown in Figure 8.10. These characteristics allow for effective phase-only control of the transmitted light, which is shown by reconfigurable beam bending and bound states in the continuum (BICs). The research emphasizes the potential applications in AR/VR spatial light modulators, LiDAR, programmable planar optics, and dynamic holography.

Yihang Chen et al. proposed Ge-clad GSST ($Ge_2Sb_2Se_4Te_1$) metasurfaces for broadband, reversible tuning of transmissive beam deflection and focusing in the long-wave infrared (LWIR) spectrum (Pan et al. 2023). Utilizing waveguide-type meta-atoms, the metasurfaces achieve broadband tunability, contrasting with previous resonant-type metasurfaces, which were limited to narrowband responses in the mid-wave infrared (MWIR). The phase transition of GSST, induced by external thermal stimuli, enables dynamic modulation of both the deflecting angle in beam steering and the focal length in achromatic metalenses, as shown in

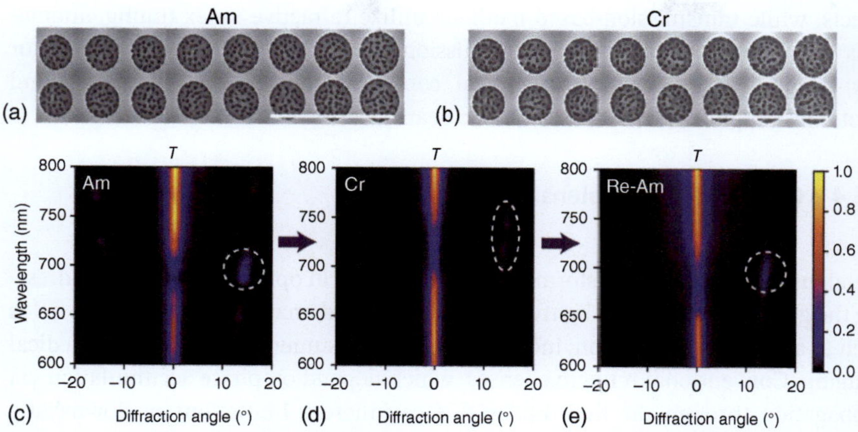

Figure 8.10 Beam steering experiments: (a, b) SEM images of the eight-element supercell for the beam bending metasurfaces made of amorphous (a) and crystallized (b) Sb_2S_3. These images were taken from the same position within the array. The scale bars indicate a length of 1 μm. (c–e) Angle-resolved normalized transmission spectra of the metasurfaces in the amorphous, crystallized, and laser-re-amorphized states. The main diffraction mode is marked with dashed circles. *Source:* Moitra et al. (2022) / John Wiley and Sons.

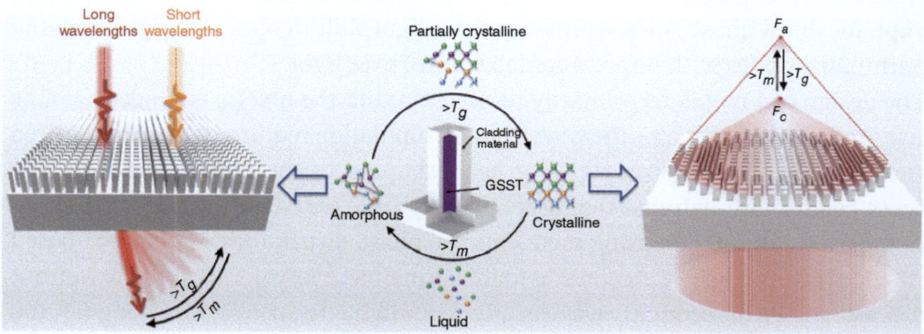

Figure 8.11 Schematic of the proposed reconfigurable GSST metasurface for broadband reversibly tunable applications in the long-wave mid-infrared (8–14 μm) band: a long-wave pass deflector with continuous beam steering (left) and an achromatic varifocal metalens (right). *Source:* Pan et al. (2023) / Optical Society of America.

Figure 8.11. Key innovations include reversible tuning, achieved by encapsulating GSST with a refractory cladding material, and the introduction of figures of merit for systematic design evaluation. The proposed deflector functions as both a beam steerer and a long-wave pass filter, offering continuous angle tuning based on the GSST crystalline fraction. Additionally, the broadband metalens enables a twofold change in focal length within the 11–14 μm range, achieving achromatic varifocal imaging from 12.5 μm to 13.5 μm.

Chalcogenide beam control techniques can be categorized into reflection-based and transmission-based approaches, each leveraging the material's unique optical properties for nonmechanical, highly efficient light steering. Reflection-based methods primarily rely on metasurfaces, nonlinear optics, and phase-change effects, while transmission-based methods utilize refractive index tuning, anomalous refraction, and nonlinear transmission. These approaches are pivotal for next-generation applications in optical communication, LiDAR, imaging, and photonic computing, where fast, tunable beam steering is essential.

8.3.4 Chalcogenide Metalens

8.3.4.1 Basics of Metalens

Metalenses represent a transformative advancement in optical technology, addressing the growing demand for lightweight, miniaturized imaging systems across fields such as aerospace exploration, industrial vision, consumer electronics, and medical imaging. Conventional refractive lenses, which depend on phase accumulation via propagation through the bulk material, face inherent limitations in downsizing due to their dependence on natural materials with finite refractive indices. While innovations like freeform lenses and microlens arrays have reduced thickness, they still struggle with cascading requirements for high imaging quality, leading to bulky architectures and alignment challenges. Diffractive lenses, which control light through constructive interference, present planar alternatives. However, they are plagued by low efficiency, dispersion, and challenges in integration. In contrast, metalenses—composed of subwavelength nanostructures (meta-atoms) that impart abrupt, localized phase shifts—provide a paradigm shift in optical design, enabling ultrathin, flat optics with unprecedented control over light.

The design of a metalens primarily revolves around the precise engineering of its phase profile, which governs the wavefront manipulation required to achieve desired optical functionalities such as focusing, beam shaping, or aberration correction. The process begins with defining the target phase profile based on the intended application. In a traditional focusing metalens, the phase distribution is obtained based on the Huygens principle. According to this principle, every point on the wavefront serves as a source of secondary wavelets. To focus light at a specific focal length f, the required phase retardation at a radial position r is calculated using the hyperboloidal phase profile:

$$\varphi(r, \omega) = \frac{n_b \omega}{c}\left(f - \sqrt{r^2 + f^2}\right) = \frac{2\pi n_b}{\lambda}\left(f - \sqrt{r^2 + f^2}\right) \tag{8.1}$$

where n_b represents the refractive index of the background medium, λ stands for the vacuum wavelength, and ω denotes the angular frequency. This equation guarantees constructive interference at the focal point by making up for the path difference between the central and peripheral parts of the lens.

However, practical metalens designs face a critical constraint: the phase profile must be discretized into finite phase levels due to the limited unit cell size of nanostructures. Instead of a continuous phase distribution, discrete phase steps covering a full 2π range are implemented. This discretization introduces a wave aberration function (WAF), which measures the difference between the ideal spherical wave-front and the envelope of the discretized phase distribution. The root–mean–square WAF (WAFrms) is determined by both the quantity of phase levels N and the intrinsic discrepancy between the aspherical target phase and the reference spherical wavefront. Increasing N reduces WAFrms, improving focusing quality, but simultaneously escalates fabrication complexity as smaller nanostructures are required. For instance, achieving high N values demands advanced nanofabrication techniques to maintain structural integrity and minimize scattering losses.

8.3.4.2 Chalcogenide Metalens

Chalcogenide materials, including GST-series compounds and the recently highlighted low-loss materials Sb_2S_3 and Se_2S_3, exhibit significant refractive index contrast, which endows them with distinctive advantages for tunable metalens applications. In the following section, we will present representative studies on metalenses fabricated using chalcogenide-based materials.

We will first introduce several studies on laser-direct-written planar lenses. Although these works fall outside the conventional definition of metalenses, they effectively exploit the phase-change characteristics of chalcogenide materials, offering unique research perspectives. Wang et al. introduce a new platform for photonic devices with optical reconfigurability by utilizing the phase-change material $Ge_2Sb_2Te_5$ (GST). Femtosecond laser pulses (85 fs, 730 nm) induce nonvolatile, reversible phase transitions in GST films, enabling high-resolution (0.59 μm) binary and grayscale optical patterning as shown in Figure 8.1a. Demonstrated devices include binary Fresnel zone plates (upper panel of Figure 8.1b), super-oscillatory lenses achieving subwavelength focusing (lower panel of Figure 8.1b). Chromatically selective lenses spatially separated 730 nm and 900 nm foci, while corrected lenses combined wavelengths into a single focus. Reconfigurability was validated via write–erase–rewrite cycles of zone plates using 0.39 nJ (crystallization) and 1.25 nJ (amorphization) pulses. Challenges include pixelation effects and partial phase transitions affecting device fidelity. This approach bridges optical data storage principles with dynamic photonics, offering a versatile framework for on-demand light–matter interaction control in integrated systems.

Zhang et al. exhibit an on-chip, nonvolatile, and reconfigurable terahertz (THz) varifocal metalens by employing $Ge_2Sb_2Te_5$ (GST) phase-change material. By integrating GST with C-shaped-slit resonator metasurfaces (Figure 8.1c), the device achieves dynamic control of surface plasmons (SPs) through optical and thermal stimuli. The design leverages GST's reversible phase transitions (amorphous/crystalline) to enable nonvolatile switching, wavefront shaping, and amplitude modulation without structural modifications. Experimental results demonstrate a reconfigurable SP metalens with a 29.9 dB extinction ratio and switchable focusing at two distinct focal points (7.4 dB and −44.4 dB contrast ratios) as shown in Figure 8.1d. Multilevel intensity modulation is achieved by tuning laser fluence, enabling partial GST crystallization. The device also exhibits repeatable switching over multiple cycles, highlighting its energy efficiency and stability (Figure 8.12).

Yin et al. demonstrate a nonmechanical bifocal zoom lensing mechanism using a hybrid plasmonic metasurface integrated with $Ge_3Sb_2Te_6$ (GST-326). This phase-change material enables reversible refractive index modulation ($\Delta n \approx 1.3$

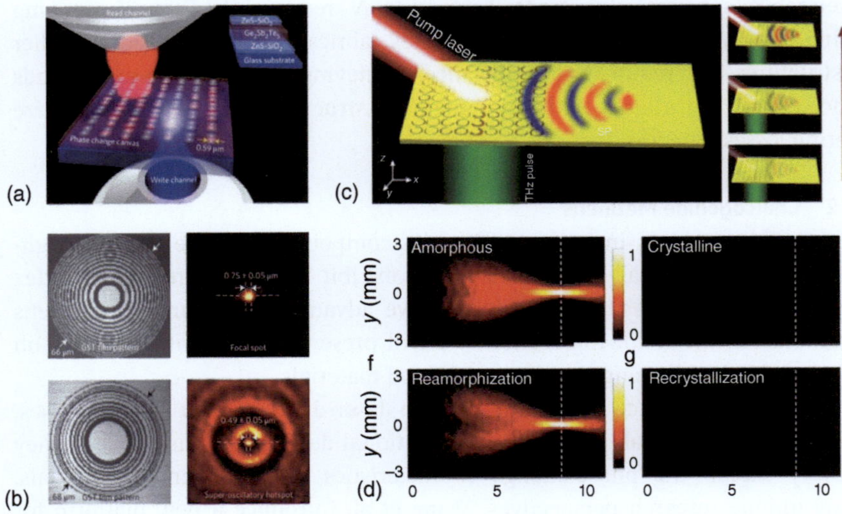

Figure 8.12 (a) A schematic diagram of the reconfigurable photonic devices within a phase-change film. (b) Top panel: An image of the Fresnel zone–plate pattern captured at a wavelength of $\lambda = 633$ nm. Bottom panel: A microscopic view of the optical hotspot formed by the focusing of the Fresnel zone plate at a wavelength of $\lambda = 730$ nm. (c) A schematic representation of the reconfigurable THz metalens. (d) The measured normalized intensity distributions and profiles in the amorphous and crystalline states, respectively. *Source:* (b) Wang et al. (2016) / Springer Nature and (c,d) Zhang et al. (2023) / John Wiley & Sons.

at 3.1 μm wavelength) through amorphous-to-crystalline transitions. The metasurface features interleaved gold nanorod arrays designed with spatially varying orientations and lengths to generate distinct phase profiles, as shown in Figure 8.13. In the amorphous state, the lens achieves a focal length of 0.5 mm, while crystallization switches the focus to 1.0 mm through selective activation of resonant nanorod subsets. Experimental characterization confirmed 5–10% optical efficiency in both states with sub-millisecond switching capability via thermal/optical excitation. This static, reconfigurable design eliminates moving parts, enabling applications in ultra-compact adaptive optics, endoscopic imaging systems, and tunable laser focusing architectures.

Shalaginov et al. present a reconfigurable all-dielectric metalens leveraging optical phase-change material (GSST) for nonmechanical tuning of focal lengths with diffraction-limited performance. The platform enables binary switching between two arbitrary phase profiles (amorphous and crystalline states) via a Huygens' meta-atom design, achieving full 2π phase control while minimizing optical loss. A novel figure-of-merit (FOM) was developed to optimize meta-atom geometries for high efficiency and low crosstalk. Experimental validation at 5.2 μm wavelength demonstrated 29.5 dB switching contrast ratio, focusing efficiencies of 23.7% (amorphous) and 21.6% (crystalline), and aberration-free imaging with resolved features matching theoretical resolution limits. The design framework supports scalability to multistate reconfigurable systems, bridging performance gaps between active metasurfaces and conventional bulk optics for applications in adaptive imaging and compact zoom lenses.

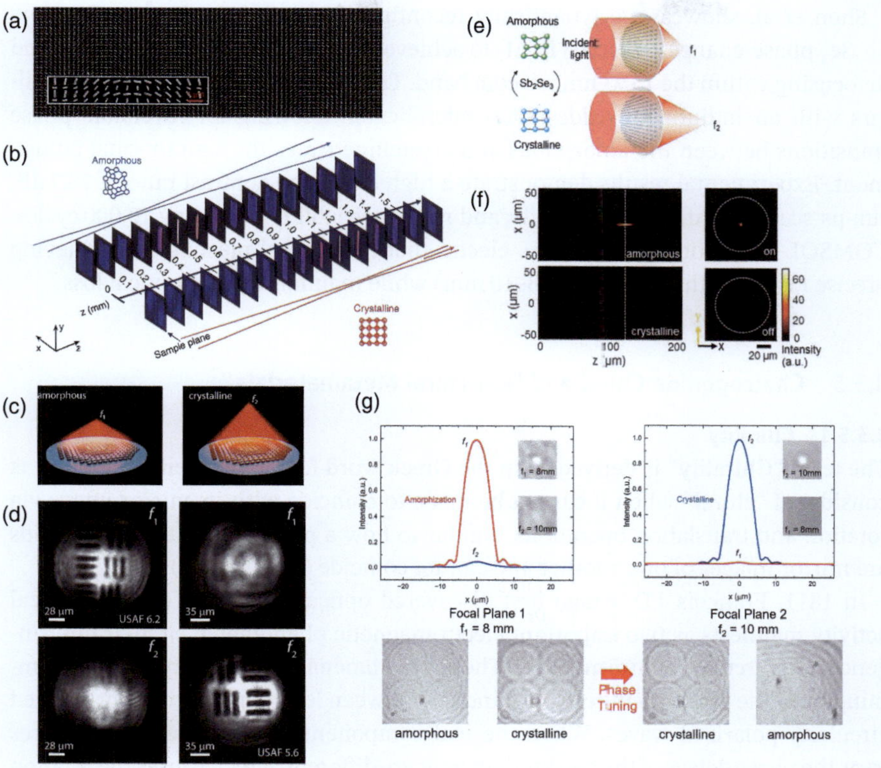

Figure 8.13 (a) SEM micrograph of a section of the manufactured cylindrical metasurface lens. (b) Camera pictures of the cylindrical bifocal lens imaged at different distances z from the metasurface. In the amorphous case, a bright line appears at $z = 0.5$ mm, whereas the image shows broad diffuse lightness at the same z position in the crystalline case. At $z = 1$ mm, the bright line focus appears in the crystalline case, whereas one can merely observe a light patch in the amorphous case. (c) Diagram of the reconfigurable varifocal metalens. (d) Clearly distinguishable lines of USAF−1951 resolution charts: the patterns have half-periods that are nearly equal to the Rayleigh resolution limits of 7 μm and 9 μm in the a-state (f_1) and c-state (f_2), respectively. (e) Schematic illustration of the single-focal metalens. (f) The measured light intensity along the light propagation direction (left) and at the focal plane (right) for amorphous Sb_2Se_3 (upper part), for crystallized Sb_2Se_3 (lower part). (g) Upper part: the normalized power intensity of the two states on focal plane 1 (left) and focal plane 2 (right). Lower part: the images of the metalens in different states at different focal planes captured by the CCD camera. *Source:* (d) Shalaginov et al. (2021) / Springer Nature / CC BY 4.0, (f) Wang et al. (2023) / John Wiley & Sons / CC BY 4.0, and (g) Shen et al. (2023) / MDPI / CC BY 4.0.

Wang and colleagues experimentally prove the functionality of a varifocal metalens that operates in the near-infrared wavelength range. They utilize Sb_2Se_3, an ultralow-loss phase-change material, for this demonstration. By leveraging its reversible amorphous-to-crystalline transition with minimal optical loss ($\Delta n \approx 1.0$, near-zero k), polarization-insensitive metalenses were designed via Huygens' metasurface principles. A single-focal metalens reached a focusing efficiency of 23% and accomplished an intensity modulation of 95% between states. A varifocal configuration exhibited tunable focal lengths from 41 μm (amorphous, upper panel of Figure 8.13f) to 123 μm (crystalline, lower panel while maintaining comparable efficiencies (5.7% versus 3%)).

Shen et al. showcase a dynamically reconfigurable metalens that employs the Sb_2Se_3 phase-change material (PCM) to achieve rapid and nonvolatile focusing and defocusing within the 1550-nm telecom band. The design integrates Sb_2Se_3 nanopillars with an indium tin oxide (ITO) microheater, allowing for reversible phase transitions between the amorphous and crystalline states through thermal adjustment. Experimental results demonstrate a high switching contrast ratio of 28.7 dB, sub-μs-scale reconfiguration speeds, and robust cyclability exceeding 10,000 cycles. COMSOL simulations validate the electro-thermal control mechanism, achieving precise focal length modulation (8–10 mm) while maintaining low optical loss.

8.3.5 Chalcogenide Chiral and Non-chiral Metamaterials

8.3.5.1 Chirality

The term "Chirality" is derived from the Greek word for hand (cheir). An object is considered "chiral" when it cannot be made to coincide with its mirror image via rotation and translation operations, similar to how a person's left and right hands are mirror images of one another and cannot coincide (Cintas 2007).

In 1811, François J.D. Arago first discovered optical activity in quartz. Optical activity manifests as two important electromagnetic phenomena: circular birefringence and circular dichroism (CD). These phenomena can be interpreted as stemming from the disparity in refractive indices between left-handed and right-handed circularly polarized waves. When the real components of their refractive indices vary, the phase delays of the two kinds of light are different, which causes the rotation of the polarization axis, referred to as circular birefringence. When the imaginary components of the refractive indices of the two electromagnetic waves are not the same, their losses are distinct, leading to different transmittances, called CD. Today, optical activity plays a role in probing the spatial arrangement of atoms in fields such as spectroscopy, analytical chemistry, crystallography, and molecular biology. Additionally, polarization rotators and circular polarizers made from chiral media have important applications in microscopy, photography, and display technologies (Rogacheva et al. 2006). Through specific structural designs, chirality can also exist in metamaterials. Chiral metamaterials (CMMs) not only display more pronounced optical activity when compared with natural chiral media, but they are also capable of generating novel properties like negative refractive indices and circular conversion dichroism (Cao et al. 2017, Mao et al. 2022).

8.3.5.2 Applications of Chalcogenide Chiral Metamaterials

Chalcogenide chiral metamaterials exhibit superior optical chirality effects compared to conventional materials, showing significantly different optical responses to left-and right-handed circularly polarized light. This characteristic enables efficient circular dichroism and facilitates precise manipulation and detection of chiral optical signals. Through the strategic modification of composition, thickness, geometric configuration, dimensions, and spatial arrangement of micro- and nanostructures within chalcogenide compounds, key optical properties (including resonance wavelength and circular dichroism intensity) can be precisely tuned to meet diverse application-specific requirements.

Distinct from traditional chiral materials, chalcogenide chiral metamaterials demonstrate reduced optical loss within specific wavelength ranges, maintaining high optical signal transmission efficiency and stable optical performance. This enhancement significantly improves the efficiency and reliability of optoelectronic devices. The integration of these materials is further facilitated by their compatibility with micro-nano fabrication technologies, enabling their preparation on compact chip scales. This compatibility promotes seamless integration with other optical and electronic components, paving the way for the development of miniaturized, multifunctional optoelectronic devices and advancing the field toward greater integration and miniaturization.

Moreover, beyond their inherent chiral optical effects, these materials can be synergistically coupled with other physical phenomena such as surface plasmon resonance and exciton effects. This coupling engenders novel physical phenomena and optical properties, expanding the horizons for light field manipulation and the innovation of functional devices. Such versatility underscores the potential of chalcogenide chiral metamaterials in driving forward the frontiers of optical science and technology.

Next, the applications of chalcogenide chiral metamaterials are explored through two aspects: tunable and switchable chiroptical responses, and tunable wavelength.

Tunable and Switchable Chiroptical Responses Chiroptical responses such as CD, manifesting themselves as the difference between the response of left-handed circularly polarized (LCP) light and right-handed circularly polarized (RCP) light, have proven to be good candidates for applications in biosensing, chiral imaging, and chiral coding. In the terahertz (THz) area, the regulation of chiral optical activity is extremely crucial. Macromolecules containing polar or ionic components absorb THz waves vigorously because of collective vibration modes. Biopolymers like DNA, proteins, and RNA, which have chiral structures, selectively absorb circularly polarized light within the THz range. These characteristics can be utilized to develop vibrational CD spectroscopy for determining molecular structures, including chirality and tertiary configurations. Beyond the significance of chirality in the THz region, the active polarization control of THz light might offer remarkable opportunities for fabricating novel and revolutionary THz devices. THz chiral metamaterials with high levels of chirality, particularly those with adjustable properties, are extremely skilled at controlling the polarization of THz waves, far outperforming other natural materials.

We demonstrated that conjugated bilayer chiral metamaterials incorporating GST-225 active elements are capable of attaining nonvolatile and continuously adjustable optical activity within the terahertz (THz) spectrum (Bao et al. 2022), as shown in Figure 8.14. A THz time-domain spectroscopic system was employed to characterize the device, which demonstrated an adjustable ellipticity (from −36° to 0°) and rotation of the plane polarization (from −32° to 0°) at approximately 0.73 THz by varying the GST-225 state from amorphous (AM) to crystalline (CR).

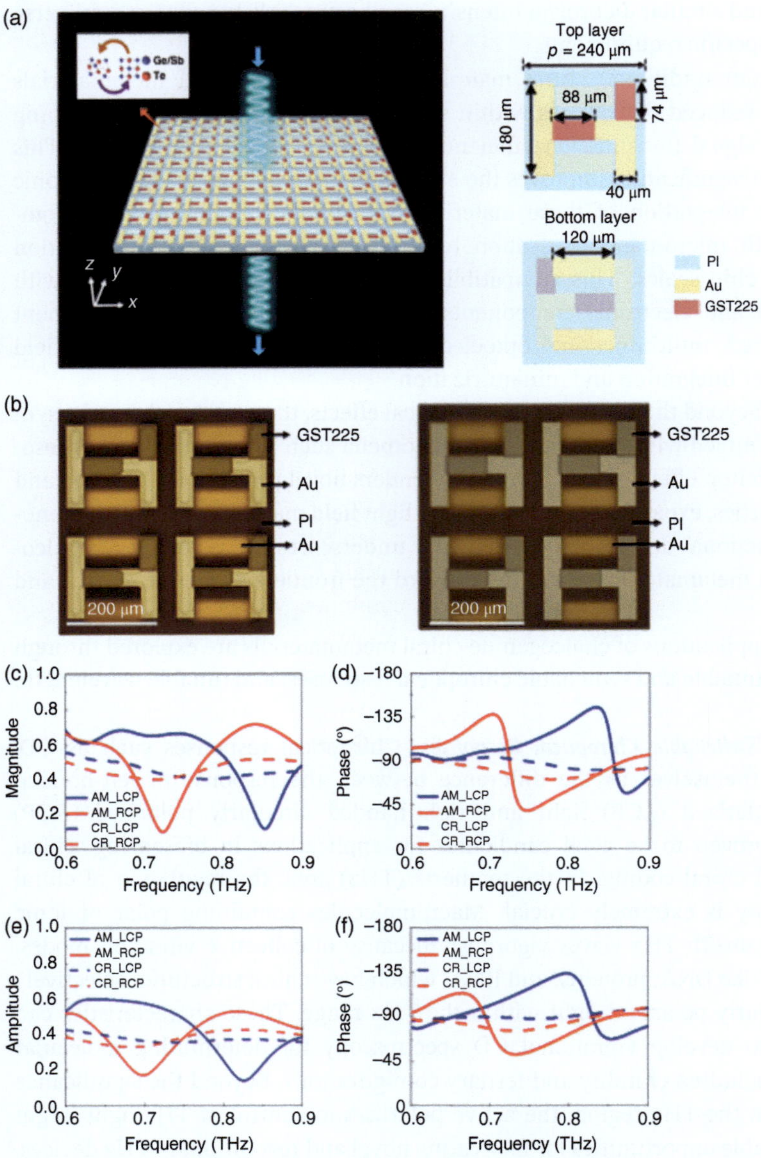

Figure 8.14 The presentation of adjustable chiral metamaterials that depend on GST-225. (a) A schematic diagram shows the bilayer metamaterials for chirality switching that utilize GST-225 strips. In the inset, the meta-atom consists of two resonators with different shapes that are patterned on either side of the PI board. (b) Images of a 2 × 2 resonator segment of the fabricated metamaterials are shown, with the left column being before crystallization and the right column after crystallization. The scale bar is 200 μm. The measured transmission (c) magnitude and (d) phase spectra of LCP (blue lines) and RCP (red lines) incident waves are displayed, where the solid lines correspond to the amorphous state and the dashed lines to the crystalline state. The simulated transmission spectra of LCP (blue lines) and RCP (red lines) incident waves, including (e) amplitude and (f) phase, are also shown. Here, the solid and dashed lines represent the amorphous and crystalline states, respectively. *Source:* Bao et al. (2022) / Springer Nature / CC BY 4.0.

Moreover, a continuously tunable chiroptical response was experimentally observed by partially crystallizing the GST-225, which can create intermediate states, having regions of both AM and CR states. It should be noted that compared with other active elements, the GST-225 has the advantage of nonvolatility, and it doesn't need any energy to maintain its structural state. Our research enables the creation of THz meta-devices that can actively control the polarization of THz waves. These devices may have applications in dynamically tunable THz circular polarizers and polarization modulators for THz emissions.

Apart from the terahertz region mentioned earlier, many biomolecules and chemical substances have chiral characteristics, and the infrared band is an important region to study the molecular vibration and rotation characteristics of these substances. The tunable chiral optical response is helpful for a more detailed study of the optical properties of chiral substances in the infrared band, such as CD, optical activity, etc., so as to deeply understand the structure, conformation and molecular interaction, and other microscopic information of chiral molecules, which offers a potent method for the fundamental research in chemistry, biology, and other disciplines.

We conducted numerical modeling of gammadion metal/phase-change material/metal three-layer planar chiral metamaterials (Cao et al. 2013), as shown in Figure 8.15. The chiral metamaterial is combined with the GST-225 phase-change material to realize a broad tuning range of circular dichroism within the mid-infrared band. A photothermal model is employed to investigate the time-dependent change of the temperature of the GST-225 layer and to demonstrate the possibility of rapidly switching the phase of GST-225 under a low incident light intensity of 0.016 mW/μm^2.

Tang et al. designed and demonstrated infrared phase-change chiral metasurfaces constructed from GST-225 material, which incorporate trapezoid-like resonators (Tang et al. 2024), as shown in Figure 8.16. This design enables the attainment of

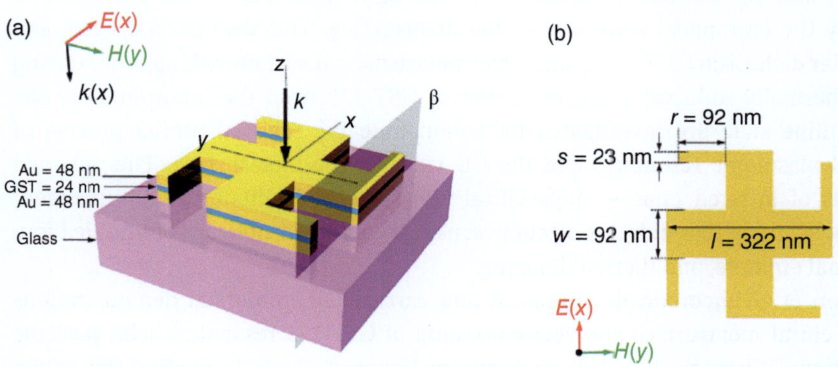

Figure 8.15 (a) A diagram illustrates the gammadion metamaterial and the polarization of the incident light. The Au film, GST-225 spacer, and another Au film have thicknesses of 48 nm, 24 nm, and 48 nm, respectively. The lattice constant in both the x and y directions is $L = 506$ nm, and the dimensions are $l = 322$ nm, $w = 92$ nm, $s = 23$ nm, and $r = 92$ nm. The entire structure is placed on a 200 μm thick BK7 silica glass. β represents a cross-section plane along the arm's edge. (b) An image of the gammadion metamaterial from a top-down perspective is presented. *Source:* Cao et al. (2015) / Optical Society of America.

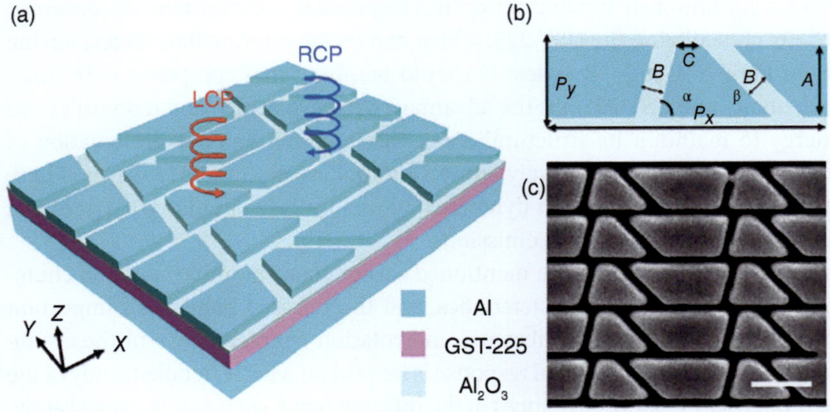

Figure 8.16 (a) Schematic diagram and SEM image of the GST-based phase-change chiral metasurface with tunable CD. (b) The dimensions of the metasurface unit cell. (c) SEM image of the chiral metasurface. Scale bar: 500 nm. *Source:* Tang et al. (2024) / Optical Society of America and Optica Publishing Group.

adjustable CD responses. The resonators configured in an asymmetric trapezoid shape are designed to support two chiral plasmonic resonance modes, where the CD values are opposite when illuminated by LCP and RCP. A CD that can be switched between negative and positive values during the phase transition of GST-225 is realized by appropriately selecting the tilted angles of two slots in the metasurface unit cell design, where the peak wavelength of the plasmonic resonance under LCP excitation in the crystalline state overlaps with the peak wavelength under RCP excitation in the amorphous state. Moreover, the electromagnetic field distributions of two chiral plasmonic resonant modes under left-circularly polarized (LCP) and right-circularly polarized (RCP) light illumination are examined to clarify the chiroptical responses of the metasurface. The absorption spectra and circular dichroism (CD) mapping of the metasurface at various baking times during the thermally induced phase transition of GST-225 from the amorphous to the crystalline state are investigated to demonstrate the thermal tuning process of chiral plasmonic resonance and the CD value of the metasurface. The obtained results of infrared phase-change chiral metasurfaces facilitate the advancement of applications like chiral molecular sensing, encrypted optical communication, thermal emitters, and thermal imaging.

Chen et al. theoretically suggested and carried out numerical demonstrations for a chiral metasurface absorber consisting of GST-225 resonators with periodic serrations (Chen et al. 2023), as shown in Figure 8.17. By leveraging the strong plasmonic resonance made possible by the gradient geometry, they achieved a significantly enhanced chiral response. Specifically, a CD value of 0.98 was achieved at $\lambda_2 = 2359$ nm, and a CD value of 0.7 was reached at $\lambda_1 = 2274$ nm. In addition, by adjusting the gradient difference in the serrated GST-225 resonator, they could control the CD intensity in multiple dimensions and almost perfectly fine-tune the chiral properties. Also, it is worth noting that the CD value can be substantially changed

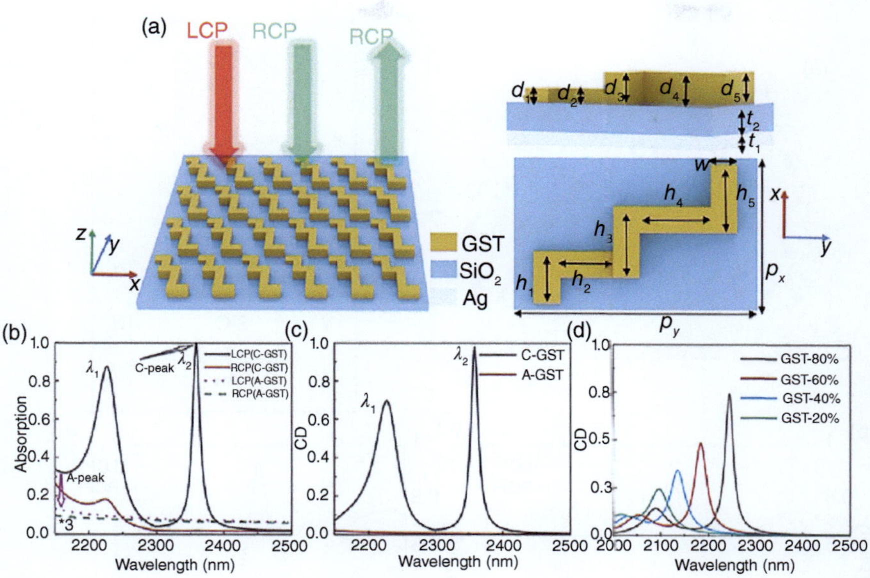

Figure 8.17 (a) Schematic illustration of the chiral metamaterial absorber. (b) Absorption spectra of GST in the crystalline and amorphous phases for LCP and RCP light. (Solid lines indicate crystalline absorption, crossed lines indicate amorphous RCP absorption, and dotted lines indicate amorphous LCP absorption). (c) Comparison of the CD spectra for crystalline and amorphous GST. (d) CD spectra of GST at different crystallization ratios. *Source:* Chen et al. (2023) / Optical Society of America.

by modifying the phase-transition characteristics of GST-225. At λ_1, the CD value can vary from 0.007 to 0.7, and at λ_2, it can range from 0.002 to 0.98, effectively representing a switch between "on" and "off" states. These research results offer novel perspectives on multifunctional chiroptics and have broad application prospects.

Ding et al. put forward an innovative metasurface absorber featuring integrated periodic GST-225 resonators (Ding et al. 2021), as shown in Figure 8.18. They then carried out numerical demonstrations to showcase the absorber's potential for reconfiguring the CD effect. Owing to the intense chiral plasmonic resonance, a significant circular dichroism (CD) value of approximately 0.75 can be obtained within a specific spectral range. Moreover, the phase transition of GST-225 resonators allows for the near-linear adjustment of the CD strength across a wide spectrum, ranging from 0.03 to 0.75. Additionally, the metasurface absorber's chirality can be reversed by independently controlling the states of the left-hand and right-hand GST-225 resonators. This feature enables the easy switching of the CD signal among the on, off, and reverse states.

Using the phase change material $Ge_3Sb_2Te_6$ (GST-326), Yin et al. presented a mid-infrared plasmonic chiral metamaterial that is tunable and switchable in concept verification experiment (Yin et al. 2015), as shown in Figure 8.19. The structure of the active chiral plasmonic dimer stack includes 50 nm of GST-326, flanked by two 10-nm protective ZnS/SiO_2 layers. These layers are located among vertically shifted, corner-stacked, orthogonal gold nanorods that are embedded in

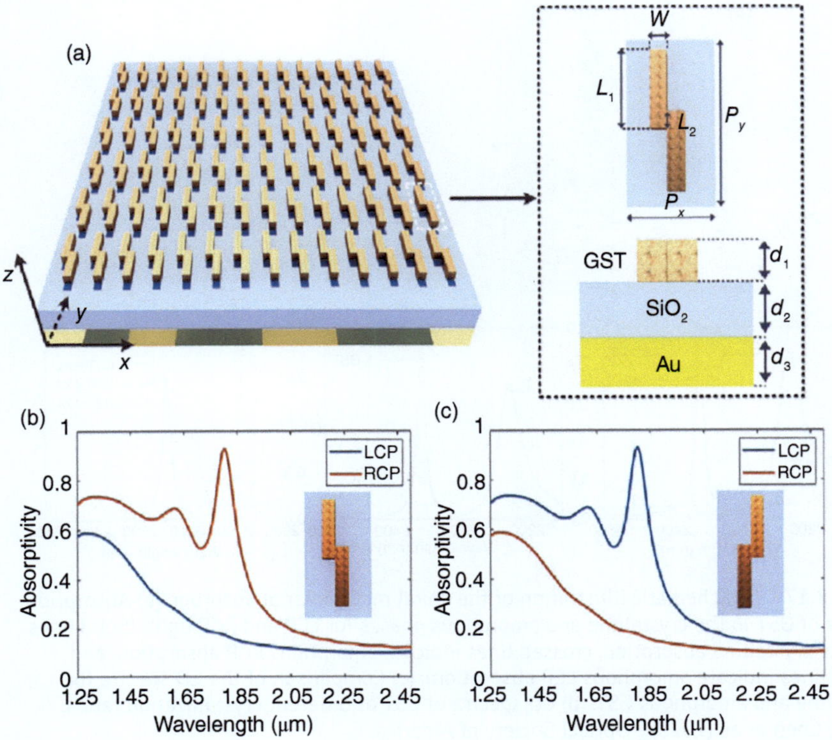

Figure 8.18 (a) A schematic depiction of the chiral metasurface absorber is shown. The inset showcases both the top-down view and cross-sectional perspectives of the unit cell. The absorption spectra of the chiral metasurface absorber are shown for (b) right-handed (RH) and (c) left-handed (LH) GST resonators when excited by circularly polarized light (CPL). *Source:* Ding et al. (2021) / Optical Society of America.

PC403. The GST layer functions as a phase-change material that can be electrically switched between the amorphous and crystalline states. A large adjustable range of circular dichroism response from $\lambda = 4.15$ to $4.90\,\mu m$ is realized, and they have proved through experiments that the combination of a passive bias chiral layer and an active chiral metamaterial can achieve switchable chirality, that is, the reversal of circular dichroism symbols, without geometric reconstruction in full-plane and layered design. Since phase change materials can switch between electricity and light, their design might pave the way for highly integrated mid-infrared polarization engineering devices that can be modulated within ultra-fast time frames.

However, the inherent chirality signal of most substances is weak, which limits its application in chiral emission and polarized light detection. On the contrary, it has been demonstrated that bound states in the continuum serve as an effective approach to boosting the intrinsic chirality. A bound state in the continuum (BIC) is a local wave that coexists with the radiative wave continuum and provides a high-quality factor (Q-factor). When the structure parameter deviates from the ideal BIC point, the BIC will transform into a quasi-bound state (QBIC) with a finite Q value. By combining QBIC with a chiral structure, a strong intrinsic CD signal can

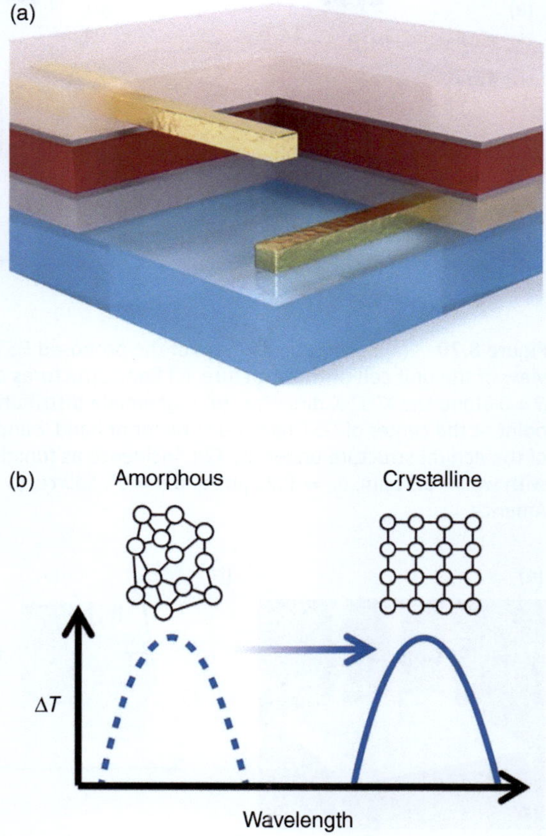

Figure 8.19 The idea of tunable chirality. (a) The active chiral plasmonic dimer stack consists of 50 nm of GST-326 (colored red), along with two 10-nm protective ZnS/SiO_2 layers (in gray). These are positioned between vertically displaced, corner-stacked, orthogonal gold nanorods that are embedded in PC403 (a light red material). To improve the visibility of the nanorods, a cross-sectional view is presented at the edge of the stack. (b) When the phase-change layer is thermally activated, it leads to a circular dichroism (CD) signal ΔT, which shifts toward longer wavelengths. *Source:* Yin et al. (2015) and Zhang et al. (2025) / American Chemical Society.

be generated. However, most of the existing studies on chiral QBIC are limited to static structures, which are unable to dynamically control their CD responses.

In 2023, Li et al. fabricated a phase-change metasurface using paired GST-225 bars. They employed this structure to showcase the dynamic regulation of the circular dichroism (CD) responses of chiral QBIC (Li et al. 2023), as shown in Figure 8.20. A refined coupled-mode theory (CMT) is formulated to evaluate the intrinsic chirality. The predictions made by this theory exhibit a remarkable level of agreement with the findings derived from finite-difference time-domain (FDTD) simulations. The maximum value of intrinsic chirality is connected to the spin-selected dipole mode. In particular, when it comes to left-circularly polarized (LCP) light, it is correlated with the coupled magnetic dipole (MD) QBIC mode. Conversely, for right-circularly polarized (RCP) light, it is associated with the decoupled electric dipole (ED) QBIC mode. By modifying the volume fraction of GST-225, the location of the chiral bound state in the continuum (BIC) can be linearly adjusted, and the corresponding chiral response can be switched.

In 2024, Chen et al. theoretically proposed an all-dielectric meta surface with a simple geometric structure, which has intrinsic chirality achieved by breaking the

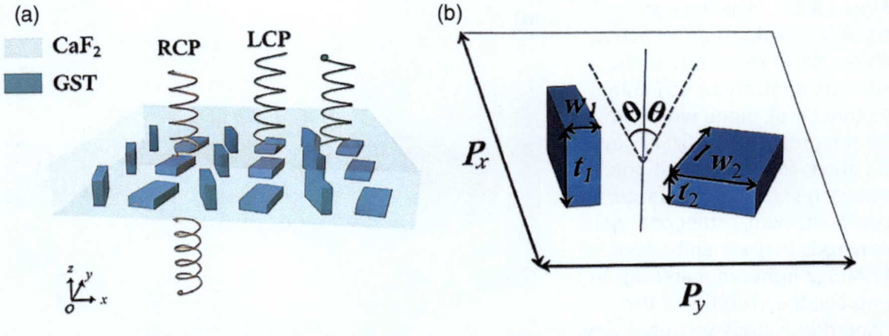

Figure 8.20 (a) Schematic diagram of the proposed GST chiral metasurface. (b) Enlarged view of the unit cell of the structure. (c) Band structures of the metasurface with $\theta = 0$ and $\delta = 0$ along the X′-Γ-X direction. (d) Eigenmode distributions of band 1 and band 2 at the Γ point at the center of GST bars. (e) Q-factor of band 1 and band 2. (f) Transmission 2D map of the achiral structure under the CPL incidence as functions of asymmetry parameter δ with $w_2 = 0.51\,\mu m$, $t_2 = 0.20\,\mu m$, and $\theta = 0°$. *Source:* Li et al. (2023) / Optical Society of America.

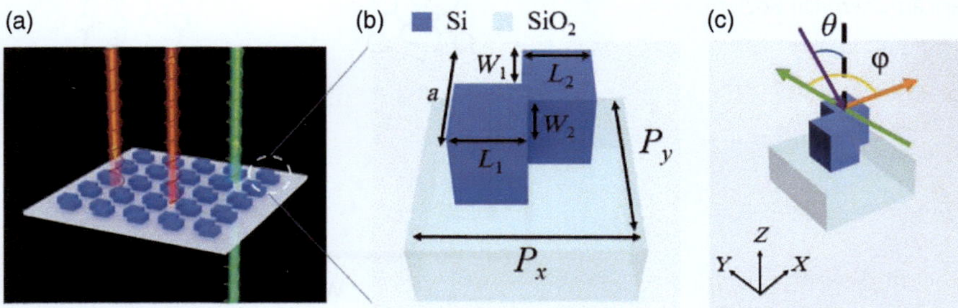

Figure 8.21 (a) Illustration of circularly polarized light acting vertically on a periodic planar metasurface. (b) The geometry of the unit cell and its related geometric parameters. (c) Schematic diagram of the incident angle θ and azimuth angle φ of polarized light obliquely incident on the metasurface. *Source:* Chen et al. (2024) / Optical Society of America.

in-plane inversion symmetry and extrinsic chirality achieved by changing the illumination symmetry, and it has near-unity circular dichroism and giant linear dichroism (Chen et al. 2024), as shown in Figure 8.21. Furthermore, the phase-change material GST-225 is incorporated into the Z-shaped Si metasurface as an extra degree of freedom to realize the manipulation of the chiral optical response. The results expand the methods and applications for achieving optical chirality, providing the possibility to realize high-performance optical devices.

Tunable Wavelength In recent years, the trend of development from passive to actively tunable chiral metasurfaces has drawn considerable attention. Actively tunable chiral metasurfaces are extremely well-suited for the dynamic control of the interaction between chiral wave-matter and circularly polarized light. GST-225, a typical kind of GST, is a nonvolatile phase-change material. When it changes from the amorphous state to the crystalline state during the phase transition, it shows a

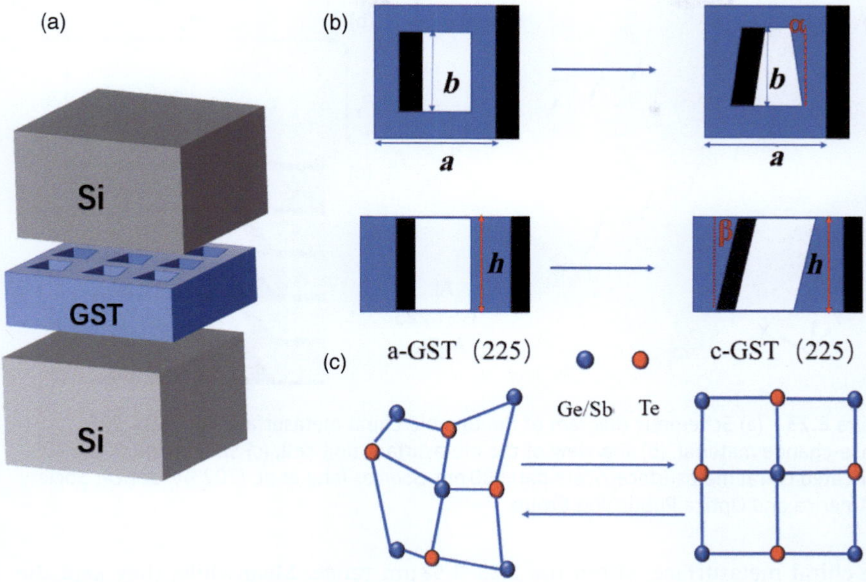

Figure 8.22 (a) A schematic representation of the slant-perturbation metasurface incorporating the GST-225 phase-change material is shown. (b) A diagram depicts the disturbance of the in-plane deformation angle α and the out-of-plane inclination angle β. The geometric parameters are as follows: $a = 1300$ nm, $b = 1000$ nm, and $h = 500$ nm. (c) An illustration depicts the phase transition of GST-225 from its amorphous state to the crystalline state. *Source:* Zhang et al. (2025) / Springer Nature / CC BY 4.0.

significant disparity in the dielectric constant. Additionally, metasurfaces based on GST can be tuned via phase transition without altering the geometric parameters of the unit cell structure. This feature gives them substantial potential in diverse applications, such as biosensing, thermal imaging, and thermal camouflage.

Zhang et al. illustrated that through the combination of the phase-change material GST-225 and chiral BIC structures, a metasurface with wavelength-tunable infrared chiral properties can be realized (Zhang et al. 2025), as shown in Figure 8.22. An infrared chiral metasurface is designed by altering the longitudinal tilt angle of the nanostructure. It features a chiral resonance with an incredibly narrow bandwidth and a CD value exceeding 0.8. Owing to the phase-change characteristics of GST-225, the chiral resonance can be adjusted in terms of wavelength without changing the structural parameters. The study also investigates the impact of the metasurface's key structural parameters on the chiral resonance wavelength and CD value. The proposed chiral BIC metasurface incorporating phase-change materials shows great promise for applications in filter devices, chiral thermal switches, infrared imaging, and tunable chiral photonics.

Tang et al. presented an infrared chiral metasurface with wavelength-tunable property, which is integrated with phase change material GST-225 and utilizes a designed parallelogram resonator unit (Tang et al. 2023), as shown in Figure 8.23. By precisely controlling the baking time at temperatures above the GST-225 phase transition temperature, they managed to adjust the resonance wavelength of

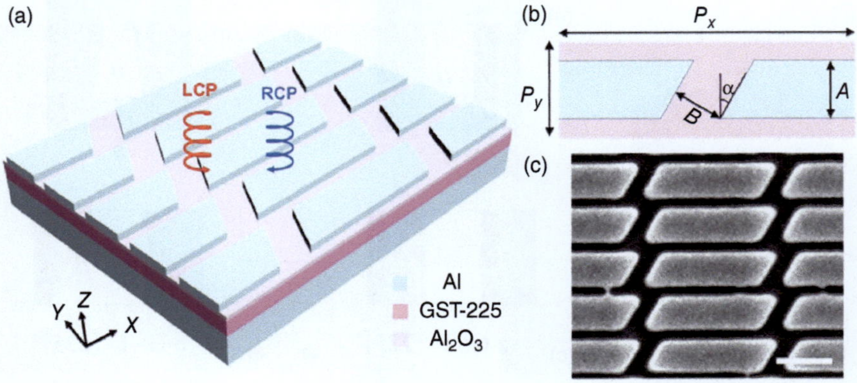

Figure 8.23 (a) Schematic diagram of the tunable chiral metasurface with GST-225 phase-change material. (b) Top view of the metasurface unit cell. (c) SEM image of the fabricated chiral metasurface. Scale bar: 500 nm. *Source:* Tang et al. (2023) / Optical Society of America and Optica Publishing Group.

the chiral metasurface within the 2.33–2.58 μm range. Meanwhile, they kept the circular dichroism of the absorption relatively stable at around 0.44. A thorough examination of the electromagnetic field and displacement current distributions under both left-circularly polarized (LCP) and right-circularly polarized (RCP) light illumination explains the chiroptical response of the designed metasurface. Moreover, the simulation of the photothermal effect shows notable temperature disparities in the chiral metasurface when illuminated by LCP and RCP light. This indicates the possibility of controlling the phase transition via circular polarization. The proposed chiral metasurfaces integrated with phase-change materials are highly promising for promoting applications in the infrared field, such as chiral thermal switching, infrared imaging, and tunable chiral photonics.

8.3.5.3 Applications of Chalcogenide Non-chiral Metamaterials

In addition to chiral metamaterials, achiral metamaterial structures can also exhibit optical rotation in the case of oblique incidence of electromagnetic waves. This phenomenon was first discovered by E. Plum et al. from the University of Southampton in the UK. They called this phenomenon external three-dimensional chirality. Subsequently, I. Sersic et al. from the FOM Institute for Atomic and Molecular Physics (AMOLF) in the Netherlands quantitatively explained this phenomenon from the perspective of the polarization tensor of the split ring resonator. In their theoretical analysis, the polarization tensor is used to establish a relationship between the incident electromagnetic field and the electric and magnetic dipole moments of the metamaterial. Then, the Green's function method is employed to derive the electric and magnetic field components of the emitted electromagnetic wave from the electric and magnetic dipole moments. Among them, the polarization tensor not only needs to satisfy the reciprocity, optical theorem of wave scattering theory, and radiation attenuation, but also needs to satisfy the symmetry, resonance peak position, and energy balance of the split ring resonator. By analyzing these conditions, the relationship between the incident electromagnetic wave and the outgoing electromagnetic wave of the split-ring resonator is ultimately determined, which confirms that the optical activity generated by the oblique incidence of the achiral structure is the inherent characteristic of the general metamaterial with electrical response and magnetic response.

The applications of chalcogenide non-chiral metamaterials are explored through two aspects: tunable and switchable chiroptical responses, and controllable valley-specific light emission.

Tunable and Switchable Chiroptical Responses Similar to chalcogenide chiral metamaterials, chalcogenide achiral metamaterials also have many applications in tunable chiroptical responses in the infrared and terahertz regions.

We conduct numerical analyses to demonstrate that when achiral metamaterials, which consist of an elliptical nanohole array (ENA) passing through a metal/phase-change material/metal multilayer stack, are tilted with respect to the incident light, multiband circular dichroism occurs (Cao et al. 2017), as shown in Figure 8.24. By transitioning GST-225 from its amorphous state to the crystalline form, the CD spectrum can be dynamically adjusted across a broad spectral range spanning from the near-infrared (NIR) to the mid-infrared (MIR). This transition allows for the on/off control of a multiband chiroptical response in the infrared spectrum. Our simulations further show that achiral multilayer stack metamaterials with strong magnetic resonance characteristics can enhance the optical chirality inside the elliptical apertures, regardless of whether GST-225 is in the amorphous or crystalline phase. The modulation of this heightened chirality might present novel opportunities for the manipulation of electromagnetic waves, such as the development of tunable circular polarizers, sophisticated chiroptical spectroscopy instruments, and highly sensitive chiral biosensors.

By integrating the nonvolatile chalcogenide semiconductor GST-225 into the achiral metamaterials design, we realized the reversible switching of exogenous chirality in the NIR region (Cao et al. 2023), as shown in Figure 8.25. When excited by a 532-nm nanosecond pulsed laser, it allows for ultrafast and reconfigurable regulation of transmittance circular dichroism (TCD) between the amorphous and crystalline states. Meanwhile, when the geometric parameters are set, the TCD can be effectively manipulated by adjusting both the rotation angle and the incident angle. Through the use of ellipsometry and analytical inversion methods, it is feasible to obtain not only linear but also circular birefringence-dichroism pairs from a single general ellipsometry measurement. This offers a comprehensive description of the polarization characteristics of non-chiral metamaterials, having an advantage over traditional circular dichroism measurements.

Shi et al. put forward a metasurface featuring tunable circular dichroism of extrinsic chirality, which is based on the phase-change material GST-225 (Shi et al. 2022), as shown in Figure 8.26. The metasurface consists of two symmetric square silver split-ring resonators and a GST-225 intermediate layer arranged periodically. When interacting with obliquely incident light, this metasurface can display electromagnetic properties similar to those of a chiral structure. Numerical simulation

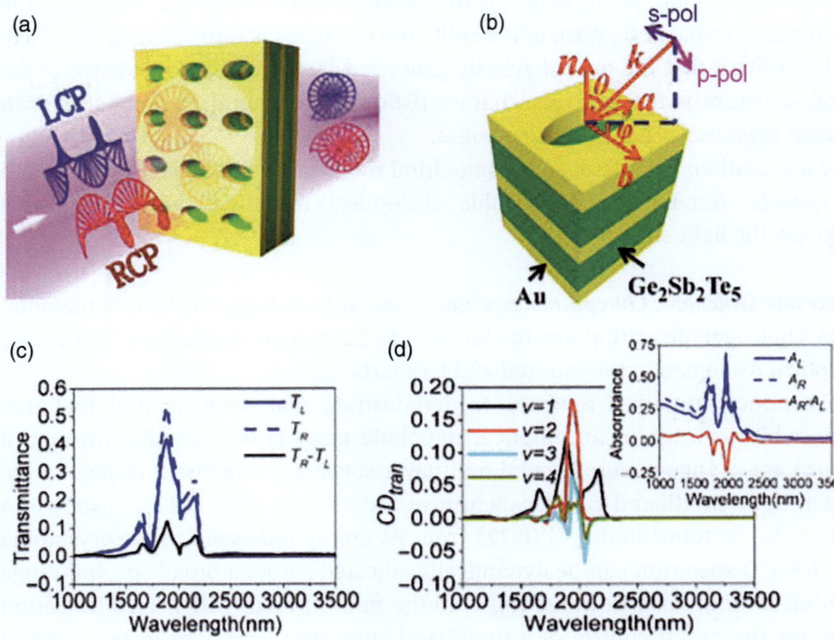

Figure 8.24 (a) A schematic diagram shows an ENA passing through an Au/$Ge_2Sb_2Te_5$/Au multilayer stack, and this structure is suspended in the air. (b) An illustration depicts the unit cell. (c) Regarding the [$(MD)_1$ M]-ENA in the amorphous state, the transmittance is shown for the LCP (blue solid line) and RCP (blue dashed line) incidences when $\theta = \varphi = 45°$. The black solid curve represents the difference in transmittance between the RCP and LCP incidences. (d) The CD_{tran} spectrum is presented for an ENA passing through (MD)vM stacks (for v up to 4) in the amorphous state at $\theta = \varphi = 45°$. The absorptances for both LCP (blue solid line), RCP (blue dashed line), and $\Delta A = A_R - A_L$ (red solid line) for the amorphous [$(MD)_2$ M]-ENA are shown in the inset. Schematic diagram of an ENA penetrating through Au/GST-225/Au multilayer stack, where the structure is suspended in air. *Source:* Cao et al. (2017) / Optical Society of America.

findings indicate that within the 50~300 THz frequency range, the metasurface attains a maximum circular dichroism value of 0.85 when the GST-225 is in the amorphous state and 0.52 when it is in the crystalline state. When the GST-225 transitions from the amorphous state to the crystalline state, a tunable frequency range of about 70 THz can be achieved. In comparison to previously reported studies, this metasurface demonstrates a higher circular dichroism and a broader tuning range. By analyzing the electric field distribution, the root cause of the circular dichroism is explained, and the effects of incidence angles and structural parameters on the circular dichroism of the metasurface are investigated. This study shows potential for applications in high-efficiency polarization modulation devices, circular polarizers, and polarization filters in the optical frequency band.

Controllable Valley-Specific Light Emission In two-dimensional transition metal dichalcogenides (TMDCs), the carriers and excitons situated at the K and K′ valleys possess a valley pseudospin, presenting a new degree of freedom that holds great

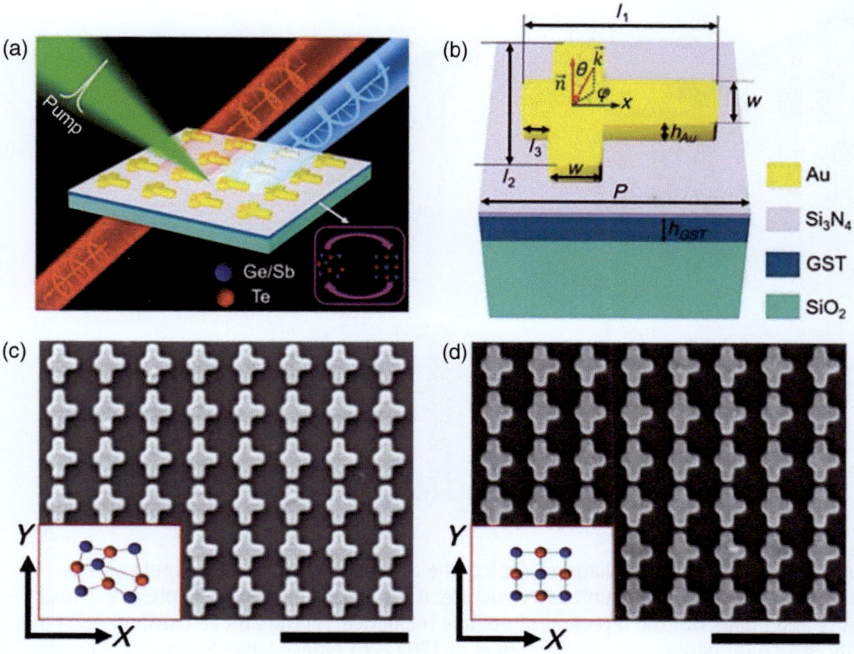

Figure 8.25 Structure of the nonchiral metasurface using chalcogenide phase-change materials (PCMs). (a) A schematic picture shows an all-optical, nonvolatile chiroptical switch based on a chalcogenide metasurface. A single nanosecond laser pulse is capable of changing a 35-nm thick GST-225 layer between its amorphous and crystalline states. This layer consists of 510,000 antisymmetric Au crosses. (b) An illustration presents the Au cross with broken symmetry. The geometrical parameters are: $l_1 = 440$ nm, $l_2 = 550$ nm, $l_3 = 70$ nm, $h_{Au} = 35$ nm, $w = 200$ nm, and $P = 700$ nm, respectively; the thickness of the GST-225 functional laminate is $h_{GST} = 35$ nm. The thickness of the Si_3N_4 diffusion barrier is 5 nm. The SEM pictures of the fabricated metasurface. (c) before and (d) after crystallizing the GST-225 laminate incorporated with the metasurface. Scale bar: 2 μm. *Source:* Cao et al. (2020) and (2023) / John Wiley & Sons.

promise for quantum information processing and storage. In these materials, the valley-polarized excitonic states selectively interact with chiral photon states through valley circular dichroism. As a result, the control and detection of valley states can be achieved by using circularly polarized light (CPL) with different helicities. This exciton–photon interaction further promotes the utilization of valley states in the creation of chiral quantum light-emitting devices. The valley-specific photoluminescence (PL) can be gauged by the degree of polarization (DOP) under circularly polarized excitation, expressed as $[I(\sigma^+) - I(\sigma^-)]/[I(\sigma^+) + I(\sigma^-)]$, where $I(\sigma^\pm)$ represents the PL intensity for right or left circular polarization. However, the magnitude of the DOP is restricted by valley-specific absorption and decoherence processes, such as intervalley scattering. These factors impede the practical application of valleytronics at room temperature.

To improve the performance of valleytronics, it has been reported that TMDCs can be integrated with various plasmonic nanostructures and metasurfaces. For

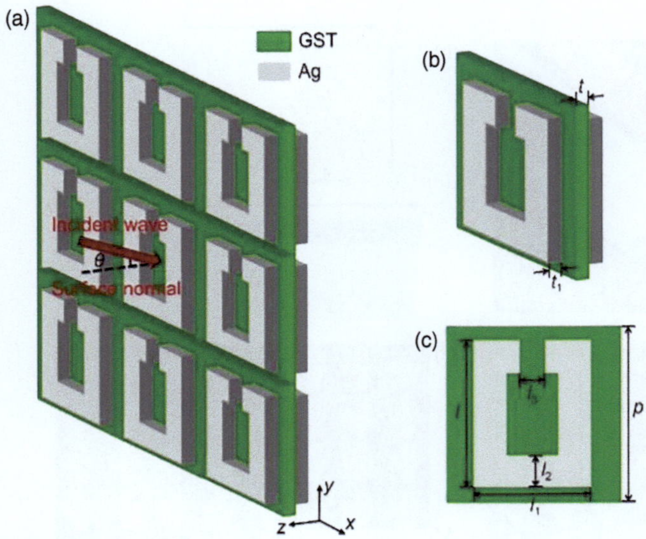

Figure 8.26 The schematic diagram depicts the designed extrinsic chiral metasurface structure array along with its unit cell structure. It also shows (a) the principle of extrinsic chirality and points out the direction of oblique incidence; (b) the unit cell structure; (c) the specific structural parameters. *Source:* Shi et al. (2022) / Chinese Laser Press.

example, plasmonic nanowire waveguides and metasurfaces have been used to increase the intensity of exciton emission and guide the emission from excitons in different valleys to different directions. This spatial segregation of emission is beneficial for extracting valley-specific information. Moreover, chiral metasurfaces and nanostructures have been applied to enhance the valley polarization of PL, allowing significant valley polarization to be preserved even at room temperature. This enhancement is due to the enhanced chiral fields produced by the metasurface, which speed up the radiative decay of chiral photon states through the chiral Purcell effect. However, the practical use of these systems in valleytronics encounters several difficulties: (i) valley polarization is enhanced when the helicity of the excitation light matches the chirality of the metasurface, but it is reduced when the helicity is reversed; (ii) the resonance of chiral structures can change the polarization of the excitation in the near-field, resulting in exciton generation in both valleys even when purely circularly polarized light is used; (iii) chiral metasurfaces can also significantly alter the polarization of exciton emission, making it more difficult to interpret valley information through DOP measurements.

Liu et al. combined MoS_2 monolayers with achiral dielectric metasurfaces made up of Si disk arrays to improve and regulate valley-specific absorption and emission (Liu et al. 2023), as shown in Figure 8.27. Through coupling with the electric and magnetic Mie modes of the metasurface, the emission intensities of neutral excitons, trions, and defect-bound excitons can be respectively boosted, and their lifetimes can be correspondingly shortened. At the same time, the spectral shape can be modified. Additionally, at 100 K, the degree of polarization (DOP) of exciton and trion emissions from the valley can be symmetrically heightened. The increase in

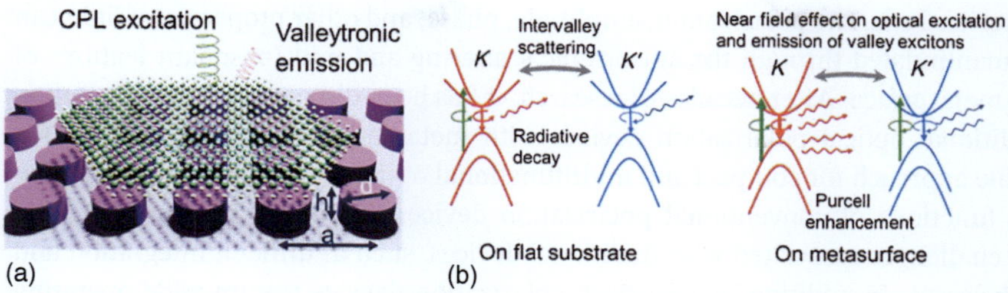

Figure 8.27 Integrating monolayer MoS_2 with achiral dielectric metasurfaces to regulate the valleytronic emission of transition metal dichalcogenides (TMDCs). (a) A schematic diagram illustrates the combination of a single-layer TMDC and a Si metasurface. (b) One schematic (left) depicts the excitation and decay processes of valley excitons with substantial intervalley scattering, while another (right) shows the near-tunable and switchable chiroptical responses and field impact of the achiral metasurface on the generation and emission of valley excitons. *Source:* Liu et al. (2023) / American Chemical Society.

DOP is attributed to both the metasurface-enhanced chiral light absorption and the metasurface-enhanced exciton emission caused by the Purcell effect. Our research, which combines Si-compatible photonic design with large-scale 2D materials, marks a substantial progress toward on-chip valleytronic applications that are nearer to room-temperature operation.

8.3.6 Chalcogenide Polarization Converter

Polarization is a fundamental property of light. It determines the oscillating directions of light during propagation. Manipulating the polarization state of light at the micro- and nanoscale has been widely used in quantum information processing, imaging, sensing, and other applications. Most of the light in nature is unpolarized or partially polarized. Thus, a polarizer is needed to separate polarized light from natural light. Polarizers typically utilize the asymmetry in light reflection, refraction, absorption, scattering, and other processes to split the incident light into two orthogonal polarized lights and then select the required one. Traditional ways of generating polarized light involve incidence based on the Brewster angle, birefringence, or the absorption dichroism of crystals. Methods for adjusting the polarization state include creating phase differences, like using wave plates and compensators, as well as the electro-optical effect and magneto-optical effect. The demand for optical systems has spurred the development of polarization devices. The advent of optical fiber polarizers and spatial light modulators has brought great convenience. Nevertheless, they still suffer from the issues of traditional polarization devices, such as large size, narrow and fixed operating bands. It remains challenging to achieve the integration of miniature photonic devices. Therefore, the design of polarization devices that are easily integrable, broadband, and tunable has become a research focus in recent years.

Metamaterials and metasurfaces are artificially fabricated ultra-thin materials. They possess electromagnetic characteristics and optical responses that natural materials lack. The polarization, amplitude, phase, and other properties of light can be manipulated through the anisotropic scattering and multi-resonant features of the metasurface. After decades of research, it has been discovered that substituting traditional optical polarization devices with metamaterials or metasurfaces is a viable approach for compact and multifunctional optical systems. Besides fulfilling the functions of conventional polarization devices, metasurfaces also overcome the challenges associated with traditional devices, such as difficult integration and high losses. In addition, metasurface polarization devices feature wide operating bandwidths, high efficiency, and versatility, among other traits. They showcase unparalleled abilities to control the polarization states of light in the aspects of polarization generation, manipulation, detection, and imaging. They can achieve polarization manipulation, including polarization selection, asymmetric transmission, and polarization conversion, and can perform multiple functions with a single device, which is not possible with conventional ones.

Polarizers play a crucial role in optical systems. They split incident light into two orthogonally polarized beams. The unit cell of a metasurface can be engineered to have distinct responses to different polarized lights, enabling the metasurface to act as a polarizer. Based on their practical functions, metasurface polarizers can be further classified into linear polarizers and circular polarizers. Metasurface linear polarizers usually generate polarization sensitivity via the shape anisotropy of the unit cell, triggering surface plasmonic resonance or Mie resonance. This enables specific polarized light to pass through, while the other orthogonally polarized light is either absorbed or reflected.

To date, the proposed dynamic metasurfaces (MSs) have made use of resonant dispersive meta-atoms. These meta-atoms inherently restrict the operational bandwidth of the corresponding optical functions. To achieve dispersionless phase control and decouple the scattered amplitude from the phase response, Pancharatnam–Berry (P–B) or geometric phase MSs have been presented as an effective approach. By spatially altering the orientation of identical meta-atoms around their optical axis, wideband gradient MSs can be achieved, and the fabrication tolerance is reduced. In this context, Yin et al. presented the first dynamic P–B-based MS composed of two interleaved meta-atoms of different sizes placed on top of a low-loss 50 nm thick $Ge_3Sb_2Te_6$ layer. When the GST layer is in the amorphous state, the resonance condition is only met for meta-atoms with longer lengths. Nevertheless, when the GST is transformed into the crystalline state, only meta-atoms with shorter lengths strongly interact with the incident light at the same wavelength. This allows for the combination of two functionalities using two parallel rows of meta-atoms with appropriate lengths interleaved in a supercell structure. By rotating the longer and shorter meta-atoms clockwise and counterclockwise, respectively, opposite phase gradients can be applied to the transmitted circularly polarized (CP) light, resulting in cross-polarized light with +1 and −1 order diffraction, respectively. Moreover, by encoding the quadratic phase distribution of cylindrical metalenses in the spatial orientation of plasmonic meta-atoms, bifocal focusing can be realized upon switching the state of GST, although the overall efficiencies are less than

10%. It is worth noting that the resonant meta-atoms in the spatially multiplexed metasurfaces (MSs) have relatively small scattering cross-sections. This leads to a reduction in the overall efficiency when compared with traditional schemes.

Zhang et al. have experimentally verified switchable spin-orbit interactions by using a hybrid P–B phase metasurface (MS). A metal–insulator–metal (MIM) structure is employed, where the insulator layer consists of a thin MgF_2 layer placed on a 600 nm thick $Ge_2Sb_2Te_5$ film. This film is sandwiched between the bottom Au substrate and the top array of Au nanoantennas. Thanks to the carefully designed meta-atoms, when the GST is in the amorphous state, the reflectance of the cross-polarized light from the MS reaches up to 60%, while the co-polarized reflectance is below 10%. Consequently, the MS shows a highly efficient half-wave plate function, with the polarization conversion ratio (PCR) exceeding 80% in the mid-IR spectral range. By annealing the a-GST active layer and changing its state to the crystalline state, the MS acts as a simple mirror with a relatively low PCR of 10%. The authors utilized this effect to dynamically control the coupling between the spin and orbital momenta of photons within a broadband wavelength range from 8.5 to 10.5 μm. As an illustration to prove the concept, three MSs for controlling the spin-orbit interactions were fabricated and characterized. These MSs enabled the spin Hall Effect, vortex beam generation, and holography in the amorphous state. The state transition from a-GST to c-GST simply disrupts the geometric phase phenomena and thus deactivates these optical functions.

More recently, de Galarreta et al. presented a reconfigurable MS for beam steering applications in the operating spectral range of 1530–1570 nm. The MS is made up of a multilayer stack of SiO_2, ITO, GST, and ITO, which is sandwiched between an Al back reflector and a top array of periodically arranged Al nanoribbons with different lateral widths. Selecting Al as the plasmonic material not only cuts down the fabrication cost but also makes the entire process compatible with CMOS technology. Moreover, the spacer layer between the nanoantennas and the PCM layer can be removed because Al diffuses less in GST compared to Au. This maximizes the interaction of the PCM layer with the near fields of the plasmonic nanoantennas. However, its relatively low melting temperature makes the amorphization process of the GST layer challenging due to possible deforming or even melting issues. It is worth noting that the MIM configuration not only supports the fundamentally enhanced GSP mode, which is marked by the antiparallel displacement current. Moreover, the subwavelength characteristic of the dielectric spacers promotes the quenching process during the amorphization. Here, the gradient phase control is accomplished by slightly deviating each nanoribbon width from the specified center resonant frequency. Beam steering with multiple reflecting angles simultaneously can also be realized by pixelating the MS. Each pixel can be specifically designed to steer the incident light at a different angle. When the GST is in its amorphous state, the MS acts as an anomalous reflector to a pre-designed angle. In the crystalline state, the incident beam does not experience a phase gradient, so the incident light is reflected specularly.

Chensheng Li put forward and experimentally verified a bilayer transmissive metasurface with four-level unit pixels, which can achieve on-demand photonic routing in the NIR region. An all-dielectric double-layer nanostructure cell, consisting of an upper GST nanoantenna and a bottom amorphous silicon (α-Si) nanofin, is used as the unit pixel for Pancharatnam–Berry (PB) phase modulation. This significantly reduces the unnecessary transmission loss compared with the metallic counterpart. The α-Si nanofins and GST nanoantennas are responsible for cross-polarization generation and amplitude modulation of cross-polarized light, respectively. To confirm the cross-polarization modulation, they first fabricated tunable metasurfaces with the same unit cell orientation. The cross-polarization transmissivity of the tunable metasurfaces in the amorphous state is approximately 76.1%. The absolute polarization conversion variation (PCV) caused by the GST phase change is about 55% at 1550 nm, indicating a large polarization conversion modulation depth. Based on this, a metasurface composed of arranged four-level phase unit pixels is proposed. Each unit pixel is expected to be independently controlled by a pulsed laser, which can well meet the requirement of on-demand nonvolatile wavefront modulation and thus photonic routing. Then they demonstrated the scheme by fabricating metasurfaces with coded unit pixels to realize desired photonic transport.

These metasurfaces are composed of periodically arranged nanofins with the same orientation. When the GST components are in the amorphous state, the polarization conversion efficiency (PCE) of the metasurface is high. This enables the transmission of most cross-polarized light at the desired wavelength, and as a result, the polarization conversion switch is "on." In stark contrast, for GST in the crystalline state, the corresponding PCE is low, with most transmitted cross-polarized light being impeded. Consequently, the polarization conversion state of the metasurface is "off." To obtain the ideal metasurface capable of polarization conversion modulation, a double-layer design is employed. One layer is dedicated to achieving high polarization conversion efficiency (PCE), while the other is accountable for an extremely sharp modulation of the optical response. The specific configuration of the unit cell is presented in Figure 8.1c. A quartz substrate is selected due to its high optical transmittance in the near-infrared (NIR) region, which offers convenience for the transmissive scheme.

8.3.7 Chalcogenide Optical Tweezers

8.3.7.1 Optical Tweezers

Optical tweezers are a technique that captures and manipulates tiny particles (such as atoms, molecules, cells, and nanoparticles) by using the light pressure of the laser. Optical tweezers technology has emerged as a crucial instrument in the domains of biophysics, nanotechnology, and quantum optics, since its invention by Arthur Ashkin in 1986.

The primary trapping force of optical tweezers stems from the light intensity gradient (gradient force), which draws particles towards the focal point of the beam (that is, the area with the highest light intensity). Particles are also influenced by the scattering force, and the direction of this scattering force aligns with the direction of light propagation.

Optical tweezers are mainly composed of four parts:

1. **Laser Source:**

   High-power continuous-wave lasers are usually used, such as Nd:YAG lasers or Ti:sapphire lasers.

2. **Optical System:**

   It includes a lens, a mirror, and an objective lens for focusing a laser beam.

3. **Sample Stage:**

   Used to place particles or samples to be manipulated.

4. **Detection System:**

   It includes a camera and a photodetector for real-time observation and measurement of particle position and motion.

Optical tweezers can manipulate and measure with nanometer precision without touching the sample, avoiding pollution and damage to the sample. In addition, optical tweezers can be manipulated, measured, and imaged at the same time, which is suitable for a variety of research scenarios.

To capture multiple objects at various locations, acousto-optic and holographic techniques utilize non-diffracting beams like Bessel and Airy beams, which preserve their transverse intensity profiles during propagation, enabling long-range object transport via radiation pressure. Traditionally, radiation pressure is expected to push objects along the beam's direction. Nevertheless, recent advancements make use of non-paraxial structured beams to draw objects towards the light source, which are referred to as optical tractor beams. When these methods are combined with optical tweezers, they are capable of generating multiple optical traps or trap arrays. Through programming spatial light modulators to modify the incident light, holographic optical tweezers can adjust the trapping positions to generate the desired field patterns on the target plane. Additionally, research on radiation pressure, resulting from momentum transfer between light and objects, has gained traction in recent years. Non-diffracting beams, such as Bessel and Airy beams, maintain their transverse intensity profiles during propagation, making them suitable for long-range object transport through radiation pressure. While radiation pressure typically pushes objects along the beam's path, recent innovations use non-paraxial structured beams to pull objects toward the source, creating optical tractor beams.

Due to the diffraction limit, the size of objects manipulated by optical tweezers or non-diffracting beams is generally restricted to the micrometer scale. To deal with sub-wavelength objects, optical forces generated by dielectric nanostructures such as waveguides, slot waveguides, and waveguide loops have become a highly significant area of interest. These waveguide structures create an exponentially decaying electric field in regions beyond the diffraction limit, enabling the trapping and transport of nanometer-scale particles along waveguide surfaces using propagation fields. A key advantage of waveguide-based manipulation is the ease of designing waveguides in specific shapes to guide objects along desired trajectories, compared to controlling the path of a laser beam. Traditionally, dielectric waveguides have been designed for single functions, such as transporting particles in the forward direction. Recently, however, there have been proposals to transport particles in the opposite direction of field propagation, particularly in the far-infrared range. When light with a frequency lower than the cutoff frequency is injected, the propagating mode changes into an evanescent mode. In this mode, the field intensity decreases along the propagation direction. Here, the gradient force exceeds the radiation pressure, drawing particles towards the light source. Nevertheless, at various wavelengths, the calculated pushing force ($\sim 4 \times 10^{-27}$ N) is notably weaker than the pulling force ($\sim 6 \times 10^{-19}$ N), which makes it difficult to demonstrate the pushing force experimentally. The next-generation "lab-on-chip" systems combine optical manipulation, microfluidics, and other technologies to develop multifunctional devices that can manipulate and analyze multiple cells on a single chip. These systems require flexible manipulation tools that can trap and move objects freely across different regions.

8.3.7.2 Chalcogenide Optical Tweezers

The chalcogenide compound GST is well-known for its application in rewritable optical discs and phase-change memory. Usually, the phase transition of GST can be triggered by laser heating within the nanosecond or microsecond range. When heated between its crystallization and melting temperatures, amorphous GST turns into a crystalline state. On the other hand, a high-intensity laser pulse can transform crystalline GST back into an amorphous state. The rapid phase transition rate of GST makes it appropriate for diverse applications in tunable and reconfigurable devices. Wang et al. have demonstrated that a femtosecond laser can bring about bidirectional phase transitions with a volume resolution as tiny as 0.02 μm^3, allowing the formation of two-dimensional reconfigurable patterns through advanced writing methods.

Zhang et al. introduced a versatile device utilizing a single rectangular GST waveguide. This waveguide can produce continuous pushing or pulling forces of comparable magnitude over distances exceeding 10 μm on gold particles (50 nm in diameter) when the GST is in its amorphous or crystalline phase, respectively, at a working wavelength of 1.75 μm. Leveraging femtosecond laser writing technology, they also demonstrated a reconfigurable optical trap array on the GST waveguide, capable of trapping small particles in predesigned areas. When a femtosecond laser is employed to convert particular areas of the amorphous GST waveguide into the crystalline phase, particles can be captured at the interfaces between the amorphous and crystalline regions. This particle-trapping setup can be removed and rearranged by altering the settings of the femtosecond laser, enabling dynamic regulation. As a result, slight adjustments to a single waveguide can achieve two crucial functions: (i) moving particles either in the positive or negative direction, and (ii) capturing particles at designated positions. These functions mimic the effects of optical pushing and pulling using non-diffracting beams and optical trap arrays, yet they provide the additional benefit of being able to manipulate objects at the nanometer scale.

8.4 Discussions

Owing to their distinctive optoelectronic and thermophysical characteristics, chalcogenide-based metamaterials have emerged as a significant research frontier in advanced functional materials. These materials show significant advantages in areas such as nanophotonics, highly sensitive sensors, and energy conversion and storage. Their core value is manifested in the accurate control of electromagnetic waves via the design of micro-nano structures. This feature endows it with significant application potential in the development of optoelectronic devices, the production of high-precision detectors, and the realm of advanced stealth technology.

Despite the promise, chalcogenide metamaterials still face challenges. For example, the lack of reliability of existing experimental techniques under extreme conditions constrains the in-depth analysis of material properties. Theoretical modeling of the coupling mechanism of metamaterials is still imperfect, which makes it difficult to reproduce some of the experimental results. The essence of these problems stems from the fact that the multi-scale effects and complex interaction mechanisms of material systems have not yet been fully revealed.

Future research can break through into three dimensions: development of novel characterization techniques to enable in situ dynamic observations and theoretical modeling across scales; exploration of non-precious metal composite systems to reduce preparation costs; and expanding the application of materials in new frequency bands such as terahertz. Multidisciplinary cross-fertilization will be the key to breaking through the bottleneck of the existing technology, and it is expected to build a complete theoretical system for the design of sulfur metamaterials by integrating the cutting-edge fields of computational materials science and quantum optics. With the advancement of material synthesis and device processing technology, sulfur compound metamaterials may achieve industrialization breakthroughs in the fields of photonic chips, all-weather photovoltaic devices, etc., and promote optoelectronic technology into a new era.

9

Perspectives

9.1 Future of Chalcogenide Nanophotonics

9.1.1 Advanced Materials for Next-Generation Photonic Devices

Nanophotonic devices can be dynamically adjusted using a range of tuning strategies and innovative materials. These methods include optical tuning via photo-carrier injection (e.g., Si), electrical tuning through carrier injection (e.g., ITO), electric field-responsive devices (e.g., LCs), chemical modulation via hydrogenation and dehydrogenation (e.g., Pd, Mg), thermal control through phase changes for both reversible (e.g., VO_2) and irreversible (e.g., chalcogenide PCMs) processes, and mechanical adjustments using flexible substrates (e.g., PDMS) (Ruiz de Galarreta C et al. 2020).

Chalcogenide phase-change materials (PCMs) are well-known for their remarkable optical, thermal, and electrical properties, making them top contenders for tunable photonic technologies. Their capacity to transition between amorphous and crystalline states provides a substantial refractive index contrast (approximately 2), enabling accurate manipulation of amplitude, phase, and wavelength. This versatility supports diverse applications, including spectral filtering, light absorption, beam steering, metalenses, and holographic devices.

However, traditional chalcogenide PCMs have certain limitations, leading to the exploration of new compositions such as Sb_2Se_3 and Sb_2S_3. Sb_2Se_3, featuring a one-dimensional crystal structure, is an outstanding photovoltaic absorber within the visible spectrum. Even though it encounters difficulties like the absence of photoluminescence because of its indirect bandgap, notable anisotropy that affects carrier mobility (Chen et al. 2017), and inherently low electrical conductivity, it has shown promise for dynamic tuning in color filter metasurfaces.

Similarly, indium antimony telluride (In_3SbTe_2) has recently emerged as a reconfigurable infrared nanophotonic PCM, exhibiting large permittivity shifts during phase transitions. Compared to conventional PCMs like GeTe, $Ag_5In_5Sb_{60}Te_{30}$, and GST225, IST thin films exhibit a significantly higher crystallization temperature (~250 °C) and an electrical contrast exceeding six orders of magnitude between crystalline and amorphous states (Devdutt Tripathi et al. 2023). These characteristics render IST a promising option for high-speed, nonvolatile optical applications.

Chalcogenide Nanophotonics, First Edition. Tun Cao, Ying Su, Yinghan Li, and Hongze Gao.

Recent studies have concentrated on PCM nanocomposites to improve the performance of materials for data storage, reconfigurable photonics, and neuromorphic computing. Alloying GST with La–Sr–Mn–O (LSMO) thin films has improved crystallization temperature (Kraft et al., 2021), thermal stability, power efficiency, and switching speed. Similarly, Sb-based alloys have demonstrated enhanced thermal stability (e.g., Sb–GeO_2), improved thermoelectric properties (e.g., Mg–Zn–Sb), and better switching performance (e.g., In–S–Sb, Zn–Sb–Bi), positioning them as crucial materials for the memory and computing devices of the next generation.

In addition to traditional PCMs, hybrid architectures that combine chalcogenides with two-dimensional (2D) materials like graphene and transition metal dichalcogenides (TMDs) have unlocked new functionalities, including enhanced optical nonlinearity and tunable photonic device applications. These hybrid structures pave the way for compact, high-performance optoelectronic components. Additionally, perovskite-chalcogenide hybrids combine the high absorption and tunable bandgap of perovskites with the strong optical nonlinearities of chalcogenides, further expanding their applications.

Combining phase-change chalcogenides with plasmonic and dielectric metasurfaces allows for reconfigurable optical functions, including dynamic beam steering and spectral tuning as needed. Topological insulator-chalcogenide composites also present a promising avenue for robust light transport immune to defects and disorder, which is beneficial for applications in optical computing and secure communication.

Advancements in material processing techniques, including molecular beam epitaxy (MBE), atomic layer deposition (ALD), and pulsed laser deposition (PLD), have significantly improved material quality. These methods allow for the fabrication of ultra-thin chalcogenide films with superior uniformity, compositional control, and enhanced interface quality—critical factors in realizing high-performance photonic devices.

9.1.2 Chalcogenides in Quantum and Nonlinear Photonics

Chalcogenide materials, with their unique optical properties, hold great promise in quantum and nonlinear photonics. These materials display robust nonlinear optical responses, rendering them appropriate for a diverse array of applications, such as frequency conversion, light generation, and quantum information processing. Nevertheless, the application of these materials in quantum and nonlinear photonics encounters several obstacles that must be overcome for extensive utilization.

9.1.2.1 Nonlinear Optical Response

Chalcogenide materials exhibit a strong nonlinear optical response, which is indispensable for applications such as optical switching, parametric amplification, and generation of entangled photons. However, nonlinear effects can lead to high optical loss, especially at high pump powers, limiting their efficiency. To improve nonlinear performance, optimizing material composition, thickness, and waveguide design is necessary. Additionally, employing techniques such as phase matching or dispersion engineering can enhance nonlinear efficiency and reduce unwanted loss.

9.1.2.2 Quantum Photonics

Chalcogenide materials have also attracted considerable attention in the field of quantum photonic applications, such as the generation and control of quantum states of light. These materials can support the generation of entangled photon pairs and provide the necessary nonlinearities for quantum computing and quantum sensing. However, challenges remain in achieving stable quantum states and reducing decoherence due to material imperfections and losses. Developing techniques to improve the purity of quantum states and to control environmental factors such as temperature and humidity will be critical for the successful deployment of chalcogenide-based quantum devices.

9.1.2.3 Challenges in Fabrication and Device Design

The fabrication of high-quality chalcogenide-based devices for quantum and nonlinear photonics is challenging due to the material's inherent susceptibility to defects and degradation under high-power or high-field conditions. Techniques such as careful defect engineering, material purification, and surface passivation can help mitigate these challenges. Furthermore, advanced fabrication techniques like electron beam lithography, nanofabrication, and integration with other photonic platforms can enable the creation of complex quantum photonic circuits.

9.1.3 Integration with Silicon Photonics

Combining chalcogenide materials with silicon photonics is a prospective method to leverage the benefits of both platforms. This combination paves the way for the creation of integrated photonic devices that are highly efficient, easily scalable, and can be well-integrated with current semiconductor technology. However, this integration comes with several challenges, including material compatibility, interface quality, and fabrication issues.

9.1.3.1 Material Compatibility

Chalcogenide materials have different thermal and mechanical properties compared to silicon, which can lead to challenges in the integration process. The mismatch in the coefficient of thermal expansion (CTE) between chalcogenides and silicon can cause strain and defects at the interface, leading to reduced device performance. Optimizing growth techniques, such as molecular beam epitaxy (MBE) or chemical vapor deposition (CVD), and employing interface passivation strategies (such as oxide or nitride encapsulation) can help address these issues.

9.1.3.2 Interface Quality and Device Integration

The quality of the interface between chalcogenide materials and silicon photonics is crucial for ensuring efficient coupling and minimizing losses. Surface defects, roughness, and the formation of unwanted phases at the interface can degrade performance. Techniques such as surface treatment, selective etching, and the development of advanced bonding techniques like wafer bonding or transfer printing can improve interface quality and facilitate seamless integration between chalcogenides and silicon photonic circuits.

9.1.3.3 Scalability and Fabrication Techniques

The ability to scale up poses a major hurdle when integrating chalcogenide materials with silicon photonics, as conventional semiconductor fabrication processes may not be fully compatible with chalcogenide materials. To enable large-scale manufacturing, new fabrication techniques that are compatible with both materials must be developed. High-throughput processes such as chemical vapor deposition (CVD), laser-assisted deposition, and hybrid fabrication approaches that combine traditional semiconductor techniques with new chalcogenide deposition methods will be essential for scaling up the production of integrated photonic devices.

9.1.3.4 Applications and Future Directions

Integrating chalcogenide materials with silicon photonics unlocks numerous application possibilities in fields like integrated quantum photonics, optical interconnects, and sensing technologies. Chalcogenides can provide tunability, nonlinear effects, and low-loss propagation, enhancing the functionality of silicon-based photonic devices. Future research should focus on improving the integration process, optimizing the performance of hybrid photonic circuits, and exploring new applications that leverage the unique properties of both chalcogenide and silicon materials.

9.2 Challenges in Chalcogenide Nanophotonics

9.2.1 Material Quality

In practical applications of chalcogenide photonic devices, material quality directly affects device performance and stability. However, due to the complex composition and susceptibility to oxidation of chalcogenide compounds, achieving high-quality and reproducible material fabrication remains challenging. The feasibility of material quality and scalable production is a key issue for chalcogenide-based devices. The uniformity of thin-film deposition, compositional stability, and defect control are critical factors that require further optimization. To address these challenges, researchers are exploring the development of high-purity precursor materials, improving deposition techniques, and optimizing defect engineering strategies. Additionally, doping or alloying strategies may help enhance the stability and optical properties of chalcogenide materials, facilitating their application in next-generation photonic systems.

9.2.1.1 Thin Film Uniformity and Reproducibility

The uniformity of chalcogenide thin films is crucial for large-area integration, as it affects refractive index, nonlinear optical response, and transmission loss. Conventional sputtering and thermal evaporation methods may introduce compositional segregation and film inhomogeneity. Therefore, optimizing substrate temperature, deposition rate, and annealing conditions is essential for improving film quality.

The use of high-purity precursor materials, along with precisely controllable deposition methods such as physical vapor deposition (PVD), CVD, and solution-based techniques, can enhance material uniformity and reproducibility.

9.2.1.2 Defect Control and Optical Loss

Structural defects (such as vacancies, grain boundaries, and stress concentration areas) increase scattering loss, reducing nonlinear efficiency and photon lifetime. Doping or alloying (e.g., Ge–Sb–Te and Se–Ti systems) can effectively reduce defect density and improve optical performance.

Low-temperature plasma-assisted deposition or nanoscale chemical modification techniques can enhance film structural uniformity, reduce internal stress, and improve stability and optical transparency.

9.2.1.3 Challenges in Scalable Fabrication

Current fabrication techniques for chalcogenide materials still face bottlenecks in wafer-level integration and large-area photonic chip manufacturing. Investigating high-throughput chemical vapor deposition (HVPE), spray deposition, and laser-induced growth methods can improve the scalability of chalcogenide material production, paving the way for industrial applications.

9.2.2 Challenges in Micro- and Nanofabrication

The micro- and nanofabrication of chalcogenide materials is critical for achieving high-performance photonic devices. However, owing to their intrinsic physical characteristics (such as low melting points, susceptibility to decomposition, and moisture sensitivity), traditional semiconductor processing techniques are difficult to apply directly. Therefore, more precise and compatible processing methods must be explored to achieve high-resolution patterning and device integration. Meanwhile, the compatibility of chalcogenide materials with standard semiconductor manufacturing processes remains a challenge for large-scale commercialization.

9.2.2.1 High-Resolution Patterning Techniques

Electron beam lithography (EBL) and ion beam etching (IBE) provide nanoscale resolution but suffer from slow processing rates and potential surface damage. Low-damage wet etching, laser direct writing, and nanoimprint lithography can improve processing accuracy and reduce defects.

Atomic layer deposition (ALD) and self-assembled template techniques enable more uniform and controllable thin film patterning, enhancing device consistency.

9.2.2.2 Process Compatibility and Integration

The compatibility between chalcogenide materials and silicon photonic processes is crucial for large-scale integration. Optimizing heteroepitaxial growth techniques and exploring interface passivation strategies (such as oxide/nitride encapsulation) can reduce interface defects between chalcogenide materials and silicon-based structures.

Advanced packaging techniques, such as wafer-level bonding and transfer printing, can facilitate the integration of reconfigurable chalcogenide photonic devices onto flexible substrates, enabling applications in wearable optics and flexible photonic chips.

9.2.2.3 Exploration of Novel Processing Methods

Researching ultrafast laser direct writing, photo-induced phase change, and thermal nanoimprint techniques can overcome the resolution and compatibility limitations of conventional etching processes, achieving high-precision, low-loss photonic structures.

The utilization of machine learning in process optimization can improve the fabrication yield and consistency of chalcogenide photonic devices.

9.2.3 Thermal Stability

The thermal stability of chalcogenide materials limits their applications in high-power laser systems and broadband-temperature photonic devices. Chalcogenide compounds are prone to phase transitions, crystallization, or chemical degradation at high temperatures. Therefore, enhancing their thermal stability requires breakthroughs in material modification, structural design, and packaging strategies.

9.2.3.1 Optimization of Material Stability

Investigating high-temperature stable chalcogenide compounds, such as Ge–Sb–Se or As–S–Se alloys, can enhance their thermal resistance and reduce crystallization tendencies.

The use of nanocomposite strategies (such as oxide or nitride doping) can improve thermal stability and enhance resistance to humidity and oxidation.

9.2.3.2 Thermally Induced Phase Transition Control

Studying the kinetics of thermal phase transitions and optimizing crystallization processes are capable of assisting in the control of nonlinear optical properties and enhancing the lifespan of reconfigurable photonic devices.

Implementing temperature-regulated structures (such as thermal insulation layers and microcavity resonance-enhanced heat dissipation structures) can reduce local temperature accumulation and improve long-term operational stability.

9.2.3.3 Packaging and Interface Stability

Using low-hygroscopicity packaging methods (such as silicon nitride protective layers and hermetic sealing) can prevent environmental degradation and improve device durability.

Optimizing thermal management designs (such as photonic crystal heat dissipation structures and thermal diffusion layers) can mitigate thermal drift issues in chalcogenide photonic chips.

9.2.4 Advances in Modulation Mechanisms

Advancements in modulation mechanisms are essential for expanding material functionalities, improving device performance, and enabling reconfigurable photonic systems in chalcogenide nanophotonics. By making use of external stimuli (like electric fields, optical fields, thermal fields, and mechanical stress), the optical, electrical, and structural characteristics of chalcogenide materials can be adjusted dynamically, which results in the creation of a variety of new-fashioned photonic devices.

9.2.4.1 Electric Field Modulation

Applying external electric fields can induce phase transitions and modify the refractive index of electro-optic chalcogenide materials, enabling tunable optical modulators and switches.

Designing chalcogenide materials with high dielectric constants can enhance electro-optic response and reduce switching power consumption.

9.2.4.2 Optical and Thermal Control

The nonlinear optical effects of chalcogenide materials allow light-induced refractive index tuning, which can be utilized for ultrafast all-optical switching applications.

Thermal tuning strategies using localized heating elements can dynamically adjust optical properties while maintaining structural integrity.

9.2.4.3 Mechanical and Strain Engineering

Mechanical strain can dynamically modulate the band structure and optical properties, enabling stretchable and flexible photonic devices.

The integration of microelectromechanical systems (MEMS) technology with chalcogenide photonic devices can enable precise control and reconfigurable optical functionalities.

The challenges in material quality, micro- and nanofabrication, thermal stability, and modulation mechanisms present significant obstacles to the widespread adoption of chalcogenide nanophotonics. However, continuous advancements in deposition techniques, defect engineering, scalable fabrication, and novel processing methods are gradually addressing these limitations. By improving material stability, optimizing fabrication processes, and integrating advanced modulation strategies, chalcogenide-based photonic devices are poised to achieve enhanced performance and broader applicability. In future research, the emphasis should be placed on the collaborative optimization of material characteristics and device structures to fully unleash the potential of chalcogenide nanophotonics in next-generation optical communication, sensing, and integrated photonic systems.

9.3 Conclusion and Outlook

Chalcogenide nanophotonics, as an emerging photonic technology, has demonstrated tremendous potential across different fields, including nonlinear optics, quantum information processing, and silicon photonic integration. Its distinctive characteristics, like a high nonlinear optical coefficient, a broad spectral transparency window, and highly tunable material characteristics, enable its wide-ranging applications in ultrafast optics, all-optical computing, optical communication, and optical sensing. Despite these advantages, several key challenges remain, including material stability, fabrication compatibility, device reliability, and large-scale integration.

In the future, advancing research in material optimization, innovative fabrication techniques, hybrid integration strategies, and intelligent control mechanisms will be critical in pushing the boundaries of chalcogenide nanophotonics. Moreover, it is anticipated that the integration of artificial intelligence and data-driven methods will have a more and more important impact on optimizing material design, device performance, and dynamic optical functionality. The synergy between AI-driven methodologies and chalcogenide nanophotonics holds promise for overcoming current limitations and unlocking new functionalities in next-generation photonic systems.

9.3.1 Material Optimization and Intelligent Design

The future of chalcogenide nanophotonics largely depends on breakthroughs in material synthesis, composition control, and property optimization. Addressing issues related to optical losses, thermal stability, and environmental durability is crucial for ensuring long-term device operation.

Developing high-performance chalcogenide materials requires engineering new high-stability, low-loss, and highly nonlinear compositions. Optimized synthesis techniques and advanced thin-film deposition methods will be essential for achieving materials with superior uniformity and enhanced processing compatibility.

AI-assisted material discovery can accelerate the prediction of new chalcogenide compounds, optimizing their bandgap, nonlinear coefficients, and thermal resistance. High-throughput computational techniques, combined with experimental validation, will facilitate the discovery of novel alloy compositions and heterostructures tailored for specific photonic applications.

The integration of chalcogenide materials into topological photonics will enable robust light transport, immune to fabrication defects and environmental perturbations. AI-driven optimization can aid in designing photonic topological insulators with tailored band structures and enhanced nonlinearity for applications in next-generation optical computing and quantum photonics.

9.3.2 Advanced Fabrication Processes and Nanoscale Device Integration

The realization of chalcogenide-based photonic devices requires precise and scalable fabrication techniques that ensure high structural integrity while maintaining low optical loss.

Techniques for high-precision nanofabrication, including electron beam lithography, nanoimprint lithography, focused ion beam milling, and femtosecond laser writing, will be crucial in developing high-resolution and uniform chalcogenide photonic structures. AI-assisted process optimization will further enhance the precision and reproducibility of device fabrication.

Heterogeneous integration with silicon photonics offers a promising route for high-performance hybrid photonic circuits. By leveraging silicon's well-established fabrication infrastructure, chalcogenide-silicon hybrid integration can lead to the development of low-power, high-speed nonlinear optical components, tunable photonic devices, and quantum photonic circuits.

The fabrication of three-dimensional photonic structures using chalcogenide materials will unlock novel functionalities. AI-driven inverse design methodologies can be employed to optimize three-dimensional photonic crystals, metasurfaces, and optical waveguides, leading to compact, high-efficiency photonic components.

9.3.3 AI-Driven Adaptive Photonic Systems

Integration of AI and Photonics: Chalcogenide nanophotonics has made significant progress in enabling novel optical functionalities and integrated photonic devices. The field continues to push the boundaries of nonlinear optics, quantum photonics, and tunable photonic systems. However, challenges remain in material synthesis, device fabrication, and long-term stability.

Application of AI in Photonic Device Design: AI is poised to revolutionize the design, optimization, and manufacturing of chalcogenide-based photonic devices. Machine learning algorithms can predict material properties, optimize device architectures, and accelerate the discovery of new chalcogenide compositions with enhanced optical characteristics. AI-driven computational techniques, such as inverse design, can help create highly efficient nanophotonic structures with tailored responses for specific applications. Furthermore, AI-assisted process control can improve fabrication precision, enhancing yield and reproducibility in large-scale manufacturing.

Phase-Change Materials and Adaptive Photonic Systems: Phase-change materials (PCMs), due to their tunable optical properties across different states, hold great potential in adaptive photonic systems. AI technologies can provide intelligent support for the design, optimization, and manufacturing of these materials. Deep learning (DL) and machine learning (ML) can assist in designing complex PCM-based nanophotonic structures, predicting their optical responses, and accelerating the discovery of novel photonic devices.

Integration of AI and PCM Design: Recent studies introduced a DL-enabled approach for discovering nanophotonic designs, focusing on material–structure coupled solutions for active metasurfaces with switchable functions. Leveraging deep neural networks (DNNs), this method allows precise prediction of optical spectra within the design space and employs surrogate optimization for inverse design of active metasurfaces. The study highlights DL's capability in crafting sophisticated nanophotonic architectures (Panda S et al. 2023).

Future of DL in PCM-Based Nanophotonics: In the future, the applications of DL in PCM-based nanophotonics can generally be classified into three crucial domains: (i) surrogate models—utilizing DNNs to forecast the optical properties of a design, integrating them as surrogates in optimization processes; (ii) autoencoders—employing DNNs to map structural geometries into a reduced-dimensional latent space, streamlining the search for optimal performance; (iii) generative adversarial approaches—leveraging generative adversarial networks (GANs) to directly generate structural designs based on specified optical responses (Devdutt Tripathi et al. 2023).

Despite the great potential of DL methods in PCM-based nanophotonics, challenges remain, including high dependency on training data, shape deviations in inverse-designed devices due to fabrication imperfections, and the difficulty of handling nonunique solutions (one-to-many relationships) in DL. Strategies like smart sampling, active learning, and transfer learning are being explored to mitigate these issues, reducing the need for extensive training data while enhancing design precision and stability. Combining DL with PCMs offers exciting possibilities for creating adaptive photonic systems, with potential applications in quantum computing, high-speed communications, and advanced optical sensing.

9.3.4 Expanding Chalcogenide Photonic Technologies for Future Applications

The unique optical properties of chalcogenide materials make them well-suited for various next-generation photonic applications. Future research will focus on harnessing these materials for transformative breakthroughs in quantum optics, biophotonics, and optical computing.

Quantum optics and photonic computing will benefit from chalcogenide-based quantum light sources, entangled photon generation, and nonlinear quantum frequency converters. These technologies are expected to play a key role in advancing quantum communication and quantum computing, with AI-assisted design enhancing the efficiency and robustness of quantum photonic circuits.

Biophotonics and medical imaging will see advancements owing to the high refractive index, strong nonlinear response, and excellent infrared transparency of chalcogenide materials. These properties make them ideal for super-resolution microscopy, deep-tissue imaging, and optical biosensing. AI-driven signal processing can further improve imaging resolution and sensitivity.

Next-generation optical storage and communication will be enabled by chalcogenide PCMs, which are widely studied for high-density optical data storage and neuromorphic photonic computing. Their integration into broadband optical networks will facilitate ultra-fast data transmission and energy-efficient optical processing.

9.3.5 The Future of Chalcogenide Nanophotonics

Looking ahead, chalcogenide nanophotonics is poised to become a key enabling technology in multiple disciplines, from classical photonics to quantum optics and AI-driven systems. To fully realize its potential, several critical areas must be addressed.

Developing new chalcogenide-based photonic materials with enhanced stability, tailored optical properties, and improved processing compatibility will be essential. Leveraging AI and ML for predictive material design, topology optimization, and intelligent optical control will drive innovation. Advancing nanofabrication and integration techniques will enable scalable manufacturing of chalcogenide-based photonic chips. Exploring interdisciplinary applications, including quantum technologies, biomedical optics, and next-generation optical computing, will expand the impact of chalcogenide photonics.

In summary, chalcogenide nanophotonics presents a wealth of opportunities and challenges. By integrating AI, advanced fabrication techniques, and novel photonic architectures, researchers can unlock new functionalities in optical information processing, photonic computing, and intelligent sensing. The convergence of AI and chalcogenide-based nanophotonics will drive unprecedented advancements, paving the way for the next generation of high-performance, adaptive, and scalable photonic systems.

References

Abdollahramezani, S., Hemmatyar, O., Taghinejad, H. et al. (2020). Tunable nanophotonics enabled by chalcogenide phase-change materials. *Nano* 9 (5): 1189–1241.

Abeles, F., Borensztein, Y., and Lopez-Rios, T. (1984). Optical properties of discontinuous thin films and rough surfaces of silver. *Advances in Solid State Physics* 24: 93–117.

Adam, J.I. and Zhang, X. (2014). *Chalcogenide Glasses: Preparation, Properties and Applications*. Woodhead: Woodhead Publishing Limited.

Ahlawat, A., Mishra, D.K., Sathe, V.G. et al. (2013). Raman tensor and domain structure study of single-crystal-like epitaxial films of $CaCu_3Ti_4O_{12}$ grown by pulsed laser deposition. *Journal of Physics: Condensed Matter* 25 (2): 025902.

Akola, J. and Jones, R.O. (2007). Structural phase transitions on the nanoscale: the crucial pattern in the phase-change materials $Ge_2 Sb_2 Te_5$ and GeTe. *Physical Review B—Condensed Matter and Materials Physics* 76 (23): 235201.

Anderson, P.W. (1975). Model for the electronic structure of amorphous semiconductors. *Physical Review Letters* 34 (15): 953–955.

Andersson, M.R., Johnsson, D.M., Thender, M. et al. (2001). The effect of the polymerisation temperature on molecular weight and photoluminescence quantum yield for a phenylsubstituted PPV. *Synthetic Metals* 119: 63.

Anonymous (2024a). Research on electro-optic modulation characteristics based on DC Kerr effect in chalcogenide glasses. *Journal of Optical Materials* 150: 115023.

Anonymous (2024b). Third-order nonlinear optical properties of chalcogenide glasses in fiber applications. *Optical Fiber Technology* 75: 103210.

Ashfold, M.N.R., Claeyssens, F., Fuge, G.M. et al. (2004). Pulsed laser ablation and deposition of thin films. *Chemical Society Reviews* 33 (1): 23–31.

Asobe, M. (1997). Nonlinear optical properties of chalcogenide glass fibers and their application to all-optical switching. *Optical Fiber Technology* 3 (2): 142–148.

Asobe, M., Kanamori, T., and Kubodera, K.I. (1993). Applications of highly nonlinear chalcogenide glass fibers in ultrafast all-optical switches. *IEEE Journal of Quantum Electronics* 29 (8).

Baldacchini, G. (2020). *Organometallic Luminescence: A Case Study on Alq3, an OLED Reference Material*. Woodhead Publishing.

Chalcogenide Nanophotonics, First Edition. Tun Cao, Ying Su, Yinghan Li and Hongze Gao.

Bao, J., Chen, X., Liu, K. et al. (2022). Nonvolatile chirality switching in terahertz chalcogenide metasurfaces. *Microsystems and Nanoengineering* 8 (1): 112.

Besra, L. and Liu, M. (2007). A review on fundamentals and applications of electrophoretic deposition (EPD). *Progress in Materials Science* 52 (1): 1–61.

Blaineau, S., Jund, P., and Drabold, D.A. (2003). Physical properties of a GeS2 glass using approximate ab initio molecular dynamics. *Physical Review B* 67 (9): 094204.

Bongiovanni, G., Botta, C., Communal, J.E. et al. (2003). Organic host–guest systems for blue emission. *Materials Science & Engineering, C: Materials for Biological Applications* 23: 909.

Boone, B.E. and Shannon, C. (1996). Optical properties of ultrathin electrodeposited CdS films probed by resonance Raman spectroscopy and photoluminescence. *The Journal of Physical Chemistry* 100 (22): 9480–9484.

Born, M. and Heisenberg, W. (1985). Zur quantentheorie der molekeln. *Original Scientific Papers Wissenschaftliche Originalarbeiten* 216–246.

Bruno, B., Zhang, X.H., Smektala, F. et al. (2004). Recent advances in chalcogenide glasses. *Journal of Non-Crystalline Solids* 345–346: 276–283.

Bruno, B., Catherine, B. et al. (2014). Chalcogenide optical fibers for mid infrared sensing. *Optical Engineering* 53 (2): 027101.

Cao, T. and Cen, M.J. (2019). Fundamentals and applications of chalcogenide phase-change material photonics. *Advanced Theory and Simulations* 2 (8): 1900094.

Cao, Y., Nie, Q.H., Xu, T.F. et al. (2010). Optical properties and structure of GeSbS-Se glass system. *Acta Photonica Sinica* 39 (7): 1535–1540.

Cao, T., Zhang, L., Simpson, R.E. et al. (2013). Strongly tunable circular dichroism in gammadion chiral phase-change metamaterials. *Optics Express* 21: 27841–27851.

Cao, T., Wei, C., Simpson, R.E. et al. (2014). Fast tuning of double Fano resonance using a phase-change metamaterial under low power intensity. *Scientific Reports* 4: 1–9.

Cao, T., Zheng, G.Z., Wang, S., and Wei, C.W. (2015). Ultrafast beam steering using gradient Au-$Ge_2Sb_2Te_5$-Au plasmonic resonators. *Optics Express* 23: 18029.

Cao, T., Li, Y., Wei, C.W. et al. (2017a). Numerical study of tunable enhanced chirality in multilayer stack achiral phase-change metamaterials. *Optics Express* 25 (9): 9911–9925.

Cao, T., Li, Y., Zhang, X. et al. (2017b). Theoretical study of tunable chirality from graphene integrated achiral metasurfaces. *Photonics Research* 5 (5): 441–449.

Cao, T., Fang, L., Cao, Y. et al. (2019a). Dynamically reconfigurable topological edge state in phase change photonic crystals. *Scientific Bulletin* 64: 814.

Cao, T., Liu, K., Lu, L. et al. (2019b). Large-area broadband near-perfect absorption from a thin chalcogenide film coupled to gold nanoparticles. *ACS Applied Materials and Interfaces* 11: 5176–5182.

Cao, T., Liu, K., Li, L. et al. (2019c). Chalcogenide–gold dual-layers coupled to gold nanoparticles for reconfigurable perfect absorption. *Nanoscale* 11 (43): 20546–20553.

Cao, T., Liu, K., Lu, L. et al. (2020a). Low-power phase transition of chalcogenide glass using Au nanoparticle plasmon resonance. *Advanced Optical Materials* 8 (6): 1901570.

Cao, T., Wang, R.Z., Simpson, R.E. et al. (2020b). Photonic Ge-Sb-Te phase change metamaterials and their applications. *Progress in Quantum Electronics* 74.

Cao, T., Li, Y., Zhang, X. et al. (2023). Reversible tuning of extrinsic 3D chirality in chalcogenide nonchiral metasurfaces. *Advanced Optical Materials* 11 (8): 2202239.

Caravati, S., Bernasconi, M., Kühne, T.D. et al. (2007). Coexistence of tetrahedral-and octahedral-like sites in amorphous phase change materials. *Applied Physics Letters* 91 (17): 171906.

Carlsson, J.-O. and Martin, P.M. (2010). Chapter 7 - chemical vapor deposition. In: *Handbook of Deposition Technologies for Films and Coatings*, 3e (ed. P.M. Martin), 314–363. Boston, William Andrew Publishing.

Ceperley, D.M. and Alder, B.J. (1980). Ground state of the electron gas by a stochastic method. *Physical Review Letters* 45 (7): 566.

Chelikowsky, J.R., Derby, J.J., Godlevsky, V.V. et al. (2001). Abinitio simulations of liquid semiconductors using thepseudopotential-density functional method. *Journal of Physics: Condensed Matter* 13 (41): R817.

Chen, Y.C., Rettner, C.T., Raoux, S. et al. (2006). Ultra-thin phase-change bridge memory device using geSb. In 2006 *International Electron Devices Meeting*, (11–13 Dec 2006). 10.1109/IEDM.2006.346910.

Chen, W., Thoreson, M.D., Ishii, S. et al. (2010). Ultra-thin, ultra-smooth, and low-loss silver films on a germanium wetting layer. *Optics Express* 18: 5124.

Chen, Y., Xiang, S., Tiefeng, X. et al. (2011). Research progress of chalcogenide glass waveguide. *Laser and Optoelectronics Progress* 48 (11): 111301.

Chen, Y.G., Kao, T.S., Ng, B. et al. (2013). Hybrid phase-change plasmonic crystals for active tuning of lattice resonances. *Optics Express* 21: 13691.

Chen, C., Bobela, D.C., Yang, Y. et al. (2017). Characterization of basic physical properties of Sb_2Se_3 and its relevance for photovoltaics. *Frontiers of Optoelectronics* 10: 18–30.

Chen, R., Fang, Z., Miller, F. et al. (2022). Opportunities and challenges for large-scale phase-change material integrated electro-photonics. *ACS Photonics* 9 (10): 3181–3195.

Chen, J., Zong, S., Liu, X. et al. (2023). Gradient-assisted metasurface absorber with dual-band chiral switching and quasi-linearly tunable circular dichroism. *Optics Letters* 48 (18): 4917–4920.

Chen, C., Yu, J., Kang, C. et al. (2024). Chiral optical modulation of Si-$Ge_2Sb_2Te_5$ composite metasurfaces driven by quasi-bound states in the continuum. *Applied Optics* 63 (29): 7682–7691.

Cheng, Z., Ríos, C., Youngblood, N. et al. (2018). Device-level photonic memories and logic applications using phase-change materials. *Advanced Materials* 30 (32): 1802435.

Chew, L.T., Dong, W.L., Lu, L. et al. (2017). Chalcogenide active photonics; proceedings of the conference on active photonic platforms IX, San Diego, CA, F (06–10 Aug 2017).

Chopra, K.L. and Das, S.R. (1983). *Thin Film Deposition Techniques and Applications*. Plenum Press.

Cintas, P. (2007). On chirality and the universal asymmetry. Reflections on image and mirror image. By Georges H. Wagnière. *Angewandte Chemie International Edition* 46 (48): 9143–9144.

Coe, S., Woo, W.K., Bawendi, M., and Bulović, V. (2002). Electroluminescence from single monolayers of nanocrystals in molecular organic devices. *Nature* 420 (6917): 800–803.

Cohen, R.W., Cody, G.D., Coutts, M.D., and Abeles, B. (1973). Optical properties of granular silver and gold films. *Physical Review B* 8: 3689–3701.

Couturier, G., Vatinel, S., Boisgard, R. et al. (2009). Quartz crystal microbalance and evaporation of sessile droplets. *Journal of Applied Physics* 106 (5): 054906.

Crow, J. D. (1989). Optical interconnect technology for multi-processor networks. In *Conference Proceedind—Lasers and Electro-Optics Society Annual Meeting-LEOS* (ed. Athale, R.) 61 (IEEE, Piscataway, NJ).

Dabbousi, B.O., Rodriguez-Viejo, J., Mikulec, F.V. et al. (1997). (CdSe) ZnS quantum dots of high luminescence: synthesis and characterization of a size series of highly luminescent materials. *The Journal of Physical Chemistry B* 1997, 101 (46): 9463–9475.

Dai, S.X. and Xu, T.F. (2020). Chalcogenide glasses for mid-infrared photonics: from materials to devices. *Advanced Optical Materials* 8 (15): 1901698.

Danto, S., Thompson, D., Wachtel, P. et al. (2013). A comparative study of purification routes for As_2Se_3 chalcogenide glass. *International Journal of Applied Glass Science* 4 (1): 31–41.

Dapkus, P.D. (1982). Metalorganic chemical vapor deposition. *Annual Review of Materials Research* 12: 243–269.

Debunne, A., Virwani, K., Padilla, A. et al. (2011). Evidence of crystallization-induced segregation in the phase change material Te-rich GST. *Journal of the Electrochemical Society* 158 (10): H965–H972.

Delaney, M., Zeimpekis, I., Du, H. et al. (2021). Nonvolatile programmable silicon photonics using an ultralow-loss Sb_2Se_3 phase change material. *Science Advances* 7 (25): eabg3500.

Delaney, M., Zeimpekis, I., Lawson, D. et al. (2020). A new family of ultralow loss reversible phase-change materials for photonic integrated circuits: Sb2S3 and Sb2Se3. *Advanced Functional Materials* 30 (36): 2002447.

Demir, U. and Shannon, C. (1996). Reconstruction of cadmium sulfide monolayers on Au (100). *Langmuir* 12 (2): 594–596.

Denier Van Der Gon, A.W., Tromp, R.M., and Reuter, M.C. (1993). Low-energy electron microscopy studies of Ge and Ag growth on Si(111). *Thin Solid Films* 236: 140–145.

Depla, D. and Mahieu, S. (2008). *Reactive Sputter Deposition*. Berlin: Springer.

Depla, D., Mahieu, S., Greene, J.E. et al. (2010). Chapter 5 - Sputter deposition processes. In: *Handbook of Deposition Technologies for Films and Coatings (Third Edition)* (ed. P.M. Martin), 253–296. Boston, William Andrew: Publishing.

Deshpandey, C. and Bunshah, R. (1991). *Evaporation Processes*. Boston: Academic.

Destriau, G. (1939). Researches upon electro-photo-luminescence. *Transactions of the Faraday Society* 35: 227–233.

Dexter, D.L. and Fowler, W.B. (1967). De-excitation rates of excited electronic states in solids. *The Journal of Chemical Physics* 47 (4): 1379–1384. https://doi.org/10.1063/1.1712091.

Ding, C., Rui, G., Gu, B. et al. (2021). Phase-change metasurface with tunable and switchable circular dichroism. *Optics Letters* 46 (10): 2525–2528.

Diouf, M., Trichlli, A., and Zghal, M. (2019). Stimulated Brillouin scattering-based slow light using singlemode As_2S_3 chalcogenide photonic crystal fiber for temperature sensing. In: *Frontiers in Optics + Laser Science APS/DLS, OSA Technical Digest*. Optica Publishing Group, paper JTu3A.63.

Dobkin, D. and Zuraw, M.K. (2003). *Principles of Chemical Vapor Deposition*. Berlin: Springer Science & Business Media.

Dong, W., Liu, H., Behera, J.K. et al. (2019). Wide bandgap phase change material tuned visible photonics. *Advanced Functional Materials* 29 (6): 1806181.

Dow, J.D. and Redfield, D. (1970). Electroabsorption in semiconductors: The Excitonic Absorption Edge. *Physical Review B* 1 (8): 3358–3371. https://doi.org/10.1103/PhysRevB.1.3358.

Edwards, A.H., Pineda, A.C., Schultz, P.A. et al. (2006). Electronic structure of intrinsic defects in crystalline germa nium telluride. *Physical Review B* 73: 045210.

Eom, J.H., Yoon, Y.G., Park, C. et al. (2006). Global and local structures of the Ge-Sb-Te ternary alloy system for a phase-change memory device. *Physical Review B: Condensed Matter and Materials Physics* 73 (21): 214202.

Esaki, L. and Tsu, R. (1970). Superlattice and negative differential conductance in semiconductors. *IBM Journal of Research and Development* January: 61–65.

Fang Z., Chen R., Zheng J., et al. (2021). Non-volatile reconfigurable silicon photonics based on phase-change materials. *IEEE Journal of Selected Topics in Quantum Electronics* 28: 1–17.

Fang, Z., Zheng, J., Saxena, A. et al. (2021). Non-volatile reconfigurable integrated photonics enabled by broadband low-loss phase change material. *Advanced Optical Materials* 9 (9): 2002049.

Farmakidis, N., Youngblood, N., Li, X. et al. (2019). Plasmonic nanogap enhanced phase-change devices with dual electrical-optical functionality. Science. *Advances* 5 (11): eaaw2687.

Fatome, J., Fortier, C., Nguyen, T.N. et al. (2009). Linear and nonlinear characterizations of chalcogenide photonic crystal fibers. *Journal of Lightwave Technology* 27: 1707.

Feigel, A., Kotler, Z., Sfez, B. et al. (2001). Chalcogenide three-dimensional photonic structures. *Proceeding of SPIE-The International Society for Optical Engineering*, 249–251. SPIE (The International Society for Optical Engineering).

Feldmann, J., Stegmaier, M., Gruhler, N. et al. (2017). Calculating with light using a chip-scale all-optical abacus. *Nature Communications* 8 (1): 1256.

Feldmann, J., Youngblood, N., Li, X. et al. (2020). Integrated 256 Cell photonic phase-change memory with 512-bit capacity. *IEEE Journal of Selected Topics in Quantum Electronics* 26 (2): 1–7.

Feldmann, J., Youngblood, N., Karpov, M. et al. (2021). Parallel convolutional processing using an integrated photonic tensor core. *Nature* 589 (7840): 52–58.

Feng, Z.C., Schurman, M., Stall, R.A. et al. (1997). Raman scattering as a characterization tool for epitaxial GaN thin films grown on sapphire by turbo disk metal-organic chemical vapor deposition. *Applied Optics* 36 (13): 2917–2922.

Feng, X., Ren, H., Xu, F. et al. (2020). Few-moded ultralarge mode area chalcogenide photonic crystal fiber for mid-infrared high power applications. *Optics Express* 28: 16658.

Fischer, A., Reisse, G., Gänsicke, F. et al. (1992). Theoretical and experimental studies of laser-induced chemical vapor-deposition of silicon thin-films. *Applied Surface Science* 54: 41–45.

Fritzsche, H. (1974). Electronic properties of amorphous semiconductors. In: *Amorphous and liquid semiconductors* (ed. J. Tauc). Boston, MA: Springer. https://doi.org/10.1007/978-1-4615-8705-7_5.

Furniss, D. and Seddon, A.B. (1999). Towards monomode proportioned ®breoptic preforms by extrusion. *Journal of Non-Crystalline Solids* 256–257: 232–236.

Gadhwal, R., Kaushik, P., and Devi, A. (2022). A review on 1D photonic crystal based reflective optical limiters. *Critical Reviews in Solid State and Materials Sciences* 48: 93.

de Galarreta, C.R., Alexeev, A.M., Au, Y.Y. et al. (2018). Nonvolatile reconfigurable phase-change metadevices for beam steering in the near infrared. *Advanced Functional Materials* 28: 1704993.

Gao, Y., Liu, Z., Sun, D.M. et al. (2015). Large-area synthesis of high-quality and uniform monolayer WS2 on reusable Au foils. *Nature Communications* 6 (1): 8569.

Gao, M., Zhang, M., Niu, W. et al. (2017). Tuning the transport behavior of centimeter-scale WTe2 ultrathin films fabricated by pulsed laser deposition. *Applied Physics Letters* 111: 031906.

Garbuzov, D.Z., Bulović, V., Burrows, P.E., and Forrest, S.R. (1996). Photoluminescence efficiency and absorption of aluminum-tris-quinolate (Alq3) thin films. *Chemical Physics Letters* 249 (5–6): 433–437.

Gholipour, B., Zhang, J., KF, M.D. et al. (2013). An all-optical, non-volatile, bidirectional, phase-change meta-switch. *Advanced Materials* 25: 3050–3054.

Gholipour, B., Karvounis, A., Yin, J. et al. (2018). Phase-change-driven dielectric-plasmonic transitions in chalcogenide metasurfaces. *NPG Asia Materials* 10: 533–539.

Goedecker, S., Teter, M., and Hutter, J. (1996). Separable dual-space Gaussian pseudopotentials. *Physical Review B* 54 (3): 1703.

Gong, Y.Q. and Zhao, X.J. (2009). Polarization-induced enhancement of second-order nonlinearity in GeSbS chalcogenide glasses. *Optics Letters* 34 (12): 1828–1830.

Gong, Y.Q., Liu, S.W., Li, X.J., and Zhao, X.J. (2008). Effects of chalcogenide glass microstructure defects on its second-order nonlinear optical properties. *The Chinese Journal of Nonferrous Metals* 18 (7): 1299–1304.

Greenham, N.C., Samuel, I.D., Hayes, G.R. et al. (1995). Measurement of absolute photoluminescence quantum efficiencies in conjugated polymers. *Chemical Physics Letters* 241 (1-2): 89–96.

Grillet, C., Lee, M.W., Gai, X. et al. (2010). Chalcogenide glass photonic crystals: progress and prospects. *Photonic and Phononic Crystal Materials and Devices X* 7609: 164.

Grillet, C, Freeman, D, Luther-Davies, B, et al. (2006). Nonlinear photonic crystals in chalcogenide films. Presented at the Photonics: Design, Technology, and Packaging II, pp. 127–139.

Gunnarsson, O., Jonson, M., and Lundqvist, B.I. (1979). Descriptions of exchange and correlation effects in inhomogeneous electron systems. *Physical Review B* 20 (8): 3136.

Guo, H., Zhang, H., Liu, L. et al. (2022). Industrial, medical and military application of flouride and chalcogenide glass fibers. In: *Mid-Infrared Fluoride and Chalcogenide*

Glasses and Fibers (ed. P. Wang et al.), 327–370. Springer Nature Singapore: Singapore.

Gustafson, T.K., Smith, P.W., Engl, W., and Sugano, T. (1987). *Photonic Switching: Proceeding of the First Topical Meeting, Incline Village, Nevada, 18–20 March 1987*, 219. Berlin: Springer.

Hablanian, M.H. (1984). Comments on the history of vacuum pumps. *Journal of Vacuum Science and Technology* 2: 118–125.

Hahn, D.V., Thomas, M.E., and Blodgett, D.W. (2005). Modeling of the frequency- and temperature-dependent absorption coefficient of long-wave-infrared (2–25 m) transmitting materials. *Applied Optics* 44 (32): 6913–6920.

Halim, M., Samuel, I.D., Rebourt, E., and Monkman, A.P. (1997). Measurement of time dependent quantum yield of poly (p-pyridine) and poly (p-phenylenevinylene). *Synthetic Metals* 84 (1–3): 951–952.

Han, J.-T., Zhang, W., Wei, F.-J. et al. (2015). Investigation of wideband slow light in $Ge_{20}Sb_{15}Se_{65}$ photonic crystal slab waveguides. *Chinese Journal of Lasers* 42: 0606002.

Han, J., Liang, F., Zhao, Y. et al. (2024). Reconfigurable topological wave routing based on tunable valley kink states and valley-polarized chiral edge states. *Optics Express* 32: 26819.

Hari, P., Su, T., Taylor, P.C. et al. (2000). Photodarkening in glassy As_2S_3. *Journal of Non-Crystalline Solids* 266: 929–932.

Hassan, P.A., Rana, S., and Verma, G. (2015). Making sense of Brownian motion: colloid characterization by dynamic light scattering. *Langmuir* 31 (1): 3–12.

Heavens, O.S. (1960). Optical properties of thin films. *Reports on Progress in Physics* 23 (1): 1.

Heavens, O.S. (1991). *Optical Properties of Thin Solid Films*. New York, NY: Dover Publications.

Hines, M.A. and Guyot-Sionnest, P. (1996). Synthesis and characterization of strongly luminescing ZnS-capped CdSe nanocrystals. *The Journal of Physical Chemistry* 100 (2): 468–471.

Ho, K., Chan, C.T., and Soukoulis, C.M. (1990). Existence of a photonic gap in periodic dielectric structures. *Physical Review Letters* 65: 3152.

Hohenberg, P. and Kohn, W. (1964). Inhomogeneous electron gas. *Physical Review* 136 (3B): B864.

Hosseini, P., Wright, C.D., and Bhaskaran, H. (2014). An optoelectronic framework enabled by low-dimensional phase-change films. *Nature* 511 (7508): 206–211.

Hubert, M, Delaizir, G, Monnier, J, et al. An innovative approach to develop highly performant chalcogenide glasses and glass-ceramics transparent in the infrared range. *Optics Express*, 2011, 19 (23): 23513–23522.

Jafari, M., Guo, L.J., and Rais-Zadeh, M. (2019). A reconfigurable color reflector by selective phase change of GeTe in a multilayer structure. *Advanced Optical Materials* 7 (5): 1801214.

Jamkhande, P.G., Ghule, N.W., Bamer, A.H. et al. (2019). Metal nanoparticles synthesis: an overview on methods of preparation, advantages and disadvantages, and applications. *Journal of Drug Delivery Science and Technology* 53: 101174.

Jia, Z., Li, J., Gao, L. et al. (2023). Dynamic light scattering: a powerful tool for in situ nanoparticle sizing. *Colloids and Interfaces* 7 (1): 15.

Jiang, C., Wang, X., Zhu, M. et al. (2016). Preparation of chalcogenide glass fiber using an improved extrusion method. *Optical Engineering* 55 (5): 056114.

Jin, L., Jarand, C.W., Brader, M.L., and Reed, W.F. (2022). Angle-dependent effects in DLS measurements of polydisperse particles. *Measurement Science and Technology* 33 (4): 045202.

John, S. (1987). Strong localization of photons in certain disordered dielectric superlattices. *Physical Review Letters* 58: 2486.

Kamble, S., Agrawal, S., Cherumukkil, S. et al. (2022). Revisiting zeta potential, the key feature of interfacial phenomena, with applications and recent advancements. *ChemistrySelect* 7 (1): e202103084.

Kanbara, H., Fujiwara, S., Tanaka, K. et al. (1997). Third-order nonlinear optical properties of chalcogenide glasses. *Applied Physics Letters* 70: 925–927.

Kane, E.O. (1957). Band structure of indium antimonide. *Journal of Physics and Chemistry of Solids* 1 (4): 249–261.

Kato, K., Kuwahara, M., Kawashima, H. et al. (2017). Current-driven phase-change optical gate switch using indium–tin-oxide heater. *Applied Physics Express* 10 (7): 072201.

Kats, M.A., Blanchard, R., Genevet, P., and Capasso, F. (2013). Nanometre optical coatings based on strong interference effects in highly absorbing media. *Nature Materials* 12 (1): 20–24.

Kazmerski, L.L. and Racine, D.M. (1975). Growth, environmental, and electrical properties of ultrathin metal films. *Journal of Applied Physics* 46: 791–795.

Keirssea, J., Boussard-Pleédel, C., Loreal, O. et al. (2003). IR optical fiber sensor for biomedical applications. *Vibrational Spectroscopy* 32 (1): 23–32.

Khurgin, J.B. (2015). How to deal with the loss in plasmonics and metamaterials. *Nature Nanotechnology* 10: 768–779.

Kiani, A., Hasko, D.G., Milne, W.I., and Flewitt, A.J. (2013). Analysis of amorphous indium-gallium-zinc-oxide thin-film transistor contact metal using Pilling-Bedworth theory and a variable capacitance diode model. *Applied Physics Letters* 102: 152102.

Kohn, W. and Sham, L.J. (1965). Self-consistent equations including exchange and correlation effects. *Physical Review* 140 (4A): A1133.

Kolobov, A.V., Fons, P., Tominaga, J. et al. (2004). Crystallization-induced short-range order changes in amorphous GeTe. *Journal of Physics: Condensed Matter* 16 (44): S5103.

Koma, A. (1992). Van der Waals epitaxy—a new epitaxial growth method for a highly lattice-mismatched system. *Thin Solid Films* 216 (1): 72–76.

Kraft, N., Wang, G., Bryja, H. et al. (2021). Phase and grain size engineering in Ge–Sb–Te–Oby alloying with la-sr-mn-o towards improved material properties. *Materials & Design* 199: 109392.

Kretschmann, E. (1972). Decay of non-radiative surface plasmons into light on rough silver films. Comparison of experimental and theoretical results. *Optics Communication* 6: 185–187.

Kuc, A., Zibouche, N., and Heine, T. (2011). Influence of quantum confinement on the electronic structure of the transition metal sulfide TS_2. *Physical Review B* 83 (24): 245213.

Lagally, M.G. and Zhang, Z. (2002). Thin-film cliffhanger. *Nature* 417 (6892): 907–909.

Lankhorst, M.H.R., Ketelaars, B.W.S.M.M., and Wolters, R.A.M. (2005). Low-cost and nanoscale non-volatile memory concept for future silicon chips. *Nature Materials* 4 (4): 347–352.

Le Person, J., Smektala, F., Chartier, T. et al. (2006). Light guidance in new chalcogenide holey fibres from GeGaSbS glass. *Materials Research Bulletin* 41: 1303.

Lee, B.S., Abelson, J.R., Bishop, S.G. et al. (2005). Investigation of the optical and electronic properties of $Ge_2Sb_2Te_5$ phase change material in its amorphous, cubic, and hexagonal phases. *Journal of Applied Physics* 97 (9).

Lee, M.W., Grillet, C., Smith, C.L. et al. (2007). Photosensitive post tuning of chalcogenide photonic crystal waveguides. *Optics Express* 15: 1277.

Lee, M.W., Grillet, C., Tomljenovic-Hanic, S. et al. (2008). Photo-induced cavities in chalcogenide photonic crystals. In *OECC/ACOFT 2008 – Joint Conference of the Opto-Electronics and Communications Conference and the Australian Conference on Optical Fibre Technology*, Sydney, NSW, Australia, 2008, pp. 1–2.

Lencer, D., Salinga, M., and Wuttig, M. (2011). Design rules for phase-change materials in data storage applications. *Advanced Materials* 23 (18): 2030–2058.

Levy, M. (1979). Universal variational functionals of electron densities, first-order density matrices, and natural spin-orbitals and solution of the v-representability problem. *National Academy of Sciences of the United States of America* 76 (12): 6062–6065.

Li, C.S., Pan, R.H., Gu, C.Z. et al. (2024). Reconfigurable micro/nano-optical devices based on phase transitions: from materials, mechanisms to applications. *Advanced Science* 11 (20): 2306344.

Li, D., Li, Z., Chen, P. et al. (2021a). Two-dimensional metal chalcogenide heterostructures: designed growth and emerging novel applications. *Advanced Materials Interfaces* 8: 2100515.

Li, K., Lin, X., Song, B. et al. (2021b). Rapid thermal evaporation for cadmium selenide thin-film solar cells. *Frontiers of Optoelectronics* 14 (4): 482–490.

Li, L., Jun Cui, T., Ji, W. et al. (2017a). Electromagnetic reprogrammable coding-metasurface holograms. Nature. *Communications* 8 (1): 197.

Li, L., Nie, W., Li, Z. et al. (2017b). All-laser-micromachining of ridge waveguides in $LiNbO_3$ crystal for mid-infrared band applications. *Scientific Reports* 7: 7034.

Li, S., Sang, T., Yang, C. et al. (2023). Phase-change metasurfaces for dynamic control of chiral quasi-bound states in the continuum. *Optics Letters* 48 (24): 6488–6491.

Li, X.J. and Liu, S.W. (2016). Microcrystalline effects on optical nonlinearity in chalcogenide glasses. *Applied Physics Letters* 108 (10): 101903.

Li, S., Qi, L. et al. (2015). Process in erbium-doped chalcogenide waveguide of optical communications. *Laser and Optoelectronics Progress* 52: 030004.

Li, X., Youngblood, N., Ríos, C. et al. (2019). Fast and reliable storage using a 5 bit, nonvolatile photonic memory cell. *Optica* 6 (1): 1–6.

Li, X., Youngblood, N., Cheng, Z. et al. (2020). Experimental investigation of silicon and silicon nitride platforms for phase-change photonic in-memory computing. *Optica* 7 (3): 218–225.

Lim, J., Yeap, S.P., Che, H.X., and Low, S.C. (2013). Characterization of magnetic nanoparticle by dynamic light scattering. *Nanoscale Research Letters* 8: 1–4.

Lin, C.M., Chuu, D.S., Yang, T.J. et al. (1997). Raman spectroscopy study of ZnSe and Zn 0.84 Fe 0.16 Se at high pressures. *Physical Review B* 55 (20): 13641.

Lin, C.G., Rüssel, C., and Dai, S.X. (2018). Chalcogenide glass-ceramics: functional design and crystallization mechanism. *Progress in Materials Science* 93: 1–44.

Liu, A., Zhu, H., Zou, T. et al. (2022). Evaporated nanometer chalcogenide films for scalable high-performance complementary electronics. *Nature Communications* 13 (1): 6372.

Liu, Y., Lau, S.C., Cheng, W.H. et al. (2023). Controlling valley-specific light emission from monolayer MoS_2 with achiral dielectric metasurfaces. *Nano Letters* 23 (13): 6124–6131.

Liu, K., Lin, Z., Han, B. et al. (2024). Non-volatile dynamically switchable color display via chalcogenide stepwise cavity resonators. *Opto-Electronic Advances* 7 (1): 230033.

Logeeswaran, V.J., Chan, M.L., Bayam, Y. et al. (2007). Ultra-smooth metal surfaces generated by pressure-induced surface deformation of thin metal films. *Applied Physics A: Materials Science and Processing* 87: 187–192.

Long, D.A. (2002). *The Raman Effect: A Unified Treatment of the Theory of Raman Scattering by Molecules*. John Wiley & Sons.

Lu, G.-W., Hong, J., Qiu, F. et al. (2020). High-temperature-resistant silicon-polymer hybrid modulator operating at up to 200 Gbit s−1 for energy-efficient datacentres and harsh-environment applications. *Nature Communications* 11 (1): 4224.

Lu, Y., Stegmaier, M., Nukala, P. et al. (2017). Mixed-mode operation of hybrid phase-change nanophotonic circuits. *Nano Letters* 17 (1): 150–155.

Lunney, J.G. and Jordan, R. (1998). Pulsed laser ablation of metals. *Applied Surface Science* 127–129: 941–946.

Lyon, L.A., Keating, C.D., Fox, A.P. et al. (1998). Raman spectroscopy. *Analytical Chemistry* 70 (12): 341–362.

Mandal, A., Cui, Y., Mcrae, L. et al. (2021). Reconfigurable chalcogenide phase change metamaterials: a material, device, and fabrication perspective. *Journal of Physics: Photonics* 3 (2): 022005.

Mansha, S., Moitra, P., Xu, X. et al. (2022). High resolution multispectral spatial light modulators based on tunable Fabry-Perot nanocavities. *Light: Science and Applications* 11 (1): 141.

Mao, L., Liu, K., Lian, M. et al. (2022). Bound states in the continuum in all-dielectric metasurface: separation of Sub-10 nm enantiomers. *Advanced Photonics Research* 3 (3): 2100280.

Martin-Monier, L., Popescu, C.C., Ranno, L. et al. (2022). Endurance of chalcogenide optical phase change materials: a review. *Optical Materials Express* 12 (6): 2145–2167.

Martins, D., Gourdon, C., Lavallard, P., and Planel, R. (1998). Photoluminescence internal quantum yield in superlattices. *Superlattices and Microstructures* 23 (2): 211–217.

Masaki, A. (1997). Nonlinear optical properties of chalcogenide glass fibers and their application to all-optical switching. *Optical Fiber Technology* 3 (2): 142–148.

Matsui, K., Kamijo-Ikemori, A., Imai, N. et al. (2016). Clinical significance of urinary liver-type fatty acid-binding protein as a predictor of ESRD and CVD in patients with CKD. *Clinical and Experimental Nephrology* 20: 195–203.

Meng, Y., Cao, T., and Long, Y. (2020). Progress in metasurfaces based on Ge-Sb-Te phase-change materials. *Journal of Applied Physics* 128: 14.

Mescia, L., Smektala, F., and Prudenzano, F.I. (2012). New trends in amplifiers and sources via chalcogenide photonic crystal fibers. *Journal of Optics* 2012: 575818.

Mie, G. (1908). Beiträge zur Optik trüber Medien, speziell kolloidaler Metallösungen. *Annalen Der Physik* 330 (3): 377–445.

Mlayah, A., Carles, R., Sayari, A. et al. (1996). Resonant Raman scattering in GaAs/AlAs superlattices: the role of electron state mixing. *Physical Review B* 53 (7): 3960.

Moitra, P., Wang, Y.Z., Liang, X.N. et al. (2023). Programmable wavefront control in the visible spectrum using low-loss chalcogenide phase-change metasurfaces. *Advanced Materials* 35: 2205367.

Monat, C., Spurny, M., Grillet, C. et al. (2011). Third-harmonic generation in slow-light chalcogenide glass photonic crystal waveguides. *Optics Letters* 36: 2818.

Morgan, A., Zakeri, R., Quint, J.K. et al. (2018). Defining the relationship between COPD and CVD: what are the implications for clinical practice? *Therapeutic Advances in Respiratory Disease* 12: 175346581775052.

Naik, G.V., Shalaev, V.M., and Boltasseva, A. (2013). Alternative plasmonic materials: beyond gold and silver. *Advanced Materials* 25: 3264–3294.

Nie, Q.H. and Wang, X.S. (2018). Infrared transmission and structural analysis of Ge-Sb-S chalcogenide glasses. *Journal of Non-Crystalline Solids* 498: 1–7.

Nosé, S. (1984). A unified formulation of the constant temperature molecular dynamics methods. *The Journal of chemical physics* 81 (1): 511–519.

Oliveira, I.A., LeonardoGomes, I., and de Souza & Vitaly Felix Rodriguez-Esquerre. (2021). Design of hybrid narrow-band plasmonic absorber based on chalcogenide phase change material in the infrared spectrum. *Scientific Reports* 11: 21919.

O'Reilly, E.P., Robertson, J., and Kelly, M.J. (1981). The structure of amorphous GeSe and GeTe. *Solid State Communications* 38 (7): 565–568.

Orjuela, J.E.A., Alfonso, E., Cubillos, G. et al. (2012). *Thin Film Growth Through Sputtering Technique and Its Applications*. Rijeka, IntechOpen: Crystallization - Science and Technology. M. Andreeta.

Palmberg, P.W., Rhodin, T.N., and Todd, C.J. (1967). Epitaxial growth of gold and silver on magnesium oxide cleaved in ultrahigh vacuum. *Applied Physics Letters* 11: 33–35.

Pan, M.Y., Fu, Y.F., Zang, Y.J. et al. (2023). Reversibly reconfigurable GSST metasurface for broadband beam steering and achromatic focusing in the long-wave infrared. *Optics Express* 31: 22554.

Panda, S., Kumar, S., Tripathi, D., and Hegde, R.S. (2023). Deep learning aids simultaneous structure-material design dis covery—a case study on designing phase change material metasurfaces. *Journal of Nanophotonics* 17: 036006.

Park, K.-H.P., Ha, J.S., and Lee, E.-H. (1997). Uniform Ag thin film growth on an Sb-terminated Si (111) surface. *ETRI Journal* 19: 71–80.

Peng, J.-P., Guan, J.Q., Zhang, H.M. et al. (2015). Molecular beam epitaxy growth and scanning tunneling microscopy study of $TiSe_2$ ultrathin films. *Physical Review B* 91 (12): 121113.

Perdew, J.P. and Yue, W. (1986). Accurate and simple density functional for the electronic exchange energy: Generalized gradient approximation. *Physical Review B* 33 (12): 8800.

Perdew, J.P. and Zunger, A. (1981). Self-interaction correction to density-functional approximations for many-electron systems. *Physical Review B* 23 (10): 5048.

Perdew, J.P., Burke, K., and Ernzerhof, M. (1996). Generalized gradient approximation made simple. *Physical Review Letters* 77 (18): 3865.

Pernice, W.H.P. and Bhaskaran, H. (2012). Photonic non-volatile memories using phase change materials. *Applied Physics Letters* 101 (17).

Petkov, K. and Ewen, P.J.S. (1999). Photoinduced changes in the linear and non-linear optical properties of chalcogenide glasses. *Journal of Non-Crystalline Solids* 249: 150–159.

Phillips, J.C. (1976). Structure and conductivity of bilinear organic compounds. *Journal of Solid State Chemistry*. https://doi.org/10.1016/0022-4596(76)90182-1.

Pi, L., Li, L., Liu, K. et al. (2019). Recent progress on 2D noble-transition-metal dichalcogenides. *Advanced Functional Materials* 29: 1904932.

Piccinotti, D., Gholipour, B., Yao, J. et al. (2019). Stoichiometric engineering of chalcogenide semiconductor alloys for nanophotonic applications. *Advanced Materials* 31 (14).

Pierson, H.O. (1999). *Handbook of Chemical Vapor Deposition: Principles*. William Andrew: Technology and Applications.

Popov, Y.O. (2005). Evaporative deposition patterns: spatial dimensions of the deposit. *Physical Review. E, Statistical, Nonlinear, and Soft Matter Physics* 71 (3 Pt 2B): 036313.

Prasad, R.D., Prasad, N.R., Prasad, N. et al. (2024). A review on scattering techniques for analysis of nanomaterials and biomaterials. *Engineered Science* 4 (33): 1332.

Prazakova, L. (2022). Advanced characterization for the development of innovative non-volatile memories. (Doctoral dissertation, Université Grenoble Alpes [2020-….]).

Prudenzano, F., Mescia, L., Allegretti, L. et al. (2010). *Optical Materials* 33: 241.

Rabe, K.M. and Joannopoulos, J.D. (1987). Theory of the structural phase transition of GeTe. *Physical Review B* 36 (12): 6631.

Ratsch, C. and Venables, J.A. (2003). Nucleation theory and the early stages of thin film growth. *Journal of Vacuum Science and Technology* 21: S96–S109.

Raty, J.Y., Gaspard, J.P., Bichara, C. et al. (2000). Re-entrant peierls distortion in IV–VI compounds. *Physica B: Condensed Matter* 1 (276): 473–474.

Redfield, D. (1963). Effect of Defect Fields on the Optical Absorption Edge. *Physical Review* 130 (3): 916–918. https://doi.org/10.1103/PhysRev.130.916.

Reed, G.T., Mashanovich, G., Gardes, F.Y., and Thomson, D.J. (2010). Silicon optical modulators. *Nature Photonics* 4: 518–526.

Reiss, P., Carayon, S., and Bleuse, J. (2003). Large fluorescence quantum yield and low size dispersion from CdSe/ZnSe core/shell nanocrystals. *Physica E: Low-dimensional Systems and Nanostructures* 1 (17): 95–96.

Ren, H., Qi, S., Hu, Y. et al. (2019). All-solid mid-infrared chalcogenide photonic crystal fiber with ultralarge mode area. *Optics Letters* 44: 5553.

Rios, C., Hosseini, P., Wright, C.D. et al. (2014). On-chip photonic memory elements employing phase-change materials. *Advanced Materials* 26 (9): 1372–1377.

Ríos, C., Hosseini, P., Taylor, R.A., and Bhaskaran, H. (2016). Color depth modulation and resolution in phase-change material nanodisplays. *Advanced Materials* 28 (23): 4720–4726.

Ríos, C., Stegmaier, M., Hosseini, P. et al. (2015). Integrated all-photonic non-volatile multi-level memory. *Nature Photonics* 9 (11): 725–732.

Ríos, C., Youngblood, N., Cheng, Z. et al. (2019). In-memory computing on a photonic platform. *Science Advances* 5 (2): eaau5759.

Rogacheva, A.V., Fedotov, V.A., Schwanecke, A.S. et al. (2006). Giant gyrotropy due to electromagnetic-field coupling in a bilayered chiral structure. *Physical Review Letters* 97 (17): 177401.

Röseler, A. (1993). IR spectroscopic ellipsometry: instrumentation and results. *Thin Solid Films* 234: 307–313.

Rossnagel, S. (2001). Sputtering and sputter deposition. In: *Handbook of Thin Film Deposition Processes and Techniques (Second Edition)* (ed. K. Seshan), 319–348. Elsevier.

Routkevitch, D., Haslett, T.L., Ryan, L. et al. (1996). Synthesis and resonance Raman spectroscopy of CdS nano-wire arrays. *Chemical Physics* 210 (1–2): 343–352.

Rudé, M., Pello, J., Simpson, R.E. et al. (2013). Optical switching at 1.55 μm in silicon racetrack resonators using phase change materials. *Applied Physics Letters* 103 (14).

Ruiz de Galarreta, C., Sinev, I., Alexeev, A.M. et al. (2020). Reconfigurable multilevel control of hybrid all-dielectric phase-change metasurfaces. *Optica* 7 (5): 476–484.

Rütten, M., Kaes, M., Albert, A. et al. (2015). Relation between bandgap and resistance drift in amorphous phase change materials. *Scientific Reports* 5: 17362.

Safarian, J. and Engh, T.A. (2013). Vacuum evaporation of pure metals. *Metallurgical and Materials Transactions A* 44 (2): 747–753.

Safi, I. (2000). Recent aspects concerning DC reactive magnetron sputtering of thin films: a review. *Surface and Coatings Technology* 127 (2): 203–218.

Sahoo, D. and Naik, R. (2022a). GSST phase change materials and its utilization in optoelectronic devices: a review. *Materials Research Bulletin* 148: 111679.

Sahoo, D. and Naik, R. (2022b). A review on the linear/nonlinear optical properties of Se doped chalcogenide thin films as potential optoelectronic applications. *Journal of Non-Crystalline Solids* 597: 121934.

Samson, Z.L., MacDonald, K.F., De Angelis, F. et al. (2010). Metamaterial electro-optic switch of nanoscale thickness. *Applied Physics Letters* 96: 14.

Sanders, L. T. (2008). Optical interconnect devices and structures based on metamaterials. United States Patent US20080212921A1.

Sanghera, J.S., Aggarwal, I.D., Busse, L.E. et al. (1995). Development of low-loss IR transmitting chalcogenide glass fibers. *In Biomedical Optoelectronic Instrumentation* 2396: 71–77.

Schrader, B. (1995). *Infrared and Raman Spectroscopy*. Weinheim: Wiley-VCH Verlag GmbH.

Schrader, B. and Moore, D.S. (1997). Laser-based molecular spectroscopy for chemical analysis-Raman scattering processes. *Pure and Applied Chemistry* 69: 1451–1468.

Shalaginov, M.Y., An, S., Zhang, Y. et al. (2021). Reconfigurable all-dielectric metalens with diffraction-limited performance. *Nature Communications* 12 (1).

Shen, X., Dai, S.X., and Xu, T.F. (2015). Ultrafast optical response and high nonlinearity in As-Se-Te chalcogenide glasses. *Optics Express* 23 (5): 6543–6552.

Shen, C., Ye, J., Peserico, N. et al. (2023). Enhancing focusing and defocusing capabilities with a dynamically reconfigurable metalens utilizing Sb_2Se_3 phase-change material. *Nanomaterials* 13 (14): 2106.

Shi, Z., Jinglin, H., Jinjin, W. et al. (2022). Design of tunable circular dichroism extrinsic chiral metasurface based on phase change material GST. *Opto-Electronic Engineering* 49 (10): 220092.

Shiang, J.J., Kadavanich, A.V., Grubbs, R.K., and Alivisatos, A.P. (1995). Symmetry of annealed wurtzite CdSe nanocrystals: assignment to the C3v point group. *The Journal of Physical Chemistry* 99 (48): 17417–17422.

Shixin, G., Haonan, R., Jingzhe, P. et al. Reconfigurable free form silicon photonics by phase change waveguides. *Advanced Optical Materials* 13: 2402997.

Siegle, H., Thurian, P., Eckey, L. et al. (1995). Spatially-resolved photoluminescence and Raman study on the GaN/substrate interface. *MRS Online Proceedings Library* 395: 577–581.

Sigmund, P. (1993). Introduction to sputtering. *Matematisk-Fysiske Meddelelser fra Det Kongelige Danske Videnskabernes Selskab* 43: 7–26.

Singh, V. (2015). *Chalcogenide Glass Materials for Integrated Infrared Photonics*. Massachusetts Institute of Technology.

Sinha, R.K., Kumar, A., and Saini, T.S.I.E.E.E.J. (2015). Analysis and design of single-mode As 2 Se 3-chalcogenide photonic crystal fiber for generation of slow light with tunable features. *IEEE Journal of Selected Topics in Quantum Electronics* 22: 287.

Slater, J.C., Wilson, T.M., and Wood, J.H. (1969). Comparison of several exchange potentials for electrons in the Cu+ ion. *Physical Review* 179 (1): 28.

Song, C.-L., Wang, Y.-L., Jiang, Y.-P. et al. (2011). Molecular-beam epitaxy and robust superconductivity of stoichiometric FeSe crystalline films on bilayer graphene. *Physical Review B* 84 (2): 020503.

Song, J.-G., Wang, Y.L., Jiang, Y.P. et al. (2013). Layer-controlled, wafer-scale, and conformal synthesis of tungsten disulfide nanosheets using atomic layer deposition. *ACS Nano* 7: 11333–11240.

Song, R.R., Deng, Q.L., Zhou, S.L., and Pu, M.B. (2021). Catenary-based phase change metasurfaces for mid-infrared switchable wavefront control. *Optics Express* 29: 23006.

Spear, W.E, Le Comber, P.G. (1976). Electronic properties of substitutionally doped amorphous Si and Ge. *Philosophical Magazine* 33 (6): 935-949. https://doi.org/10.1080/14786437608221926.

Stafe, M., Marcu, A., and Puscas, N.N. (2014). *Pulsed Laser Ablation of Solids*. Berlin: Springer.

Stegmaier, M., Ríos, C., Bhaskaran, H. et al. (2017). Nonvolatile all-optical 1 × 2 switch for chipscale photonic networks. *Advanced Optical Materials* 5 (1): 1600346.

Stuart, R.V. (2012). *Vacuum Technology, Thin Films, and Sputtering: An Introduction*. New York: Academic.

Su, N. (2002). A molecular dynamics method for simulations in the canonical ensemble. *Molecular Physics* 100 (1): 191–198.

Sun, Z., Zhou, J., and Ahuja, R. (2006). Structure of phase change materials for data storage. *Physical Review Letters* 96 (5): 055507.

Sun, Y., Dai, S., Zhang, P. et al. (2015). Fabrication and characterization of multimaterial chalcogenide glass fiber tapers with high numerical apertures. *Optics Express* 23 (18): 23472–23483.

Suthar, B. (2015). Tuning of guided mode in two dimensional chalcogenide based photonic crystal waveguide. *Optik* 126: 3429.

Suzuki, K. and Baba, T. (2010). Nonlinear light propagation in chalcogenide photonic crystal slow light waveguides. *Optics Express* 18: 26675.

Suzuki, K., Hamachi, Y., and Baba, T. (2009). Fabrication and characterization of chalcogenide glass photonic crystal waveguides. *Optics Express* 17: 22393.

Talapin, D.V., Rogach, A.L., Kornowski, A. et al. (2001). Highly luminescent monodisperse CdSe and CdSe/ZnS nanocrystals synthesized in a hexadecylamine–trioctylphosphine oxide–trioctylphospine mixture. *Nano Letters* 1 (4): 207–211.

Tanaka, K. (2002). Two-photon absorption spectroscopy of As_2S_3 glass. *Applied Physics Letters* 80: 177–179.

Tanaka, K. (2018). Applications of chalcogenide glasses in photonic devices. *Optical Materials Express* 8 (9): 2065–2078.

Tang, H., Stan, L., Czaplewski, D.A. et al. (2023). Wavelength-tunable infrared chiral metasurfaces with phase-change materials. *Optics Express* 31 (13): 21118–21127.

Tang, H., Stan, L., Czaplewski, D.A. et al. (2024). Infrared phase-change chiral metasurfaces with tunable circular dichroism. *Optics Express* 32 (11): 20136–20145.

Tauc, J., Grigorovici, R., and Vancu, A. (1966). Optical properties and electronic structure of amorphous germanium. *Physica Status Solidi B* 15 (2): 627–637.

Townsley, C.M., Davies, J.J., Wolverson, D. et al. (1996). Spin-flip Raman-scattering studies of compensating donor centers in nitrogen-doped zinc selenide grown by molecular-beam epitaxy. *Physical Review B* 53 (16): 10983.

Toyozawa, Y. (1959). A proposed model for the explanation of the urbach rule. *Progress of Theoretical Physics* 22 (3): 455–457.

Toyozawa, Y. (1964). Interband effect of lattice vibrations in the exciton absorption spectra. *Journal of Physics & Chemistry of Solids* 25 (1): 59–71. https://doi.org/10.1016/0022-3697(64)90162-3.

Treacy, D.J., Taylor, P.C., and Klein, P.B. (1979). Photodarkening and photostructural effects in glassy As2Se3. *Solid State Communications* 32 (6): 423–427.

Tripathi, D. and Hegde, R. (2023). Optimally designed tunable phase change material-based narrowband perfect absorber. *Journal of Nanophotonics* 17 (1): 016004.

Tripathi, D., Vyas, H.S., Kumar, S. et al. (2023). Recent developments in chalcogenide phase change material-based nanophotonics. *Nanotechnology* 34 (50).

Troles, J., Brilland, L., Smektala, F. et al. (2007). *9th International Conference on Transparent Optical Networks* (1–5 July 2007). IEEE (Institute of Electrical and Electronics Engineers).

Tyndall, J. (1869). On the blue color of the sky, the polarization of skylight, and polarization of light by cloudy matter generally. *Journal of the Franklin Institute* 88 (2): 114–120.

Uozumi, Y. and Katoda, T. (1997). Effects of stress on generation of P-line defects near insulator-silicon interface. *Applied Surface Science* 2 (117): 624–628.

Verger, F., Nazabal, V., Colas, F. et al. (2013). RF sputtered amorphous chalcogenide thin films for surface enhanced infrared absorption spectroscopy. *Optical Materials Express* 3 (12): 2112–2131.

Vetterhöffer, J., Wagner, J., and Weber, J. (1996). Isolated hydrogen molecules in GaAs. *Physical Review Letters* 77 (27): 5409.

Vina, L., Calleja, J.M., Cros, A. et al. (1996). Role of excitons in double Raman resonances in GaAs quantum wells. *Physical Review B* 53 (7): 3975.

Vlasenko, N.A. and Popkov, I.A. (1960). Study of the electroluminescence of a sublimed ZnS-Mn phosphor. *Optics and Spectroscopy* 8: 39.

Von Klitzing, K. (1986). The quantized hall effect. *Reviews of Modern Physics* 126 (1–3).

Vook, R.W. (1984). Nucleation and Growth of thin films. *Optical Engineering* 23: 343–348.

Wang, R. and Cao, T. (2020). Reconfigurable slow light in phase change photonic crystal waveguide. *Journal of Applied Physics* 128.

Wang, X.-W., Zhang, W., Zhang, L. et al. (2013). Precise analysis of formation and suppression of intensity transmittance fluctuations of glan-taylor prisms. *Shanghai Institute of Optics and Fine Mechanics* 50 (5): 052302–052302.

Wang, X., Zhang, W., Zhang, L. et al. (2013). Research progress in fabrication of chalcogenide glass photonic crystal waveguides. *Laser & Optoelectronics Progress* 50 (12): 5.

Wang, Q., Rogers, E.T.F., Gholipour, B. et al. (2016). Optically reconfigurable metasurfaces and photonic devices based on phase change materials. *Nature Photonics* 10: 60–65.

Wang, M., Lee, J.S., and Aggarwal, S. (2023). Varifocal metalens using tunable and ultralow-loss dielectrics. *Advanced Science* 10 (6).

Weber, A. (1979). *Raman Spectroscopy of Gases and Liquids*. Berlin Heidelberg: Springer.

Wełnic, W., Pamungkas, A., Detemple, R. et al. (2006). Unravelling the interplay of local structure and physical properties in phase-change materials. *Nature Materials* 5 (1): 56–62.

Wełnic, W., Botti, S., Reining, L., and Wuttig, M. (2007). Origin of the optical contrast in phase-change materials. *Physical Review Letters* 98 (23): 236403.

Westwood, W.D. (1989). Physical vapor deposition. In: *Microelectronic Materials and Processes* (ed. R.A. Levy), 133–201. Netherlands: Dordrecht, Springer.

Wu, X.L., Yan, F., Zhang, M.S., and Feng, D. (1995). Influence of illumination of Raman spectra of porous silicons. *Physics Letters A* 205 (1): 117–120.

Wu, F., Zhang, J.Z., Kho, R., and Mehra, R.K. (2000). Radiative and nonradiative lifetimes of band edge states and deep trap states of CdS nanoparticles determined by time-correlated single photon counting. *Chemical Physics Letters* 330 (3–4): 237–242.

Wu, C., Deng, H., Huang, Y.-S. et al. (2024). Freeform direct-write and rewritable photonic integrated circuits in phase-change thin films. *Science Advances* 10 (1): eadk1361.

Wu, C., Yu, H., Li, H. et al. (2019). Low-loss integrated photonic switch using subwavelength patterned phase change material. *ACS Photonics* 6 (1): 87–92.

Wu, C., Yu, H., Lee, S. et al. (2021). Programmable phase-change metasurfaces on waveguides for multimode photonic convolutional neural network. *Nature Communications* 12 (1): 96.

Wu, X.W., Gong, C.J., Wu, K. (2023). Research progress on nonlinear optical crystals in the chalcogenide system. *Foundry Technology* 44 (1): 9–14.

Wu, Y., Duan, K., Wang, Q. et al. (2025). Fabrication of large-mode area multi-core mid-infrared photonic crystal fiber with bending resistance. *Optics and Laser Technology* 186: 112648.

Wuttig, M., Bhaskaran, H., and Taubner, T. (2017). Phase-change materials for non-volatile photonic applications. *Nature Photonics* 11 (8): 465–476.

Wuttig, M., Lüsebrink, D., Wamwangi, D. et al. (2007). The role of vacancies and local distortions in the design of new phase-change materials. *Nature Materials* 6 (2): 122–128.

Wuttig, M., Schöen, C.F., Löetfering, J. et al. (2023). Revisiting the nature of chemical bonding in chalcogenides to explain and design their properties. *Advanced Materials* 35 (20).

Xiao, S.Y., Wang, T., Liu, T.T. et al. (2020). Active metamaterials and metadevices: a review. *Journal of Physics D: Applied Physics* 53 (50): 503002.

Xin, C., Qing, Z., Liang, F. et al. (2011). Fabriction of 100 nm mask by laser interference lithography. *High Power Laser and Particle Beams* 23 (3): 806–810.

Yablonovitch, E. (1987). Inhibited spontaneous emission in solid-state physics and electronics. *Physical Review Letters* 58: 2059.

Yablonovitch, E., Gmitter, T., and Leung, K.-M. (1991). Photonic band structure: the face-centered-cubic case employing nonspherical atoms. *Physical Review Letters* 67: 2295.

Yang, C.Y., Lee, J.M., Paesler, M.A. et al. (1987). Quantitative measurements of structural changes associated with photo-darkening in a-As_2S_3 films. *Journal of Non-Crystalline Solids* 97: 1119–1122.

Yang, Z. and Ramanathan, S. (2015). Breakthroughs in photonics 2014: phase change materials for photonics. *IEEE Photonics Journal* 7 (3): 1–5.

Yan-Tao, X.U., Hai-Tao, G.U.O. et al. (2015). Preparation and applications of low-loss As-S chalcogenide glass fibers. *Journal of Inorganic Materials* 30 (1): 97–101.

Yau, L.D. and Sah, C.T. (1972). Measurement of trapped-minority-carrier thermal emission rates from Au, Ag, and Co traps in silicon. *Applied Physics Letters* 21: 157–158.

Yin, X., M, S., Michel, A.K. et al. (2015). Active chiral plasmonics. *Nano Letters* 15 (7): 4255–4260.

Yu, P.Y. and Cardona, M. (2010). *Fundamentals of Semiconductors, Physics and Materials Properties*. Springer Verlag.

Yu, N., Genevet, P., Kats, M.A. et al. (2011). Light propagation with phase discontinuities: generalized laws of reflection and refraction. *Science* 334: 333.

Yutaka, T. (1959). On the dynamical behavior of an exciton. *Progress of Theoretical Physics Supplement* 253–266. https://doi.org/10.1143/PTPS.12.111.

Zakery, A. and Elliott, S.R. (2003). Optical properties and applications of chalcogenide glasses: a review. *Journal of Non-Crystalline Solids* 330: 1–12.

Zhang, Y. and Chen, X. (2019). Optical properties of chalcogenide glasses for infrared applications. *Journal of Non-Crystalline Solids* 521: 119–125.

Zhang, Z. and Lagally, M.G. (1997). Atomistic processes in the early stages of thin-film growth. *Science* 276 (5311): 377–383.

Zhang, Y., Zhang, L., Zhou, C. et al. (2013). Review of chemical vapor deposition of graphene and related applications. *Accounts of Chemical Research* 46 (10): 2329–2339.

Zhang, H., Zhou, L., Lu, L. et al. (2019). Miniature multilevel optical memristive switch using phase change Material. *ACS Photonics* 6 (9): 2205–2212.

Zhang, C.H., Zhang, M., Xie, Y. et al. (2020a). Wavelength-selective 2 × 2 optical switch based on a $Ge_2Sb_2Te_5$-assisted microring. *Photonics Research* 8: 1171–1176.

Zhang, H.-Z., Xu, C.-Y., Nan, H.-Y. et al. (2020b). Research progress of two-dimensional transition metal dichalcogenide phase transition methods. *Acta Physica Sinica* 69 (24): 246101.

Zhang, Y.F., Fowler, C., Liang, J.H. et al. (2021). Electrically reconfigurable non-volatile metasurface using low-loss optical phase-change material. *Nature Nanotechnology* 16: 661.

Zhang, S., Chen, X., Liu, K. et al. (2023). On-chip non-volatile reconfigurable THz varifocal metalens. *Laser and Photonics Reviews* 17 (11): 2300482.

Zhang, T., Liu, J., Gu, L. et al. (2025). Wavelength-tunable infrared metasurfaces with chiral bound states in the continuum. *Scientific Reports* 15 (1): 128.

Zhao, X.J. and Dai, S.X. (2017). Nonlinear optical properties of chalcogenide glasses for infrared photonics. *Progress in Materials Science* 89: 1–40.

Zhao, J., Zhang, J., Jiang, C. et al. (2004). Electroluminescence from isolated Cdse/Zns quantum dots in multilayered light-emitting diodes. *Journal of Applied Physics* 96 (6): 3206–3210.

Zhao, M., Yang, Z., Zhang, R. et al. (2021). High Q chalcogenide photonic crystal nanobeam cavities. *IEEE Photonics Technology Letters* 33: 525.

Zheng, J., Khanolkar, A., Xu, P. et al. (2018). GST-on-silicon hybrid nanophotonic integrated circuits: a non-volatile quasi-continuously reprogrammable platform. *Optical Materials Express* 8 (6): 1551–1561.

Zheng, J., Fang, Z., Wu, C. et al. (2020a). Nonvolatile electrically reconfigurable integrated photonic switches using phase-change materials. In 2020 *Conference on Lasers and Electro-Optics* (CLEO), (10–15 May 2020). IEEE (Institute of Electrical and Electronics Engineers).

Zheng, J., Zhu, S., Xu, P. et al. (2020b). Modeling electrical switching of nonvolatile phase-change integrated nanophotonic structures with graphene heaters. *ACS Applied Materials and Interfaces* 12 (19): 21827–21836.

Zhou, Z.H., Hashimoto, T., Nasu, H., and Kamiya, K. (1998). Two-photon absorption and nonlinear refraction of lanthanum sulfide-gallium sulfide glasses. *Journal of Applied Physics* 84 (5): 2380–2384.

Zhu, M., Cojocaru-Mirédin, O., Mio, A.M. et al. (2018). Unique bond breaking in crystalline phase change materials and the quest for metavalent bonding. *Advanced Materials* 30 (18): 1706735.

Zhu, W., Fan, Y., Li, C. et al. (2020). Realization of a near-infrared active Fano-resonant asymmetric metasurface by precisely controlling the phase transition of $Ge_2Sb_2Te_5$. *Nanoscale* 12: 8758–8767.

Index

Chalcogenide Nanophotonics, First Edition. Tun Cao, Ying Su, Yinghan Li, and Hongze Gao.